Signals and Communication Technology

For further volumes:
http://www.springer.com/series/4748

Hermann Rohling
Editor

OFDM

Concepts for Future Communication Systems

Springer

Editor
Hermann Rohling
Institut für Nachrichtentechnik
Technische Universität Hamburg-Harburg
Eißendorfer Str. 40
21073 Hamburg
Germany
e-mail: rohling@tu-harburg.de

ISSN 1860-4862

ISBN 978-3-642-17495-7 e-ISBN 978-3-642-17496-4

DOI 10.1007/978-3-642-17496-4

Springer Heidelberg Dordrecht London New York

Cover design: eStudio Calamar, Belin/Figueres

Printed on acid-free paper

Springer is part of Springer Science+Business Media (www.springer.com)

The contents of this book is based on the work of the Priority Program TakeOFDM (SPP1163), funded by the German Research Foundation (Deutsche Forschungsgemeinschaft, DFG).

Preface

Dear readers, dear friends,

The Orthogonal Frequency-Division Multiplexing (OFDM) digital transmission technique has several advantages in broadcast and mobile communications applications. Therefore, the German Research Foundation (DFG) funded a so-called priority program "Techniques, Algorithms, and Concepts for Future OFDM Systems" (TakeOFDM), which started in 2004. The main objective of this research program is to study the specific research topics in a collaborative work between experts and young scientists from different universities.

In broadcast applications like Digital Audio Broadcast (DAB), Digital Video broadcast (DVB-T), Digital Radio Mondiale (DRM), and single-cell WLAN systems, the OFDM transmission technique is already mature and operational for several years. However, in wireless and wireline communications there is still need for further research and optimization. The OFDM transmission technique has gained a lot of attention in research, and it has proven to be a suitable choice in the design of digital transmission concepts for mobile applications, such as the Long Term Evolution (LTE) standard, which is the system proposal for the fourth generation (4G) of mobile communication systems. The OFDM transmission method, specific medium access techniques, cellular networks with full coverage, and multiple antenna systems have been playing an important and ever-growing role in the 4G development.

Since 2004, more than 15 different universities in over 40 specific research projects have been contributing to the TakeOFDM program, covering a large variety of detailed aspects of OFDM and all related system design aspects. This TakeOFDM research project led to several PhD theses and is a basis for the young generation of excellent scientific researchers and staff. The results have been exchanged in several workshops, conferences, and direct cooperations. Besides topics regarding the physical layer, such as coding, modulation, and non-linearities, a special emphasis was put on system aspects and concepts, in particular focusing on cellular networks and multiple antenna techniques. The challenges of link adaptation, adaptive resource allocation, and interference mitigation in such systems were addressed extensively. Moreover, the domain of cross-layer design, i.e., the combination of physical layer aspects and issues of higher layers, was considered in detail.

This book summarizes the main results and gives an overview of the combined research efforts which have been undertaken in the past 6 years within the TakeOFDM priority program. As a coordinator of the program, I would like to express my gratitude to the German Research Foundation (DFG) for their generous funding, continuous support and fruitful collaboration throughout the lifetime of the TakeOFDM project. Moreover, my sincere thanks go to all scientists for their excellent work on a high scientific level and their valuable contributions over the last years.

The aim of this book is to give a good insight into these efforts, and provide the reader with a comprehensive overview of the scientific progress which was achieved in the framework of TakeOFDM. I am convinced that also in the future, these results will facilitate and stimulate further innovation and developments in the design of modern communication systems, based on the powerful OFDM transmission technique.

Prof. Dr. Hermann Rohling Hamburg, June 2010

Contents

1 Introduction

H. Rohling, Hamburg University of Technology, Germany

In the evolution of mobile communication systems, approximately a 10 years periodicity can be observed between consecutive system generations. Research work for the current 2nd generation of mobile communication systems (GSM) started in Europe in the early 1980s, and the complete system was ready for market in 1990. At that time, the first research activities had already been started for the 3rd generation (3G) of mobile communication systems (UMTS, IMT-2000) and the transition from second generation (GSM) to the new 3G systems was observed around 2002. Compared to today's GSM networks, these UMTS systems provide much higher data rates, typically in the range of 64 to 384 kbit/s, while the peak data rate for low mobility or indoor applications is 2 Mbit/s. With the extension of High Speed Packet Access (HSPA), data rates of up to 7.2 Mbit/s are available in the downlink. The current pace, which can be observed in the mobile communications market, already shows that the 3G systems will not be the ultimate system solution. Consequently, general requirements for a fourth generation (4G) system have been considered in the process of the "Long Term Evolution" (LTE) standardization. These requirements have mainly been derived from the types of service a user will require in future applications. Generally, it is expected that data services instead of pure voice services will play a predominant role, in particular due to a demand for mobile IP applications. Variable and especially high data rates (100 Mbps and more) will be requested, which should also be available at high mobility in general or high vehicle speeds in particular (see Fig. 1.1). Moreover, asymmetrical data services between up- and downlink are assumed and should be supported by LTE systems in such a scenario where the downlink carries most of the traffic and needs the higher data rate compared to the uplink.

To fulfill all these detailed system requirements, the OFDM transmission technique applied in a wide-band radio channel has been chosen as an air interface for the downlink in the framework of LTE standardization due to its flexibility and adaptivity in the technical system design. From the above considerations, it already becomes apparent that a radio transmission system for LTE must provide a great flexibility and adaptivity at different levels, ranging from the highest layer (requirements of the application) to the lowest layer (the transmission medium, the physical layer ,i.e., the radio channel) in the ISO-OSI stack. Today, the OFDM transmission technique is in a completely matured stage to be applied for wide-band communication systems integrated into a cellular mobile communications environment.

H. Rohling (ed.), *OFDM: Concepts for Future Communication Systems*,
Signals and Communication Technology, DOI: 10.1007/978-3-642-17496-4_1,

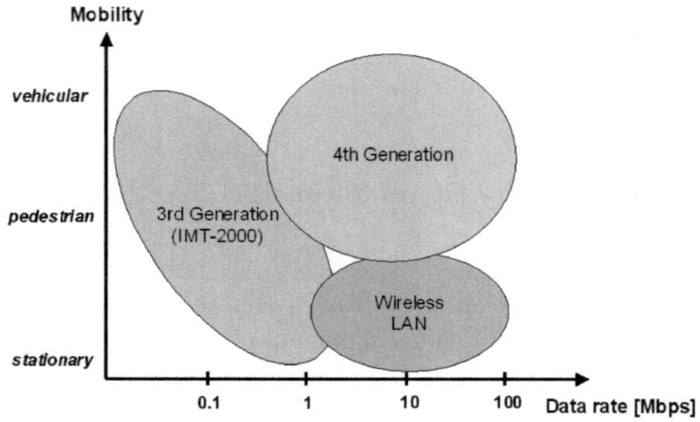

Figure 1.1: General requirements for 4G mobile communication systems

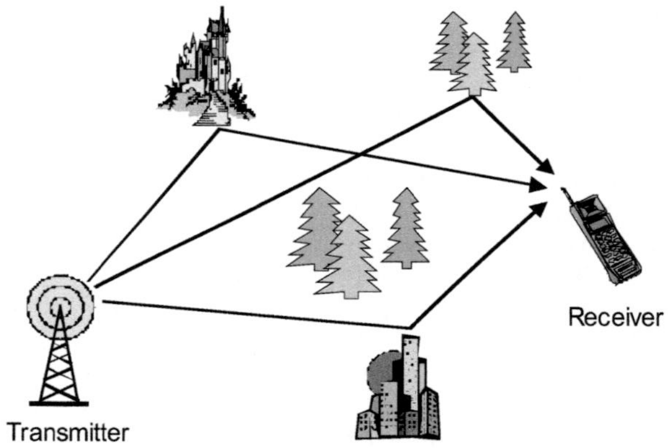

Figure 1.2: Multipath propagation scenario

1.1 Radio Channel Behavior

The mobile communication system design is in general always dominated by the radio channel behavior [19] [25]. In a typical radio channel situation, multi-path propagation occurs (Fig. 1.2) due to the reflections of the transmitted signal at several objects and obstacles inside the local environment and inside the observation area. The radio channel is analytically described unambiguously by a linear (quasi) time-invariant (LTI) system model and by the related channel impulse response $h(\tau)$ or, alternatively, by the channel transfer function $H(f)$. An example for these channel characteristics is shown in Fig. 1.3, where $h(\tau)$ and $H(f)$ of a so-called Wide-Sense Stationary, Uncorrelated Scattering-channel (WSSUS) are given.

Due to the mobility of the mobile terminals, the multipath propagation situation will be continuously but slowly changed over time which is described analytically by a time-variant channel impulse response $h(\tau, t)$ or alternatively by a frequency-

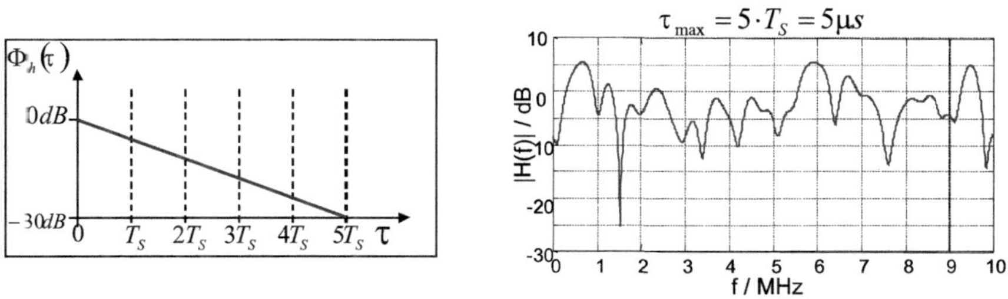

Figure 1.3: Power delay profile (left) and channel transfer function of a WSSUS Channel

selective and time-dependent radio channel transfer function $H(f,t)$ as it is shown as an example in Fig. 1.5. All signals on the various propagation paths will be received in a superimposed form and are technically characterized by different delays and individual Doppler frequencies which finally lead to a frequency-selective behavior of the radio channel (see Fig. 1.4). The other two system functions, the Delay Doppler function, $\nu(\tau, f_D)$ and the Frequency Doppler function, $U(f, f_D)$ can be used as an alternative description of the radio channel behavior. The Delay Doppler function $\nu(\tau, f_D)$ describes the variation of the channel impulse response related to certain values of the Doppler frequency f_D. This means, the channel delays change due to alteration of the relative speed between a mobile terminal and the base station. The Frequency Doppler function $U(f, f_D)$ models the same effects for the channel behavior in the frequency domain.

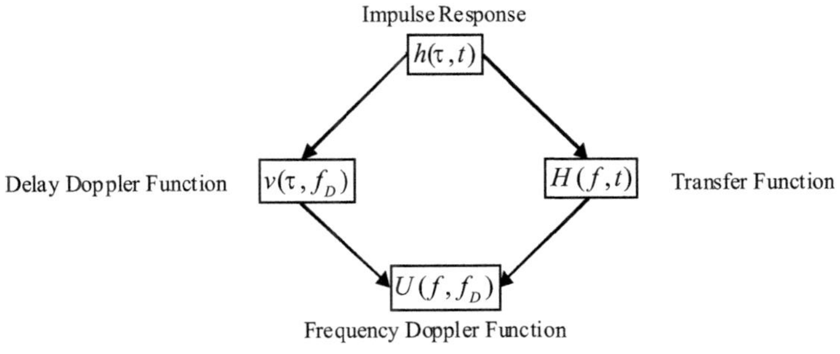

Figure 1.4: Relationships between different system functions

The radio channel can roughly and briefly be characterized by two important system parameters: the maximum multipath delay τ_{max} and the maximum Doppler frequency f_{Dmax} which are transfered into the coherence time T_C and the coherence bandwidth B_C of the radio channel:

$$T_C = \frac{1}{f_{Dmax}}, B_C = \frac{1}{\tau_{max}} \tag{1.1}$$

Over time intervals significantly shorter than T_C, the channel transfer function can be assumed to be almost stationary. Similarly, for frequency intervals significantly smaller than B_C, the channel transfer function can be considered as nearly constant. Therefore, it is assumed in this chapter that the coherence time T_C is much larger compared to a single OFDM symbol duration T_S and the coherence bandwidth B_C is much larger than the distance Δf between two adjacent subcarriers:

$$B_C \gg \Delta f, \Delta f = \frac{1}{T_S}, T_S \ll T_C \tag{1.2}$$

This condition should always be fulfilled in well-dimensioned OFDM systems and in realistic time-variant and frequency-selective radio channels.

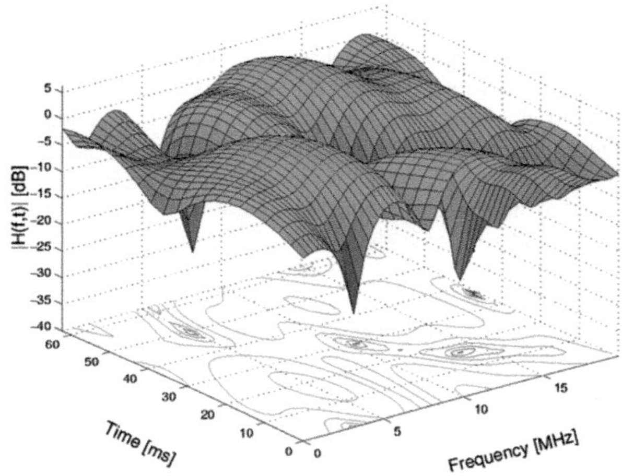

Figure 1.5: Frequency-selective and time-variant radio channel transfer function

There are always technical alternatives possible in new system design phases. However, future mobile communication systems will in any case require extremely large data rates and therefore large system bandwidth. If conventional single carrier (SC) modulation schemes with the resulting very low symbol durations are applied in this system design, very strong intersymbol interference (ISI) is caused in wide-band applications due to multi-path propagation situations. This means, for high data rate applications, the symbol duration in a classical SC transmission system is extremely small compared to the typical values of maximum multi-path delay τ_{max} in the considered radio channel. In these strong ISI situations, a very powerful equalizer is necessary in each receiver, which needs high computation complexity in a wide-band system. These constraints should be taken into consideration in the system development phase for a new radio transmission scheme. The computational complexity for the necessary equalizer techniques to overcome all these strong ISI in a SC modulation scheme increases quadratically for a given radio channel with increasing system bandwidth and can be extremely large in wide-band applications. For that reason alternative transmission techniques for broadband applications are of high interest.

Alternatively, OFDM can efficiently deal with all these ISI effects, which occur in multi-path propagation situations and in broadband radio channels. Simultaneously, the OFDM transmission technique needs much less computational complexity in the equalization process inside each receiver. The performance figures for an OFDM based new air interface and for next generation of mobile communications are very promising even in frequency-selective and time-variant radio channels.

1.2 Basics of the OFDM Transmission Technique

If a high data rate is transmitted over a frequency-selective radio channel with a large maximum multi-path propagation delay τ_{max} compared to the symbol duration, an alternative to the classical SC approach is given by the OFDM transmission technique. The general idea of the OFDM transmission technique is to split the total available bandwidth B into many narrowband sub-channels at equidistant frequencies. The sub-channel spectra overlap each other but the subcarrier signals are still orthogonal. The single high-rate data stream is subdivided into many low-rate data streams for the sub-channels. Each sub-channel is modulated individually and will be transmitted simultaneously in a superimposed and parallel form. An OFDM transmit signal therefore consists of N adjacent and orthogonal subcarriers spaced by the frequency distance Δf on the frequency axis. All subcarrier signals are mutually orthogonal within the symbol duration of length T_S, if the subcarrier distance and the symbol duration are chosen such that $T_S = \frac{1}{\Delta f}$. For OFDM-based systems, the symbol duration T_S is much larger compared to the maximum multipath delay τ_{max}. The k-th unmodulated subcarrier signal is described analytically by a complex valued exponential function with carrier frequency $k\Delta f$, $\tilde{g}_k(t), k = 0, \ldots, N-1$.

$$\tilde{g}_k(t) = \begin{cases} e^{j2\pi k\Delta f t} & \forall t \in [0, T_S] \\ 0 & \forall t \notin [0, T_S] \end{cases} \tag{1.3}$$

Since the system bandwidth B is subdivided into N narrowband sub-channels, the OFDM symbol duration T_S is N times larger than in the case of an alternative SC transmission system covering the same bandwidth B. Typically, for a given system bandwidth, the number of subcarriers is chosen in such a way that the symbol duration T_S is sufficiently large compared to the maximum multi-path delay τ_{max} of the radio channel. Additionally, in a time-variant radio channel, the Doppler spread imposes restrictions on the subcarrier spacing Δf. In order to keep the resulting Inter-Carrier Interference (ICI) at a tolerable level, the system parameter of the subcarrier spacing Δf must be large enough compared to the maximum Doppler frequency f_{Dmax}. In [18] the appropriate range for choosing the symbol duration T_S as a rule of thumb in practical systems is given as (cf. Eq. (1.2)):

$$4\tau_{max} \leq T_S \leq 0.03\frac{1}{f_{Dmax}} \tag{1.4}$$

The duration T_S of the subcarrier signal $\tilde{g}_k(t)$ is additionally extended by a cyclic prefix (so-called guard interval) of length T_G, which is larger than the maximum multi-path delay τ_{max} in order to avoid ISI completely which could occur in multi-path channels in the transition interval between two adjacent OFDM symbols [17].

$$g_k(t) = \begin{cases} e^{j2\pi k\Delta f t} & \forall t \in [-T_G, T_S] \\ 0 & \forall t \notin [-T_G, T_S] \end{cases} \tag{1.5}$$

The guard interval is directly removed in the receiver after the time synchronization procedure. From this point of view, the guard interval is a pure system overhead. The total OFDM symbol duration is therefore $T = T_S + T_G$. It is an important advantage of the OFDM transmission technique that ISI can be avoided completely or can be reduced at least considerably by a proper choice of OFDM system parameters.

The orthogonality of all subcarrier signals is completely preserved in the receiver even in frequency-selective radio channels, which is an important advantage of OFDM. The radio channel behaves linearly and in a short time interval of a few OFDM symbols even time-invariant. Hence, the radio channel behavior can be described completely by a Linear and Time Invariant (LTI) system model, characterized by the impulse response $h(t)$. The LTI system theory gives the reason for this important system behavior that all subcarrier signals are orthogonal in the receiver even when transmitting the signal in frequency-selective radio channels. All complex-valued exponential signals (e.g., all subcarrier signals) are eigenfunctions of each LTI system and therefore eigenfunctions of the considered radio channel, which means that only the signal amplitude and phase will be changed if a subcarrier signal is transmitted over the linear and time-invariant radio channel.

The subcarrier frequency is not affected at all by the radio channel transmission, which means that all subcarrier signals are even orthogonal in the receiver and at the output of a frequency-selective radio channel. The radio channel disturbs only amplitudes and phases individually, but not the subcarrier frequency of all received sub-channel signals. Therefore all subcarrier signals are still mutually orthogonal in the receiver. Due to this important property, the received signal which is superimposed by all subcarrier signals can be split directly into the different sub-channel components by a Fourier transform and each subcarrier signal can be restored by a single-tap equalizer and demodulated individually in the receiver.

At the transmitter side, each subcarrier signal is modulated independently and individually by the complex valued modulation symbol $S_{n,k}$, where the subscript n refers to the time interval and k to the subcarrier signal number in the considered OFDM symbol. Thus, within the symbol duration time interval T the time-continuous signal of the n-th OFDM symbol is formed by a superposition of all N simultaneously modulated subcarrier signals.

$$s_n(t) = \sum_{k=0}^{N-1} S_{n,k} g_k(t - nT) \tag{1.6}$$

The total time-continuous transmit signal consisting of all OFDM symbols sequentially transmitted on the time axis is described by Eq. (1.7):

$$s(t) = \sum_{n=0}^{\infty} \sum_{k=0}^{N-1} S_{n,k} e^{j2\pi k \Delta f(n-nT)} \mathrm{rect}\left(\frac{2(t-nT)+T_G-T_S}{2T}\right) \tag{1.7}$$

The analytical signal description shows that a rectangular pulse shaping is applied for each subcarrier signal and each OFDM symbol. Due to the rectangular pulse shaping, the spectra of all the considered subcarrier signals are sinc-functions which are equidistantly located on the frequency axis, e.g., for the k-th subcarrier signal, the spectrum is described in Eq. (1.8):

$$G_k(f) = T \cdot \mathrm{sinc}\left[\pi T(f - k\Delta f)\right] \tag{1.8}$$

The typical OFDM spectrum shown in Fig. 1.6 consists of N adjacent sinc-functions, which are shifted by Δf in the frequency direction.

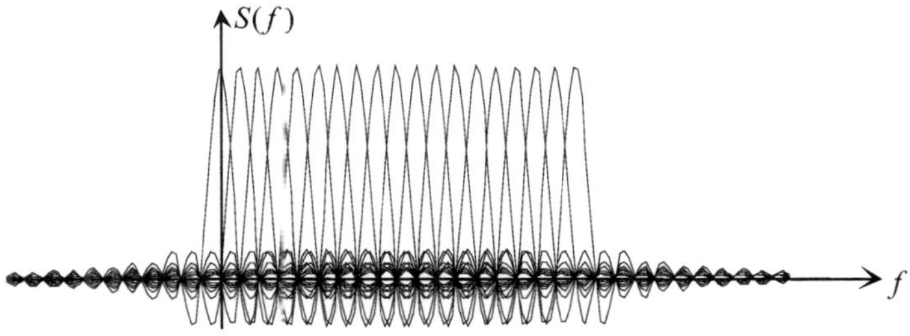

Figure 1.6: OFDM spectrum which consists of N equidistant sinc functions

The spectra of the considered subcarrier signals overlap on the frequency axis, but the subcarrier signals are still mutually orthogonal, which means the transmitted modulation symbols $S_{n,k}$ can be recovered by a simple correlation technique in each receiver if the radio channel is assumed to be ideal in a first analytical step:

$$\frac{1}{T_S}\int_0^{T_S} g_k(t) g_l^*(t) dt = \begin{cases} 1 & k = l \\ 0 & k \neq l \end{cases} \tag{1.9}$$

$$S_{n,k} = \frac{1}{T_S}\int_0^{T_S} s_n(t) g_k^*(t) dt = \frac{1}{T_S}\int_0^{T_S} s_n(t) e^{-j2\pi k \Delta f t} dt \tag{1.10}$$

where $g_k^*(t)$ is the conjugate complex version of the subcarrier signal $g_k(t)$. Eq. (1.11) shows the correlation process in detail:

$$\begin{aligned} Corr &= \frac{1}{T_S}\int_0^{T_S} s_n(t) g_k^*(t) dt = \frac{1}{T_S}\int_0^{T_S} \sum_{m=0}^{N-1} S_{n,m} g_m(t) g_k^*(t) dt \\ &= \sum_{m=0}^{N-1} S_{n,m} \frac{1}{T_S}\int_0^{T_S} g_m(t) g_k^*(t) dt = \sum_{m=0}^{N-1} S_{n,m}\delta_{m,k} = S_{n,k} \end{aligned} \tag{1.11}$$

In practical applications, the OFDM transmit signal $s_n(t)$ is generated as a time-discrete signal in the digital baseband. Using the sampling theorem while considering the OFDM transmit signal inside the bandwidth $B = N\Delta f$, the transmit signal must be sampled with the sampling interval $\Delta t = 1/B = 1/N\Delta f$. The individual samples of the transmit signal are denoted by $s_{n,i}, i = 0, 1, \ldots, N-1$ and can be calculated as follows (see Eq. (1.7)):

$$
\begin{aligned}
s(t) &= \sum_{k=0}^{N-1} S_{n,k} e^{j2\pi k \Delta f t} \\
s(i\Delta t) &= \sum_{k=0}^{N-1} S_{n,k} e^{j2\pi k \Delta f i \Delta t} \\
s_{n,i} &= \sum_{k=0}^{N-1} S_{n,k} e^{j2\pi i k/N} \qquad (1.12)
\end{aligned}
$$

Equation (1.12) above exactly describes the inverse discrete Fourier transform (IDFT) applied to the complex valued modulation symbols $S_{n,k}$ of all subcarrier signals inside a single OFDM symbol.

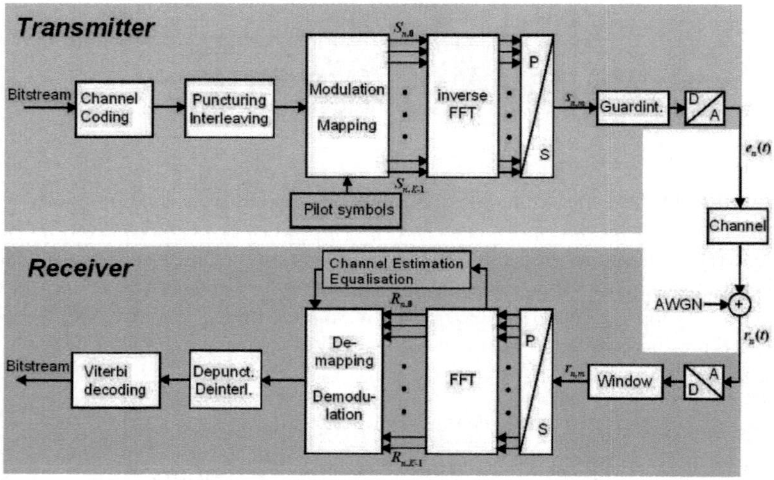

Figure 1.7: OFDM system structure in a block diagram

The individually modulated and superimposed subcarrier signals are transmitted in a parallel way over many narrowband sub-channels. Thus, in each sub-channel, the symbol duration is large and can be chosen much longer than the maximum multi-path delay of the radio channel. In this case, each sub-channel has the property to be frequency non-selective.

Figure 1.7 shows the general OFDM system structure in a block diagram. The basic principles of the OFDM transmission technique have already been described in several publications, e.g., [19], [27]. In the very early and classical multi-carrier system considerations like [20], [26], narrowband signals have been generated independently, assigned to various frequency bands, transmitted, and separated by analog filters at the receiver. The new and modern aspect of the OFDM transmission technique is that the various subcarrier signals are generated digitally and jointly by an IFFT in the transmitter and that their spectra strongly overlap on the frequency axis. As a result, generating the transmit signal is simplified and the bandwidth efficiency of the system is significantly improved.

The received signal is represented by the convolution of the transmitted time signal with the channel impulse response $h(t)$ and an additive white Gaussian noise term:

$$r_n(t) = s_n(t) * h_n(t) + n_n(t) \tag{1.13}$$

Due to the assumption that the coherence time T_C will be much larger than the symbol duration T_S, the received time-continuous signal $r_n(t)$ can be separated into the orthogonal subcarrier signal components even in frequency-selective fading situations by applying the correlation technique mentioned in Eq. (1.10):

$$R_{n,k} = \frac{1}{T_S} \int_0^{T_S} r_n(t) e^{-j2\pi k \Delta f t} dt \tag{1.14}$$

Equivalently, the correlation process at the receiver side can be applied to the time-discrete receive signal at the output of an A/D converter and can be implemented as a DFT, which leads to Eq. (1.15):

$$R_{n,k} = \frac{1}{N} \sum_{i=0}^{N-1} r_{n,i} e^{-j2\pi ik/N} \tag{1.15}$$

In this case $r_{n,i} = r_n(i\Delta t)$ describes the i-th sample of the received time-continuous base-band signal $r_n(t)$ and $R_{n,k}$ is the received complex-valued symbol at the DFT output of the k-th subcarrier.

If the OFDM symbol duration T is chosen much smaller than the coherence time T_C of the radio channel, the time-variant transfer function of the radio channel $H(f,t)$ can be considered to be constant within the time duration T of each modulation symbol $S_{n,k}$ for all subcarrier signals. In this case, the effect of the radio channel in multi-path propagation situations can be described analytically by only a single multiplication of each subcarrier signal $g_k(t)$ with the complex transfer factor $H_{n,k} = H(k\Delta f, nT)$. As a result, the received complex valued symbol R_n, k at the DFT output can be described analytically as follows:

$$\begin{aligned} r_n(t) &= s_n(t) * h_n(t) + n_n(t) \\ r_{n,i} &= s_{n,i} * h_{n,i} + n_{n,i} \\ R_{n,k} &= S_{n,k} \cdot H_{n,k} + N_{n,k} \end{aligned} \tag{1.16}$$

where $N_{n,k}$ describes an additive noise component for each specific subcarrier generated in the radio channel. Eq. (1.16) shows the most important advantage of applying the OFDM transmission technique in practical applications. It encompasses the complete signal transfer situation of the OFDM block diagram including IDFT, guard interval, D/A conversion, up- and down-conversion in the RF part, frequency-selective radio channel, A/D conversion and DFT process in the receiver, neglecting non-ideal behavior of any system components.

The transmitted Symbol $S_{n,k}$ can be recovered, calculating the quotient of the received complex valued symbol and the estimated channel transfer factor $\tilde{H}_{n,k}$:

$$S_{n,k} = \frac{R_{n,k} - N_{n,k}}{H_{n,k}}, \tilde{S}_{n,k} = R_{n,k}\frac{1}{\tilde{H}_{n,k}} \tag{1.17}$$

It is obvious that this one tap equalization step of the received signal is much easier compared to a single carrier system for high data rate applications. The necessary IDFT and DFT calculations can be implemented very efficiently using the Fast-Fourier-Transform (FFT) Algorithms such as Radix 2, which reduces the system and computation complexity even more. It should be pointed out that especially the frequency synchronization at the receiver must be very precise in order to avoid any Inter Carrier Interferences (ICI). Algorithms for time and frequency synchronization in OFDM based systems will be described in a later chapter of this book.

Besides the complexity aspects, another advantage of the OFDM technique lies in its high degree of flexibility and adaptivity. Division of the available bandwidth into many frequency-nonselective subbands gives additional advantages for the OFDM transmission technique. It allows a subcarrier-specific adaptation of transmit parameters, such as modulation scheme (PHY mode) and transmit power (e.g., Water Filling) in accordance to the observed and measured radio channel status. In a multi-user environment the OFDM structure offers additionally an increased flexibility for resource allocation procedures as compared to SC systems [21]. The important system behavior that all subcarrier signals are mutually orthogonal in the receiver makes the signal processing and the equalization process realized by a single tap procedure very simple and leads to a low computation complexity.

1.3 OFDM Combined with Multiple Access Schemes

A very high degree of flexibility and adaptivity is required for new mobile communication systems and for the 4G air interface. The combination between multiple access schemes and OFDM transmission technique is an important factor in this respect. In principle, multiple access schemes for the OFDM transmission technique can be categorized according to OFDM-FDMA, OFDM-TDMA, OFDM-CDMA and OFDM-SDMA [1] [22]. Clearly, hybrid schemes can be applied which are based on a combination of the above techniques. The principles of these basic multiple access schemes are summarized in Fig. 1.8, where the time-frequency plane is depicted and the user specific resource allocation is distinguished by different colors.

These access schemes provide a great variety of possibilities for a flexible user specific resource allocation. In the following, one example for OFDM-FDMA is briefly sketched (cf. [7]). In the case that the magnitude of the channel transfer function is known for each user the subcarrier selection for an OFDM-FDMA scheme can be processed in the BS for each user individually which leads to a multi user diversity (MUD) effect. By allocating a subset of all subcarriers with the highest SNR to each user the system performance can be improved. This allocation technique based on

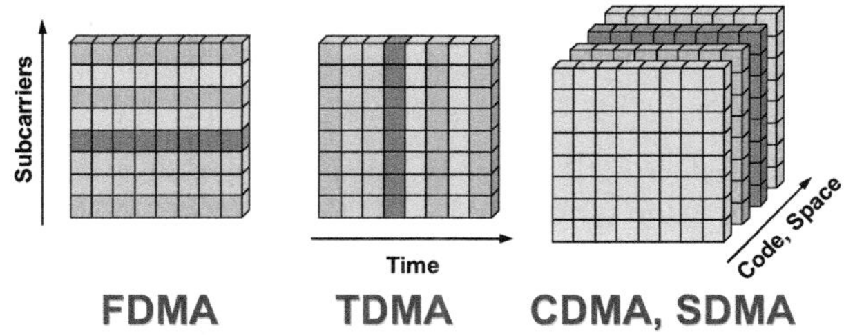

Figure 1.8: OFDM transmission technique and some multiple access schemes

the knowledge of the channel transfer function shows a large performance advantage and a gain in Quality of Service (QoS). Nearly the same flexibility in resource allocation is possible in OFDM-CDMA systems. But in this case the code orthogonality is destroyed by the frequency-selective radio channel resulting in multiple access interferences (MAI), which reduces the system performance.

Bibliography

[1] R. Grünheid, H. Rohling, "Performance comparison of different multiple access schemes for the downlink of an ofdm communication system," In *Proc. IEEE VTC*, 1997.

[2] T. May, H. Rohling, V. Engels, "Performance Analysis of 64-DAPSK and 64-QAM Modulated OFDM Signals," *IEEE Transactions on Communications*, Vol. 46, No. 2, Feb. 1998 pp. 182 - 190.

[3] H. Rohling, R. Grünheid, T. May, K. Brüninghaus, "Digital Amplitude Modulation," *Encyclopedia of Electrical and Electronics Engineering*, John Wiley and Sons 1999.

[4] H. Rohling, T. May, K. Brüninghaus, R. Grünheid, "Broad-Band OFDM Radio Transmission for Multimedia Applications," *Proceedings of the IEEE*, Vol. 87, No. 10, Oct 1999.

[5] B. Chen, R. Grünheid, H. Rohling, "Scheduling Policies for Joint Optimization of DLC and Physical Layer in Mobile Communication Systems," *Proc. of the 13th IEEE International Symposium on Personal, Indoor and Mobile Radio Communications (PIMRC 2002)*, Lisbon, Portugal, September 2002.

[6] R. Grünheid, H. Rohling, J. Ran, E. Bolinth, R. Kern, "Robust Channel Estimation in Wireless LANs for Mobile Environments," *Proc. VTC 2002 Fall*, Vancouver, September 2002.

[7] E. Costa, H. Haas, E. Schulz, D. Galda, H. Rohling, "A low complexity transmitter structure for the OFDM-FDMA uplink," In *Proc. IEEE VTC*, 2002.

[8] M. Lampe, T. Giebel, H. Rohling, W. Zirwas, "PER Prediction for PHY Mode Selection in OFDM Systems," *Proc. Globecom 2003*, San Francisco, USA, December 2003.

[9] M. Lampe, T. Giebel, H. Rohling, W. Zirwas, "Signalling-Free Detection of Adaptive Modulation in OFDM Systems," *Proc. PIMRC 2003*, Peking, China, September 2003.

[10] H. Rohling, D. Galda, "An OFDM based Cellular Single Frequency Communication Network," *Proceedings of WWRF*, Beijing, February 23-24, 2004.

[11] H. Rohling, R. Grünheid, "Cross Layer Considerations for an Adaptive OFDM-Based Wireless Communication System," *Wireless Personal Communications*, pp. 43 – 57, Springer 2005.

[12] M. Stemick, H. Rohling, "OFDM-FDMA Scheme for the Uplink of a Mobile Communication System," *Wireless Personal Communications* (2006), June 2006.

[13] R. Grünheid, H. Rohling, K. Brüninghaus, U. Schwark, "Self-Organised Beamforming and Opportunistic Scheduling in an OFDM-based Cellular Network," *Proc. VTC 2006 Spring*, Melbourne.

[14] H. Busche, A. Vanaev, H. Rohling, "SVD based MIMO Precoding on Equalization Schemes for Realistic Channel Estimation Procedures," *Frequenz* 61 (2007), 7-8, July/August 2007.

[15] C. Fellenberg, H. Rohling, "Spatial Diversity and Channel Coding Gain in a noncoherent MIMO-OFDM transmission system," *Frequenz Journal of RF-Engineering and Telecommunications*, November/December 2008.

[16] C. Fellenberg, H. Rohling, "Quadrature Amplitude Modulation for Differential Space-Time Block Codes," *Wireless Personal Communications*, vol. 50, no. 2, pp.247-255, July (II) 2009.

[17] A. Ruiz A. Peled, "Frequency domain data transmission using reduced computational complexity algorithms," In *Proc. IEEE ICASSP*, pp. 964-967, 1980.

[18] M. Aldinger, "Multicarrier COFDM scheme in high bitrate radio local area networks," In *Proc. Of Wireless Computer Networks*, 1994.

[19] P. A. Bello, "Characterization of randomly time-variant linear channels," *IEEE Transactions on Communications*, Dec. 1964.

[20] R.W. Chang, "Synthesis of band-limited orthogonal signals for multichannel data transmission," *Bell. Syst. Technical Journal*, Vol. 45:pp. 1775–1796, 1966.

[21] L. Hanzo et al, *OFDM and MC-CDMA for Broadband Multi-User Communications, WLANs and Broadcasting*, Wiley, 2003.

[22] S. Kaiser, *Multi-Carrier CDMA Mobile Radio Systems - Analysis and Optimization of Detection, Decoding and Channel Estimation*, VDI-Verlag, 1998.

[23] K. J. R. Liu, A. K. Sadek, W. Su, A. Kwasinski, *Cooperative Communications and Networking*, Cambridge University Press, 2009.

[24] H. Schulze, C. Lueders, *Theory and Applications of OFDM and CDMA: Wideband Wireless Communications*, John Wiley and Sons, 2005.

[25] M. Pätzold, *Mobile Fading Channels*, Wiley, 2002.

[26] B.R. Saltzberg, "Performance of an efficient parallel data transmission system," *IEEE Transactions on Communications*, 1967.

[27] P. M. Ebert, S.B. Weinstein, "Data Transmission by Frequency-Division Multiplexing Using the Discrete Fourier Transform," *IEEE Transactions on Communication Technology*, Vol. Com-19, no. 5, pp. 628–634, 1971.

2 Channel Modeling

M. Narandžić, A. Hong, W. Kotterman, R. S. Thomä, Ilmenau University of Technology, Germany
L. Reichardt, T. Fügen, T. Zwick, Karlsruhe Institute of Technology (KIT), Germany

In the OFDM signalling concept, the wide-band radio-communication channel is effectively utilized as a collection of narrow-band channels. Basic system parameters like the number of subcarriers N and symbol duration T are selected to mitigate the key channel impairments: Inter-Symbol-Interference (ISI) induced by frequency selectivity and loss of subcarrier orthogonality due to time selectivity [51]. Therefore, a proper channel model is required both for system design and for performance evaluation. Additionally, when Channel-State-Information (CSI) is available during system operation, transmission characteristics, such as the signal constellation or the allocated power, could be adaptively adjusted at transmitter per subcarrier in order to maximize total throughput. Further improvements of the spectral efficiency could be obtained by simultaneous transmission and/or reception from/by multiple antenna elements. Additionally to time and frequency, this concept known as MIMO (*Multiple-Input-Multiple-Output*) exploits the spatial propagation dimension or, more specific, multiplicity of energy propagation paths. Since the reachable spectral efficiency is tightly related to the signal correlation across the antenna array [15], the proper representation of correlation levels becomes essential for the analysis of MIMO systems. In order to obtain an antenna-independent representation of the channel that implicitly comprises correlation properties, *geometry-based* models are generally used.

In this chapter, the necessary concepts for representation of the multidimensional radio-channel are summarized. Data collected during multidimensional channel sounding and post-processed by high-resolution parameter estimation algorithms offer the most detailed insight into radio-propagation mechanisms. In that way, joint space-time-frequency representations being consistent with measurements can be obtained (Section 2.1). On the other hand, when an appropriate description of the EM environment is available (in the form of databases defining geometry and material properties), the EM field could be predicted by use of the Geometrical or Uniform Theory of Diffraction (GTD/UTD), as explained in Section 2.2. Note that for tuning and verification of ray tracing/launching procedures, sounding experiments are still required. For system design and performance evaluation, site-independent modeling with lower complexity is preferred. For this purpose, stochastic characterization of different radio-environment classes could be combined with geometry based propagation aspects. This results in the class of *Geometry-based Stochastic* (GbS) channel models that are described in Section 2.3. In order to properly reproduce space-time channel evolution, this class of empirical models uses stochastic characterization of

H. Rohling (ed.), *OFDM: Concepts for Future Communication Systems*,
Signals and Communication Technology, DOI: 10.1007/978-3-642-17496-4_2,

Large-Scale Parameters (LSPs) as explained in Subsection 2.3.1. The radio channels corresponding to specific propagation/deployment scenarios are given as examples of listed general modeling classes. The characteristics of a radio-link that is established between vehicle and stationary or moving objects, analyzed by ray-tracing tools, are presented in Subsection 2.2.2. Subsection 2.3.2 introduces GbS model for relay-links, based on a stochastic representation of channel LSPs. Specific aspects of spatially-distributed transmission corresponding to cooperative downlink are given in Subsection 2.3.3. Due to some inherent weaknesses regarding the representation of spatio-temporal evolution, so called *analytical* models were not considered.

2.1 Joint Space-Time-Frequency Representation

The multidimensional channel transfer function can be equivalently expressed using the system functions [4] in either *faded domains* ($\mathbf{r}$-space, t-time, f-frequency) or *resolved domains* ($\boldsymbol{\Omega}$-directions, ν-Doppler shift, τ-delay) [27]. The physical models being discussed here only use resolved domains for channel analysis and synthesis, equivalent to characterization by constituent Multi-Path-Components (MPCs). Then, the point-to-point propagation channel (i.e. link) is represented as an antenna response to a set of MPCs (usually conveniently grouped into *clusters* [9], [20]):

$$H(\mathbf{r}_{Tx}, \mathbf{r}_{Rx}, t, f) = \sum_i \mathbf{F}^T_{Tx}(\Omega^{Tx}_i)\boldsymbol{\alpha}_i \mathbf{F}_{Rx}(\Omega^{Rx}_i) e^{j2\pi(\tau_i f + \nu_i t)}. \tag{2.1}$$

Interaction between antennas and MPCs is through the complex, polarimetic antenna response

$$\mathbf{F}_\mathbf{r}(\Omega) = [F_\theta(\Omega) F_\varphi(\Omega)]^T \cdot e^{j\mathbf{k}(\Omega)(\mathbf{r}-\mathbf{r_0})}, \tag{2.2}$$

where F_θ and F_φ represent projections onto corresponding unitary vectors of the spherical coordinate system. The exponential term in (2.2) defines the phase shift of MPCs coming from direction Ω w.r.t. the phase center at $\mathbf{r_0}$. The given representation covers all spatial degrees of freedom: transversal movement and antenna rotation, as well as any array geometry for the MIMO case. A single MPC corresponds to a homogeneous plane wave that within narrow frequency bandwidth can be characterized by the following parameters:

$$\mathbf{p} = [\Omega^{Tx}, \Omega^{Rx}, \boldsymbol{\alpha}, \tau, \nu], \tag{2.3}$$

where $\boldsymbol{\Omega}_T$ and $\boldsymbol{\Omega}_R$ describe Directions-of-Departure (DoD) and Arrival (DoA), respectively. Due to the inability (in the general case) to represent the Power-Directional-Spectrum as a product of marginal spectra on departure and arrival, joint characterization of DoD and DoA is to be used - as suggested by the *double-directional* modeling concept [50], [38]. The complex 2-by-2 matrix $\boldsymbol{\alpha} \in \mathcal{C}^{2\times 2}$ is used to jointly describe MPC magnitude, MPC phase, and cross-polarization effects.

The necessary parameters, normally for a large number of MPCs, could be estimated from appropriate multidimensional channel sounding data. These data are gathered during wide-band measurement experiments with specially designed antenna arrays and real-time channel sounding devices [49], [26], [7], [8].

2.1.1 Multidimensional Channel Sounding

In a broader sense, multidimensional sounding comprises investigations into the spatio-temporal structure of a radio channel, aiming to resolve not only the temporal delay of incoming waves (signal components) but also their angular directions at transmission and at reception as well as their polarizations. Especially the combination of angular resolution and polarimetric state is potentially very costly and laborious to record and process at its full extent. Many antenna elements are necessary for high-resolution results, both to fully cover the angular domain and to create the required apertures. Providing coverage in a particular direction demands that antenna elements still have sufficient sensitivity in that direction. Aperture, required for resolution, means that (sensitive) elements are to be spread over space. A popular shortcut like using single-polarized antenna elements leads to biased results [32]. Additionally, for accurate parameter estimation, calibration of every antenna element in the measurement array is mandatory, providing complex radiation pattern $\mathbf{F}_{\mathbf{r}}(\Omega)$ of (2.2), required to estimate parameters of resolved MPCs in (2.1), in order to relate these to observed faded dimensions. Restricting Ω to the azimuthal cut, another popular saving, also means to risk grossly distorted estimates [32].

Characterization of propagation delay requires nearly instantaneous measurements, meaning the time needed for a measurement over bandwidth or over the full delay span should be considerably shorter than the time it takes the channel to change. Pseudo-random noise sequences, multi-sine tone bursts, or fast frequency-sweeps can be used, each with its own advantages and disadvantages. If the repetition rate is high enough, also the Doppler spectrum or time variability can be determined without aliasing. The temporal and spatial dimension have to be measured jointly, but measuring all antenna elements simultaneously and all transmit-receive combinations in parallel is deemed technically infeasible (exception: the 16×4 parallel sounding in [41]). Therefore, the antenna combinations are multiplexed, making use of one and the same temporal sounding unit. The multiplexing units themselves are still a technical challenge, due to requirements on switching speed, damping losses, feed-through, frequency transfer, delay, and power handling (especially on the transmit side). Seen these imperfections, the multiplexing units should be calibrated too. Synchronization of transmit and receive side, which are often too far apart for synchronization through a cable connection, requires two free-running clocks of very high stability; typically Rubidium or Cesium standards.

So, what is needed? A dedicated channel sounder with calibrated dedicated multiplexing equipment both at transmit and receive side, calibrated dedicated antennas, stable (atomic) clocks, and a high-speed data logger. As an example for the latter, the COST2100 urban reference scenario "Ilmenau" had to be measured at a modest trawling speed of 3 m/s, in order not to exceed the maximum sustained data transfer rate of 1.2 Gbit/s, the product of snapshot rate, number of transmit-receive combinations, impulse response length, and number of bits per time sample [48].

2.1.2 Extraction of Parameters for Dominant MPCs

The estimation procedure of MPC parameters from channel sounding data requires the use of so called *high-resolution algorithms* , like, e.g., Maximum Likelihood

Estimation [60], ESPRIT (Estimation of Signal Parameters via Rotational Invariance Techniques) [44], [52], SAGE (Space-Alternating Generalized Expectation-maximization) [14], or RIMAX [53], [46], [32]. An alternative to measurements is the extraction of model parameters by means of ray tracing (Section 2.2).

Both methods could provide reliable (reality matching) parameters for only a limited number of MPCs: the measurement-based estimation due to the limited number of space-time-frequency observations [46] and the limited precision of antenna calibration [32], and ray-tracing due to the limited precision of the radio-environment model. The remaining part, usually associated with diffuse scattering, is typically characterized by stochastic means both during parameter estimation [46], [32] and ray generation [10].

2.2 Deterministic Modeling

Deterministic models are used for site-specific channel modeling; they consist of an environment model and a wave propagation model. The environment model describes position, geometry, material composition and surface properties of the wave propagation relevant objects and obstacles (e.g. trees, houses, vehicles, walls, etc.). The well-known Maxwell equations [3] always form the basis for all investigations of electromagnetic fields. In practical applications an analytic solution of the Maxwell equations, due to the computation time, is not possible. Also numeric approximation methods, like, e.g., the Parabolic Equation Method (PEM) or the Finite Difference Time Domain Method (FDTD) [21], [57], [59] fail for efficiency reasons with problems, which are larger than some wavelengths in the examined frequency range. Substantially less complexity and computing time is achievable with geometric-optical models [30], [5], [56], [2], [25], [35], [17], [16]. These models are based on iterated approaches, which use the border behavior of electromagnetic fields for high frequencies [37]. The use of these procedures makes substantial simplifications of the description of the wave propagation possible. This allows to compute electrically very large problems very efficient and exactly.

2.2.1 Relevant GTD/UTD Aspects

The modern geometrical optics (GO) is an important representative of these iterated procedures, and it forms the basis for the uniform geometrical theory of diffraction (UTD). The validity of the GO does not alone depend on the frequency. A further condition is, that the scattering objects contained in the propagation vicinity are large in relation to the wavelength. Additionally the surface texture is not allowed to change over a wavelength. Further the material properties of the propagation medium must be constant within the range of a wavelength [37]. This is fulfilled in good approximation for frequencies above 1 GHz.

Due to its flexibility and accuracy geometric-optical models are already today in use. They are able to calculate, a place-dependent prognosis of the full-polarimetric field strength and/or receiving power in the regarded propagation area. Besides this a complete narrow- and wide-band description of the mobile channel is possible, why they find increased use in system simulations [12], [36].

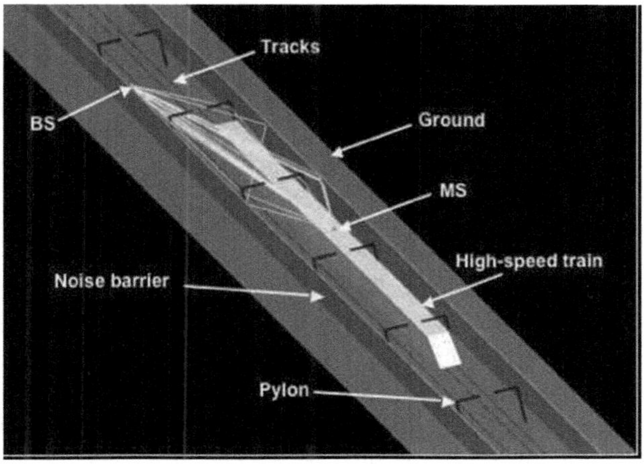

Figure 2.1: With Ray-tracing calculated wave propagation in a high-speed train scenario [29].

2.2.2 Vechicle2X Channel Modeling

The realistic channel representation at *very high participant velocities* in combination with high data rate transmission and MIMO-OFDM techniques can be obtained by a ray-optical description of the multi-path propagation. In the context of the key program *TakeOFDM* of the Deutsche Forschungsgesellschaft (DFG) such a channel model for high-speed train communication was developed (Fig. 2.1) [29]. A detailed description of the vehicle's vicinity is essential for a proper modeling of the wave propagation. This includes the track, on which the vehicles are driving, and the environment adjacent to the track. E.g. in the surrounding of train tracks possible objects are noise barriers, trees, signs, bridges, and pylons, whereas in urban or suburban areas buildings are more probable. A new map generator has been developed for this ray-tracing simulator. With this, it is possible to import standard CAD (Computer Aided Design) data with the STL (Standard Triangulation Language) format. In the map generator the electrical parameters like the permittivity ϵ_r, permeability μ_r and the standard deviation of the surface roughness σ are assigned to the objects and it is possible to shift, scale or rotate them. Furthermore it is possible to define velocities for the objects to create a time series of snapshots of the scenario to simulate the time-variant behavior of the channel. For the channel simulations each object can be equipped with a receiver and a transmitter. The position of the corresponding antennas as well as the antenna pattern and orientation can be chosen arbitrarily. An accurate description of the multi-path wave propagation in the aforementioned scenarios is required to produce realistic time series of Channel Impulse Responses (CIRs).

At the Institut für Hochfrequenztechnik und Elektronik a three-dimensional ray-tracing algorithm has been developed and implemented [36]. The results of the applied ray tracing algorithms have been verified by measurements in different sce-

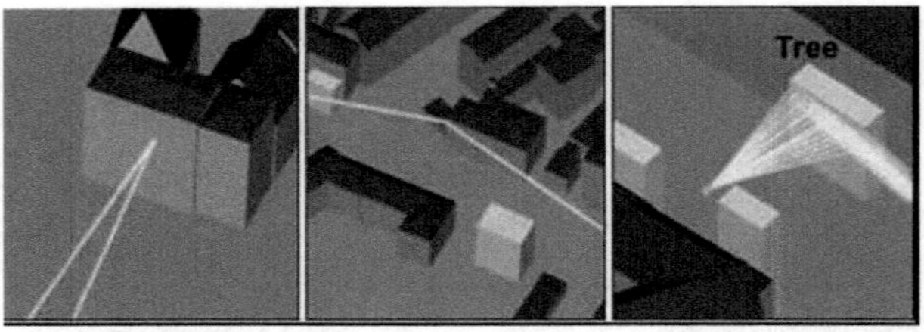

Figure 2.2: Considered multi-path effects: reflection (left), diffraction (middle), and scattering (right).

narios and have shown to reach a very high accuracy [18], [29]. Ray-optics are based on the assumption, that the wavelength is small compared to the dimensions of the modeled objects in the simulation scenario. If this is the case, different multi-path components, characterized by different types of propagation phenomena (e.g. reflection, diffraction, scattering (Fig. 2.2)), can be considered. Each multi-path is represented by a ray, which may consecutively experience several different propagation phenomena. As propagation phenomena multiple reflections, multiple diffractions and single scattering are taken into account. Mixed propagation paths containing reflections and diffractions are possible as well. The modified Fresnel reflection coefficients, which account for slightly rough surfaces, are used to model the reflections. Diffractions are described by the Uniform Theory of Diffraction (UTD) and the corresponding coefficients for wedge diffraction. To describe scattering, e.g., from trees, the surface of scattering objects is subdivided into small squared tiles. Depending on the energy, which is incident on the surface of the objects, each tile gives rise to a Lambertian scattering source. The adjustment of ray-optical models to the reality takes place via the exact modeling of the environment and the physical wave propagation. This means that measurements are not needed for the alignment of model parameters but only for the verification of the model. Investigations for the accuracy of deterministic channel models are subject of numerous publications [30], [28], [24], [11], [33], [2], [45], [47], [36], [29].

A realistic evaluation of the behavior of a communication system is however only possible if a multiplicity of spatial scanning points are used in the system simulation. Due to the complexity of geometric-optical models a substantial computing and expenditure time must be taken into account. The main advantage in contrast to other channel models is that spatially-colored multi-user interference, one of the most limiting factors for the achievable performance in multi-user MIMO-systems, is inherently considered [19].

2.3 Stochastic Driving of Multi-Path Model

When designing a wireless transmission system, it is useful to evaluate its performance over at least a minimum number of channel realizations. These could be generated by deterministic propagation models described in the previous section, however, their high computational complexity prohibits the intensive link or system level simulations required during system design. Thus, procedures with a lower computational complexity that could emulate a whole class of radio-propagation environments (i.e. *propagation scenario*) are preferred. These requirements have led to the *Geometry-based Stochastic* (GbS) channel models where generated multipath components are not directly related to any particular (or very detailed) radio-environment. Instead, the channel realizations are determined as realizations of a multidimensional random process that characterizes all aspect of physical plane-wave propagation.

The stochastic generation of multipath can be done in several different forms. We would distinguish two classes according to the use of the scattering (or interacting) objects during the physical model synthesis. E.g. it is possible to place *interacting objects* in a 2D/3D coordinating system, and to perform their abstraction in the form of multipath clusters as in the COST 273 model [8]. By assigning *visibility regions* [1] to each of the clusters, a simplified ray-tracing engine is obtained. The randomness in this approach is attained by random selection of visibility regions and the intra-cluster structure. An alternative would be to fully remove scatterers from the model synthesis. In this case multipath components are no longer related to particular scatterers, but are generated in the so called *parametric domain* instead. This term relates to the parameters of multipath components as given by (2.3). Typical representatives are the 3GPP Spatial-Channel-Model [54], the channel model developed in the WINNER project [31], and the reference model for evaluation of IMT-Advanced radio interface technologies [34].

2.3.1 Usage of the Large-Scale Parameters for Channel Characterization

The consequence of the environment abstraction introduced by parametric domain synthesis is that the evolution of a space-time model can not be implicitly given by relative distances of scattering objects. Instead, the channel dynamic is represented by correlated realizations (over space-time) of so called *Large-Scale Parameters* (LSPs). The term LSPs is used to denote a group of channel parameters that typically experience notable change only over distances exceeding several wavelengths. The relative MPC positions in parametric space (2.3) define the *MPC structure* , that can be described by the power distribution over resolved channel dimensions. Since (dis)appearance of a small portion of MPCs have minor effect on the marginal (e.g. delay and directional) spread parameters, they could be exploited for abstraction of large-scale channel behavior. The main role of the LSPs is, therefore, to describe the joint distribution of the MPC power over different domains (direction, polarization, delay, Doppler, etc.) as observed at the same instant and additionally to describe space-time channel evolution. The set of relevant LSPs established within the SCM/WINNER models is listed in Table 2.1[1].

Using the concept of correlated random LSPs it is easy to repeat stochastic properties of parameters being observed during channel sounding and therefore this enables the straightforward scenario-based representation. By performing the measurement

[1] Please, note that the Doppler shift is not explicitly parametrized, but for a given velocity vector it will be implicitly determined by the directions of departure and arrival

Table 2.1: Large-Scale Parameters of SCM/WINNER model.

LSP Name	Acronym	Power distribution...
Shadow Fading	SF	around mean transmission loss
Delay Spread	DS	over delay domain
Directional (Angular) Spread	AS	over angular domain:
		- at departure and arrival
		- over azimuth and elevation
Narrowband K-factor	K	btw. LoS and NLoS clusters
Cross polarization Ratio	XPR	btw. co- and cross-polar MPCs

experiment with particular antenna deployment in a given scenario it is possible to define empirical multipath model. This process is illustrated in Fig. 2.3.

a Multidimensional channel sounding,

b High-resolution estimation of joint MPC parameters,

c Statistic characterization of LSPs and their space-time dependencies,

d Guided random positioning of MPC in parameter space, according to random realization of multivariate LSP process,

e Determination of antenna array response to given multi-path structure.

Figure 2.3: Generation of empirical, scenario-based multipath channel model

LSPs Viewed as Correlated Multivariate Random Process

General methods for generation of random variables (RVs) with targeted first-order (i.e. probability distribution) and second-order (auto-correlation over time) statistics have been suggested in literature [6], [13]. These methods reproduce statistical behavior of a random process w.r.t. its realization over time, by using a transformation of the Gaussian autoregressive process. In order to avoid complex matching of correlations between original and transformed domain the LSPs are first mapped into new variables (*transformed LSPs*) having Gaussian distributions and the subsequent analysis of LSP inter-dependence is performed in transformed domain [54], [31]. For LSP P_i with cumulative distribution function (cdf) F_i, the necessary mapping[2] could be determined in the form of $P_i = F_i^{-1}(\Phi(Q_i))$, where Q_i designates the transformed

[2]The solution of an inverse problem, [43]

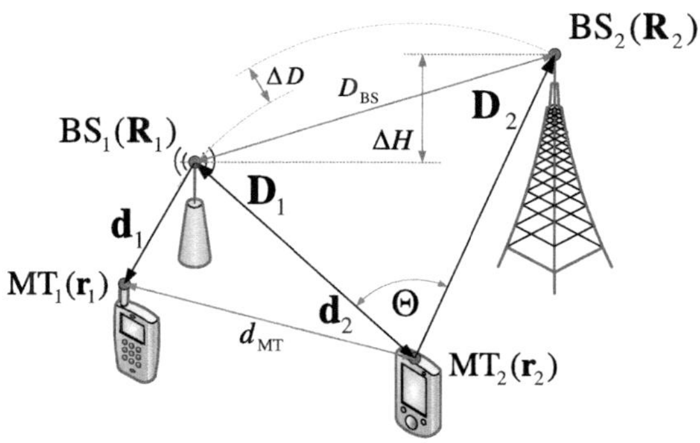

Figure 2.4: System layout defined by positions of communication terminals.

LSP with normal cdf Φ. Using a linear transformation

$$\boldsymbol{Q} = \boldsymbol{C\xi} + \boldsymbol{b} \tag{2.4}$$

of the standard multivariate normal process $\boldsymbol{\xi}$ with distribution $\mathcal{N}_{\xi}(\mathbf{0}_{M\times 1}, \mathbf{I}_{M\times M})$ a process $\mathbf{Q} = [Q_1, Q_2, \ldots, Q_M]^T$ with the targeted covariance matrix $\mathbf{CC^T}$ could be easily reproduced.

Dependence of Covariance Matrix on System Layout

The channel model of a system with K coexisting links should generate $K \cdot M$ correlated LSPs, where M is the number of LSP's per each link. The corresponding full covariance matrix $\mathbf{CC^T}$ would have, for each time instant, size $M \cdot K \times M \cdot K$. This matrix characterizes the correlations between all LSPs describing all coexisting links, however its proper synthesis is not trivial due to strong dependence upon the system layout. The problem can be addressed by proper decomposition of the transformation matrix, $\mathbf{C}$, according to *link-level*[3] and *layout-level* correlations . The link-level correlations correspond to cross-correlations of the LSPs characterizing the same link, and according to the proposed simplification they will not change over space-time. On the other hand, the layout-level correlations explicitly depend on the relative position of the terminals at both link ends. Depending on at which link's end a terminal displacement occurs, it is possible to distinguish *intra-site* and *inter-site* correlations (Fig. 2.4). Since two different links with single common end could not simultaneously exhibit both correlation types, the intra-site ($\mathbf{R}_i = \mathbf{R}_j$) and inter-site ($\mathbf{r}_i = \mathbf{r}_j$) correlations could be conveniently combined for given system layout. These correlations are typically expressed in the form of layout-dependent correlation coefficient $\rho_{XY}(\mathbf{L}) = \frac{\sigma_{XY}}{\sqrt{\sigma_{XX}\sigma_{YY}}}$, where $\sigma_{XY} = E[(X - E[X])(Y - E[Y])]$ denotes covariance between LSPs X and Y. The geometry parameters $\mathbf{L}$ are determined from the vectors defining the relative position of mobile terminals $(\mathbf{d}_i, \mathbf{d}_j) = (\mathbf{r}_i -$

[3]A single link realization, when compared to itself, could be considered as a special case in system layout, where there is neither displacement of the mobile terminal nor of the base station.

$\mathbf{R}, \mathbf{r}_j - \mathbf{R}$) or base stations $(\mathbf{D}_i, \mathbf{D}_j) = (\mathbf{R}_i - \mathbf{r}, \mathbf{R}_j - \mathbf{r})$ w.r.t. to single common position (Fig. 2.4). The set of relevant parameters $\mathbf{L}$ for intra-site correlations could be reduced to Euclidean distance between mobile terminals $d_{\mathrm{MT}} = ||\mathbf{r}_i - \mathbf{r}_j||$ [31]. The characterization of inter-site correlations , however, requires a more complex parameter space $\mathbf{L} = [\Theta, \Delta D, D_{\mathrm{BS}}, \Delta H]^T$ [40] being defined at Fig. 2.4.

2.3.2 Relaying

In wireless communication systems, the nodes with a relaying capability are integrated into conventional networks in order to provide a ubiquitous coverage with high data rates, especially in the areas with a high shadowing [42]. In relay networks, intermediate Relay-Stations (RSs) are introduced into the communication between a base station and a mobile terminal. If station labeled as BS_1 has relay functionality, than the Fig. 2.4 can be interpreted as an example of the basic three-station structured relay network [55]. The purpose of intermediate RS (BS_1) would be to forward received signals from BS_2 toward mobile terminal MT_1, and vice versa [58]. The introduction of intermediate RSs results in a meshed topology of relay networks, and brings new challenges in channel modeling. Moreover, characterizing and modeling of the relationship between meshed links, is one of the most crucial points in the channel modeling of relay networks . The correlation properties between meshed links can be captured in the form of the intra- and inter-site correlations of large scale parameters [22], as discussed in previous subsection. The observed correlation properties for relay measurements in Ilmenau inner city, could be summarized as follows [23]:

1. The de-correlation distance (used to characterize intra-site correlations) of SF, DS as well as XPR decrease with a reduced BS height. This confirms that even the intra-site correlation could exhibit more complex layout dependence.
2. The inter-site correlation of LSPs is high when two BSs/RSs are near to each other but a MT is far away from both.
3. The larger the difference in the height of two BSs/RSs, the lower the inter-site correlation.
4. The inter-site correlation decreases for larger angular separation of BSs, Θ. Figure 2.5 shows the experimental results for inter-site correlation coefficient of XPR. Note, that measured correlation does not decrease monotonically neither with angular nor distance separation ΔD.

2.3.3 Cooperative Downlink

One of the main goals behind the physical modeling is to make the channel representation as independent from the system aspects as possible. However, when characterizing the cooperative downlink (e.g. $(-\mathbf{D}_1, -\mathbf{D}_2)$ from Fig. 2.4) it is not possible to disregard the influence of the receiver's limited dynamic range on *perceived* LSPs of the cooperative links [39]. Namely, the perception of power spreading expressed

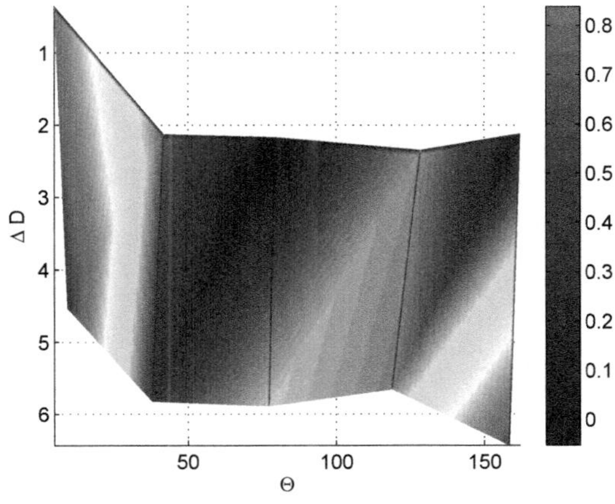

Figure 2.5: Dependence of XPR correlation coefficient from network layout parameters, $\rho(\Theta, \Delta D)$.

by DS or AS depends on the effective dynamic range of the particular radio-link, as shown in Fig. 2.6. Consequently, the characterization of inter-site correlations between cooperative links requires previous adjustment of effective dynamic ranges[4]. These will depend on the total power received from all cooperative links, and in general they will be lower for the weaker links. If *peak power level differences*, ΔP are statistically characterized for particular multi-link configurations and targeted scenario, they can be included into model as an additional LSP [39]. During model synthesis the randomly generated values of ΔP will define the effective dynamic ranges, and the parameters of spread-related LSP distributions should be modified accordingly.

One of the implications of *receiver perceived* channel representation is that reciprocity (normally assumed in channel modeling) will not be preserved. In a mesh network, the link between two communication sinks, each having other spatially distributed links too, will be experienced differently by the two sinks because of the unequal influence of the additional (distributed) links at both sides.

[4]When measured separately each link will be characterized according to the dynamic range of the measurement equipment, what is not relevant for reception of simultaneous signals.

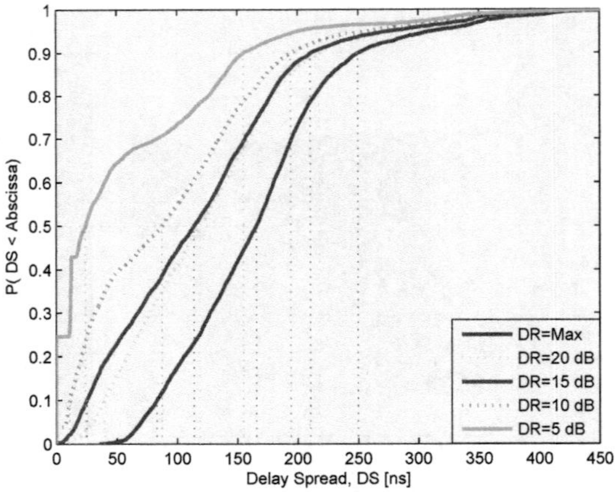

Figure 2.6: Dependence of DS empirical distribution from effective Dynamic-Range (DR).

Bibliography

[1] H. Asplund, A. A. Glazunov, A. F. Molisch, K. I. Pedersen, and M. Steinbauer, "The COST 259 Directional Channel Model – Part II: Macrocells," *IEEE Transactions on Wireless Communications*, Vol. 5, No. 12, pp. 3434-3450, December 2006.

[2] G.E. Athanasiadou, A.R. Nix, and J.P. McGeehan, "A Microcellular Ray-Tracing Propagation Model and Evaluation of its Narrow-Band and Wide-Band Predictions," *IEEE Journal on Selected Areas in Communications*, Vol. 18, pp. 322-335, March 2000.

[3] C.A. Balanis, *Advanced Engineering Electromagnetics*, John Wiley & Sons, New York, 1989.

[4] P. A. Bello, "Characterization of Randomly Time-Variant Linear Channels," *IEEE Trans. Commun. Sys.*, CS-I Vol. 1, No. 4, pp. 360-393, Dec. 1963.

[5] H.L. Bertoni, L.R. Honcharenko, L.R. Macel, and H.H. Xia, "UHF Propagation Prediction for Wireless Personal Communications," *Proceedings of the IEEE*, Vol. 82, pp. 1333-1359, September 1994.

[6] C. Cario and B. Nelson, "Auto-Regressive to Anything: Time Series Input Processes for Simulation," *Operations Research Letters*, Vol. 19, pp. 51-58, 1996.

[7] L.M. Correia, editor, *Wireless Flexible Personalized Communications* -

COST259: European Commission in Mobile Radio Research, Wiley, New York, 2001.

[8] L.M. Correia, editor, *Mobile Broadband Multimedia Networks - Techniques, Models and Tools for 4G, COST 273 Final Report*, Elsevier, Oxford, 2006.

[9] N. Czink, *The Random-Cluster Model*, PhD thesis, Technische Universität Wien, 2007.

[10] D. Didascalou, M. Döttling, N. Geng, and W. Wiesbeck, "An Approach to Include Stochastic Rough Surface Scattering into Deterministic Ray-Optical Wave Propagation Modeling," *IEEE Transactions on Antennas and Propagation*, Vol. 51, No. 7, pp. 1508–1515, July 2003.

[11] D.L. Didascalou, *Ray-Optical Wave Propagation in Arbitrarily Shaped Tunnels*, PhD thesis, Forschungsberichte aus dem Institut für Höchstfrequenztechnik und Elektronik der Universität Karlsruhe (TH), 2000.

[12] M.W. Döttling, *Strahlenoptisches Wellenausbreitungsmodell und Systemstudien für den Sattelitenmobilfunk*, PhD thesis, Forschungsberichte aus dem Institut für Höchstfrequenztechnik und Elektronik der Universität Karlsruhe (TH), Karlsruhe, Germany, 2000.

[13] S. Fitzgerald, J. Placeb, and A. van de Liefvoort, "Generating correlated matrix exponential random variables," *Advances in Engineering Software*, Vol. 37, pp. 75-84, 2006.

[14] B.H. Fleury, M. Tschudin, R. Heddergott, D. Dahlhaus, and K. Ingeman Pedersen, "Channel Parameter Estimation in Mobile Radio Environments using the SAGE Algorithm," *IEEE Journal on Selected Areas in Communications*, Vol. 17, No. 3, pp. 434-450, March 1999.

[15] G. J. Forschini and M. J. Gans, "On limits of wireless communications in a fading environment when using multiple antennas," *Wireless Personal Communications*, Vol. 6, No. 3, pp. 311-335, 1998.

[16] T. Fügen, *Richtungsaufgelöste Kanalmodellierung und Systemstudien für Mehrantennensysteme in urbanen Gebieten* , PhD thesis, Universität Karlsruhe, Karlsruhe, Germany, 2009.

[17] T. Fügen, J. Maurer, T. Kayser, and W. Wiesbeck, "Capability of 3D Ray Tracing for Defining Parameter Sets for the Specification of Future Mobile Communications Systems," *IEEE Transactions on Antennas and Propagation, Special Issue on Wireless Communications*, Vol. 54, No. 11, pp. 3125-3137, November 2006.

[18] T. Fügen, J. Maurer, T. Kayser, and W. Wiesbeck, "Verification of 3D Raytracing with Non-Directional and Directional Measurements in Urban Macrocellular Environments," in *Proceedings of the 63th IEEE Vehicular Technology Conference, VTC2006-Spring*, Melbourne, Australia, May 2006.

[19] T. Fügen, J. Maurer, C. Kuhnert, and W. Wiesbeck, "A Modelling Approach for Multiuser MIMO Systems Including Spatially-Colored Interference," in *Proceedings of the IEEE Global Telecommunications Conference, GLOBECOM 2004*, Vol. 2, pp. 938-942, Dallas, TX, December 2004.

[20] T. Fügen, J. Maurer, W. Sörgel, and W. Wiesbeck, Characterization of Multipath Clusters with Ray-Tracing in Urban MIMO Propagation Environments at 2 GHz. in *IEEE International Symposium on Antennas and Propagation*, Vol. 3B, pp. 410-413, Washington DC, USA, July 2005.

[21] N. Geng, *Modellierung der Ausbreitung elektromagnetischer Wellen in Funksystemen durch Lösung der parabolischen Approximation der Helmholtz-Gleichung*, PhD thesis, Forschungsberichte aus dem Institut für Höchstfrequenztechnik und Elektronik der Universität Karlsruhe (TH), 1996.

[22] A. Hong, M. Narandžić, C. Schneider, and R. S. Thomä, "Estimation of the Correlation Properties of Large Scale Parameters from Measurement Data," in *IEEE International Symposium on Personal, Indoor and Mobile Radio Communications (PIMRC)*, Athens, Greece, Sep. 2007.

[23] A. Hong and R. S. Thomä, "Experimental Study on the Impact of the Base Station Height on the Channel Parameters," in *ITG Workshop on Smart Antennas*, Berlin, Germany, Feb. 2009.

[24] R. Hoppe, G. Wölfle, and F.M. Landstorfer, "Accelerated Ray Optical Propagation Modeling for the Planning of Wireless Communication Networks," in *Proceedings of the IEEE Radio and Wireless Conference, RAWCON 99*, pp. 159-162, Denver, CO, USA, August 1999.

[25] M.F. Iskander and Z. Yun, "Propagation Prediction Models for Wireless Communication Systems,", *IEEE Transactions on Microwave Theory and Techniques*, Vol. 50, pp. 662-673, March 2002.

[26] M.A. Jensen and J.W. Wallace, "MIMO Wireless Channel Modeling and Experimental Characterization," in A.B. Gershman and N.D. Sidiropoulos, editors, *Space-Time Signal Processing for MIMO Communications*, pp. 1-39. John Wiley & Sons, 2005.

[27] R. Kattenbach, "Statistical Modeling of Small-Scale Fading in Directional Radio Channels," *IEEE J. Select. Areas Communications*, Vol. 20, pp. 584-592, Apr. 2002.

[28] Seong-Cheol Kim, B.J. Jr. Guarino, T.M. III Willis, V. Erceg, S.J. Fortune, R.A. Valenzuela, L.W. Thomas, J. Ling, and J.D. Moore, "Radio Propagation Measurements and Prediction Using Three-Dimensional Ray Tracing in Urban Environments at 908 MHz and 1.9 GHz," *IEEE Transactions on Vehicular Technology*, Vol. 48, No. 3, pp. 931-946, May 1999.

[29] S. Knörzer, *Funkkanalmodellierung für OFDM-Kommunikation bei Hochgeschwindigkeitszügen*, PhD thesis, Universität Karlsruhe, Karlsruhe, Germany, 2009.

[30] T. Kürner, D.J. Cichon, and W. Wiesbeck, "Concepts and Results for 3D Digital Terrain-Based Wave Propagation Models: An Overview," *IEEE Journal on Selected Areas in Communications*, Vol. 11,pp. 1002-1012, September 1993.

[31] P. Kyösti, J. Meinilä, L. Hentilä, X. Zhao, T. Jämsä, C. Schneider, M. Narandžić, M. Milojević, A. Hong, J. Ylitalo, V.-M. Holappa, M. Alatossava, R. Bultitude, Y. de Jong, and T. Rautiainen, "IST-4-027756 WINNER II Deliverable 1.1.2. v.1.2, WINNER II Channel Models," Technical report, IST-WINNERII, September 2007.

[32] M. Landmann, *Limitations of Experimental Channel Characterisation*, PhD thesis, Technische Universität Ilmenau, Ilmenau, Germany, March 2008.

[33] H.-J. Li, C.C. Chen, T.-Y. Liu, and H.-C. Lin, "Applicability of Ray-Tracing Technique for the Prediction of Outdoor Channel Characteristics,". *IEEE Transactions on Vehicular Technology*, Vol. 49, No. 6, pp. 2336-2349, November 2000.

[34] ITU-R M.2135, "Guidelines for evaluation of radio interface technologies for imt-advanced," Technical report, International Telecommunication Union, Radiocommunication Sector (ITU-R), 2008.

[35] J. Maurer, *Strahlenoptisches Kanalmodell für die Fahrzeug-Fahrzeug-Funkkommunikation*, PhD thesis, Forschungsberichte aus dem Institut für Höchstfrequenztechnik und Elektronik der Universität Karlsruhe (TH), 2005.

[36] J. Maurer, T. Fügen, and W. Wiesbeck, "Physical Layer Simulations of IEEE802.11a for Vehicle-to-Vehicle Communications," in *Proceedings of the 62th IEEE Vehicular Technology Conference, VTC2005-Fall*, Vol. 3, pp. 1849-1853, Dallas, TX, September 2005.

[37] D.A. McNamara, C.W.I. Pistorius, and J.A.G. Malherbe, *Introduction to the Uniform Geometrical Theory of Diffraction*, Artech House, Boston, 1990.

[38] A. F. Molisch, H. Asplund, R. Heddergott, M. Steinbauer, and T. Zwick, "The COST259 Directional Channel Model – Part I: Overview and Methodology,". *IEEE Transactions on Wireless Communications*, Vol. 5, No. 12, pp. 3421-3433, December 2006.

[39] M. Narandžić, W. Kotterman, M. Käske, C. Schneider, G. Sommerkorn, A. Hong, and R. S. Thomä, "On a Characterization of Large-Scale Parameters for Distributed (Multi-Link) MIMO - the Impact of Power Level Differences," in *European Conference on Antennas and Propagation (EuCAP)*, Barcelona, Spain, Apr. 2010.

[40] M. Narandžić, P. Kyösti, J. Meinilä, L. Hentilä, M. Alatossava, T. Rautiainen, Y.L.C. de Jong, C. Schneider, and R. S. Thomä, "Advances in WINNER Wideband MIMO System-Level Channel Modelling," in *Proc. of the 2nd Europian Conference on Antennas and Propagation (EUCAP)*, Edinburgh, UK, Nov. 2007.

[41] J.Ø. Nielsen, J.B. Andersen, P.C.F. Eggers, G.F. Pedersen, K. Olesen, E.H. Sørensen, and H. Suda, "Measurements of Indoor 16×32 Wideband MIMO Channels at 5.8 GHz," in *ISSSTA2004*, pp. 864–868, 2004.

[42] R. Pabst et al., "Relay based deployment concepts for wireless and mobile broadband radio," *IEEE Communication Magazine*, Vol. 42, pp. 80-89, Sep. 2004.

[43] A. Papoulis, *Probability, Random Variables, and Stochastic Processes*, McGraw Hill, 3rd edition, 1991.

[44] A.J. Paulraj, R. Roy, and T. Kailath, "Estimation of Signal Parameters via Rotational Invariance Techniques - ESPRIT," in *Proceedings of the 19th Asilomar Conference on Circuits, Systems and Computers*, pp. 83-89, Nov. 1985.

[45] T. Rautiainen, G. Wölfle, and R. Hoppe, "Verifying Path Loss and Delay Spread Predictions of a 3D Ray Tracing Propagation Model in Urban Environment," in *Proceedings of the 56th IEEE Vehicular Technology Conference, VTC2002-Fall*, pp. 1264-1268, Vancouver, BC, Canada, September 2002.

[46] A. Richter, *Estimation of Radio Channel Parameters: Models and Algorithms*, PhD thesis, Technische Universität Ilmenau, Ilmenau, Germany, 2005.

[47] J.-P. Rossi and Y. Gabillet, "A Mixed Ray Launching/Tracing Method for Full 3-D UHF Propagation Modeling and Comparison With Wide-Band Measurements," *IEEE Transactions on Antennas and Propagation*, Vol. 50, No. 4, pp. 517-523, April 2002.

[48] C. Schneider, G. Sommerkorn, M. Narandžić, M. Käske, A. Hong, V. Algeier, W. Kotterman, R.S. Thomä, and C. Jandura, "Multi-User MIMO Channel Reference Data for Channel Modelling and System Evaluation from Measurements," in *International IEEE Workshop on Smart Antennas (WSA 2009)*, Berlin, Germany, February 2009.

[49] L. Schumacher, L.T Berger, and J. Ramiro-Moreno, "Propagation Characterization and MIMO Channel Modeling for 3G," in S. Chandran, editor, *Adaptive Antenna Arrays; Trends and Applications*, pp. 377-393. Springer, 2004.

[50] M. Steinbauer, A.F. Molisch, and E. Bonek, "The Double-Directional Radio Channel," *IEEE Transactions on Antennas and Propagation*, Vol. 43, No. 4, pp. 51-63, August 2001.

[51] G. L. Stüber, *"Principles of Mobile Communications,"* Kluwer Academic Publishers, 2002.

[52] R.S. Thomä, D. Hampicke, A. Richter, G. Sommerkorn, A. Schneider, U. Trautwein, and W. Wirnitzer, "Identification of Time-Variant Directional Mobile Radio Channels," *IEEE Transactions on Instrumentation and Measurement*, Vol. 49, No. 2, pp. 357-364, April 2000.

[53] R.S. Thomä, M. Landmann, and A. Richter, "RIMAX - a Maximum Likelihood Framework for Parameter Estimation in Multidimensional Channel Sounding," in *2004 International Symposium on Antennas and Propagation*, Sendai, Japan, August 2004.

[54] 3GPP TR 25.996 V6.1.0, "Spatial Channel Model for MIMO Simulations," Technical report, Third Generation Partnership Project (3GPP), Sept. 2003.

[55] E. C. van der Meulen, "Three-Terminal Communication Channels," *Advances in Applied Probability*, Vol. 3, pp. 120-154, 1971.

[56] G.A.J. van Dooren, *A Deterministic Approach to the Modelling of Electromagnetic Wave Propagation in Urban Environments*, PhD thesis, University of Eindhoven, The Netherlands, 1994.

[57] Y. Wang, S. Safavi-Naeini, and S.K. Chaudhuri, "A Hybrid Technique Based on Combining Ray Tracing and FDTD Methods for Site-Specific Modeling of Indoor Radio Wave Propagation," *IEEE Transactions on Antennas and Propagation*, Vol. 48, No. 5, pp. 743-753, May 2000.

[58] H. Wijaya, *Broadband Multi-Hop Communication in Homogeneous and Heterogeneous Wireless Lan Networks*, PhD thesis, Technische Hochschule Aachen, Germany, Feb. 2005.

[59] K.I. Ziri-Castro, W.G. Scanlon, and N.E. Evans, "Prediction of Variation in MIMO Channel Capacity for the Populated Indoor Environment using a Radar Cross-Section-Based Pedestrian Model,". *IEEE Transactions on Wireless Communications*, Vol. 4, No. 3, pp. 1186-1194, May 2005.

[60] M. Ziskind, I. Wax, "Maximum Likelihood Localization of Multiple Sources by Alternating Projection," *IEEE Transactions on Acoustics, Speech, and Signal Processing*, Vol. 36, pp. 1553-1560, October 1988.

3 Link-Level Aspects

3.1 OFDM Data Detection and Channel Estimation

P. A. Hoeher, University of Kiel, Germany

3.1.1 Introduction

OFDM is a popular multicarrier transmission technique employing orthogonal subcarriers. It has been successfully applied in data modems, audio and video broadcasting systems, wireless local area networks, cellular radio, and is a suitable candidate for next generation wireless and wireline systems. Inherent advantages of OFDM include its ease of implementation (due to FFT processing), its robustness against multipath fading (due to a guard interval/cyclic extension), and its bandwidth efficiency (due to the ability of adaptive power and bit loading) [1]. Multi-user MIMO systems based on OFDM are currently under intensive investigation. However, the bit error performance degrades if orthogonality can not be maintained. Reasons for this include fast fading, phase jitter, frequency offset, delay spread exceeding the guard interval, and nonlinear distortions. All these effects cause crosstalk between the subcarriers. In contrast to multicarrier transmission techniques with non-orthogonal subcarrier signals (such as multicarrier systems with Gaussian pulse shaping), *all* subcarriers interfere with each other. The problems are especially severe in the presence of a "dirty RF" receiver design [2]. As a consequence, optimum reception is prohibitive if orthogonality is lost. Particularly challenging are issue like data detection, channel estimation, and synchronization. These topics have been addressed in our working group. In this chapter, we tackle two specific topics: Data detection in the presence of nonlinearities, and a contribution towards graph-based joint soft-output data detection and soft channel estimation.

3.1.2 Data Detection in the Presence of Nonlinear Distortions

Nonlinear distortions are mainly caused by the high power amplifier (HPA), since OFDM is characterized by a large peak-to-average-power ratio in conjunction with an orthogonal design. It is well known from numerous papers that due to nonlinear distortions (i) the bit error probability degrades and (ii) the spectral mask is difficult to maintain, see, e.g., [3–10], since the power density spectrum is significantly widened. Hence, in most current OFDM systems either expensive, highly linear amplifiers are being used and/or a large power back-off is chosen to maintain

H. Rohling (ed.), *OFDM: Concepts for Future Communication Systems*,
Signals and Communication Technology, DOI: 10.1007/978-3-642-17496-4_3,

quasi-orthogonality. Particularly in future mobile terminals, it is desirable to apply low-cost amplifiers operating at a fairly small power back-off in order to maintain power efficiency. Therefore, the influence of nonlinear distortion is inevitable unless some form of compensation is done.

Compensation techniques can be classified into techniques applied at the transmitter side and at the receiver side. *Predistortion* is a popular compensation technique applied at the transmitter side [11–15]. The main idea of predistortion is to shape the transmitted data symbols ("data predistortion") or the input signal of the HPA amplifier ("signal predistortion") so that the output signal of the HPA is less distorted. Predistortion does not reduce the information rate and does not increase the transmit power. Due to predistortion, the power density spectrum of the transmit signal improves. The bit error performance also improves, but only slightly since clipping can not be avoided. Therefore, in most publications on predistortion a large power back-off is assumed.

An alternative to predistortion are peak-to-average power reduction (PAPR) techniques applied at the transmitter side [16]. PAPR can be achieved by clipping & filtering, channel coding, interleaving, or by dropping or loading some carriers, among other techniques. As opposed to predistortion, PAPR either comes at the expense of transmit signal power increase, bit error rate increase, data rate loss, and so on.

At the receiver side, linear as well as nonlinear equalization/detection techniques can be applied. With nonlinear equalization/detection techniques the bit error performance can be enhanced significantly, see, e.g., [17–20], even in the presence of a small power back-off. However, the out-of-band radiation is not affected, of course. The main challenge is the derivation of cost-efficient baseband algorithms.

In this work, we *simultaneously* use a memoryless polynomial-based (signal) predistorter at the transmitter side and a low-cost nonlinear detector at the receiver side [21]. The predistorter reduces the out-of-band power, whereas the nonlinear detector improves the bit error rate. Since maximum-likelihood detection is prohibitive, reduced-state symbol detection is considered. The predistorter is not only useful for spectral shaping, but also in order to reduce the computational complexity of the nonlinear detector and to provide more robustness concerning an incomplete knowledge of the characteristics of the nonlinearity at the receiver side. All processing is done at complex baseband. A related (simultaneous) concept has been proposed in [18, 20] for OFDM systems and in [11, 15] for single-carrier systems. It is demonstrated that by means of predistortion the computational complexity of the nonlinear detector can be significantly reduced. For a QPSK/OFDM system, the performance loss with respect to the linear case can be made negligible even in the presence of a small power back-off, without needing iterative processing.

Transmission Model

Throughout this work, the equivalent discrete-time channel model in complex baseband notation is used. Vectors are written in bold face.

An OFDM signal can be calculated by means of an inverse discrete Fourier transform (IDFT):

$$\mathbf{s}[n] = \mathrm{IDFT}_N\{\mathbf{a}[n]\}, \tag{3.1}$$

where N is the number of subcarriers, $\mathbf{a}[n]$ is the nth data vector of length N, and n is the time index after serial/parallel (S/P) conversion. According to the central limit theorem, the quadrature components of the OFDM signal are Gaussian distributed, i.e., the amplitude is Rayleigh distributed. Therefore, in the presence of a nonlinear HPA a large power back-off is needed in order to avoid crosstalk between all subcarriers ("intercarrier interference"), unless predistortion, peak-to-average power reduction, or nonlinear detection is applied.

In this work, a memoryless, quasi time-invariant nonlinearity is assumed. If we denote the modulated OFDM signal as $s[k] \doteq A[k] \exp(j\phi[k])$, where k is the time index before serial/parallel conversion, $A[k]$ the amplitude of the transmit signal and $\phi[k]$ the phase, the output signal of the HPA can be modeled as

$$s_{HPA}[k] = g(A[k]) \exp(j[\phi[k] + \Phi(A[k])]). \tag{3.2}$$

The real-valued functions $g(A[k])$ and $\Phi(A[k])$ are called AM/AM and AM/PM conversion, respectively. In order to provide a fair comparison among transmission schemes with different nonlinearities, the same output back-off (OBO) is considered.

Predistortion

Memoryless predistortion has been investigated in many papers as a potential solution to decrease the nonlinear distortion caused by a HPA, see, e.g., [11]- [15]. Naturally, this technique tries to invert the nonlinearity of the HPA. The output samples of the predistorter can be written as

$$s_p[k] = f(A[k]) \exp(j[\phi[k] + \Psi(A[k])]), \tag{3.3}$$

where $f(A[k])$ and $\Psi(A[k])$ are the AM/AM and AM/PM conversion of the predistorter, respectively. The combination of a given memoryless HPA and the corresponding predistorter results in

$$s_{HPA}[k] = g(f(A[k])) \exp(j[\phi[k] + \Psi(A[k]) + \Phi(f(A[k])]). \tag{3.4}$$

Ideal predistortion is characterized as

$$\begin{aligned} g(f(A[k])) &= \begin{cases} \alpha\, A[k] & \text{if } \alpha\, A[k] \le A_0 \\ A_0 & \text{otherwise} \end{cases} \\ \Psi(A[k]) + \Phi(f(A[k])) &= 0, \end{aligned} \tag{3.5}$$

where α is a real-valued constant ($\alpha > 0$). In this case, the combination of the HPA and the corresponding predistorter (i.e., the overall transmitter-side nonlinearity) is equivalent with the so-called soft envelope limiter.

Throughout this work we assume that the AM/PM conversion of the HPA is negligibly small and does not have to be compensated, i.e., $\Psi(A[k]) = 0$. The AM/AM conversion of the predistorter is modeled by a polynomial as

$$f(A[k]) = f_1\, A[k] + f_2\, A^2[k] + \cdots + f_L\, A^L[k] \doteq \mathbf{f}\, \mathbf{A}^T[k], \tag{3.6}$$

where L is the order of the polynomial, $\mathbf{f} \doteq [f_1, f_2, \ldots, f_L]$, and $\mathbf{A}[k] \doteq [A[k], A^2[k], \ldots, A^L[k]]$. To find the coefficient set, $\mathbf{f}$, we apply the least mean square algorithm proposed in [13], which minimizes the mean squared error between the input and output amplitudes of the combined predistorter and HPA:

$$J(\mathbf{f}) \doteq E\left\{\left(\underbrace{g(\mathbf{f}\,\mathbf{A}^T[k])}_{|s_{HPA}[k]|} - \alpha\, A[k]\right)^2\right\}. \tag{3.7}$$

In (3.7), averaging is done over time. The coefficient set can be calculated recursively according to

$$\begin{aligned} \mathbf{f}[k+1] &= \mathbf{f}[k] - \mu \nabla_{\mathbf{f}} J(\mathbf{f}[k]) \\ &= \mathbf{f}[k] + \mu \mathbf{A}[k] g'(\mathbf{f}[k]\mathbf{A}^T[k]) \big(|s_{HPA}[k]| - \alpha A[k]\big), \end{aligned} \tag{3.8}$$

where $\nabla_{\mathbf{f}}$ denotes the gradient, $g'(.)$ is the derivative of $g(.)$ and μ a (small) positive step size. A suitable choice for the initial coefficient set is $\mathbf{f}[0] \doteq [1, 0, \ldots, 0]$. The steady-state coefficient set is denoted as $\mathbf{f}_\infty \doteq \lim_{k\to\infty} \mathbf{f}[k]$. Convergence is obtained after a few thousand iterations. Since the AM/AM conversion is quasi time-invariant, computations may be done off-line or from time-to-time only, leading to a negligible computational complexity.

A drawback of this particular adaptation algorithm is the fact that $g'(.)$ and hence $g(.)$ has to be known a priori. Since $g'(.)$ is well-behaved, it can easily be approximated, however. At least one alternative technique exists, where $g'(.)$ does not have to be calculated [14].

Predistortion can only compensate the smooth nonlinearity before the saturation point. The bit error performance in conjunction with predistortion can not be better than that of a linear transmission scheme in conjunction with a soft envelope limiter, because the predistorter can not invert clipping. Therefore, we apply an additional nonlinear detector in order to improve the bit error performance further. The proposed nonlinear detector is a simplified version of the maximum-likelihood receiver. It is particularly useful if the output back-off is small, i.e., if the power efficiency is high.

Nonlinear Detection

In the remainder, the transmitted signal is assumed to be distorted by additive white Gaussian noise. The received samples can then be written as

$$r[k] = s_{HPA}[k] + w[k]. \tag{3.9}$$

Conceptionally, the maximum-likelihood (ML) receiver for the transmission scheme under investigation computes all possible OFDM signals. These signal hypotheses are passed through the nonlinear function $g_{NL}(.)$ representing the HPA (eventually including the predistorter). The signal hypothesis causing the smallest squared

Euclidean distance with respect to the received samples is finally selected [19]:

$$\hat{\mathbf{a}}_{ML}[n] = \underset{\tilde{\mathbf{a}}[n]}{\operatorname{argmin}} \left\{ \| \mathbf{r}[n] - g_{NL}\left(\mathrm{IDFT}_N\{\tilde{\mathbf{a}}[n]\}\right) \|^2 \right\}. \tag{3.10}$$

If the overall nonlinearity, $g_{NL}(.)$, is perfectly known at the ML receiver, the bit error performance does not degrade compared to the corresponding linear system. Unfortunately, the computational complexity of the ML receiver is $\mathcal{O}(M^N)$, where M is the cardinality of the symbol alphabet and N the number of subcarriers. Even for a moderate number of subcarriers, the computational complexity is prohibitive.

This motivates us to derive a reduced-complexity receiver, providing an adjustable trade-off between complexity and performance. The simplest version corresponds to the conventional OFDM receiver ignoring nonlinear distortions, whereas the most complex version corresponds to the ML receiver, assuming that the overall nonlinearity is known at the receiver.

The following two effects motivate the receiver structure under investigation:

- In the presence of severe nonlinear distortions, some subcarriers are more distorted than others, even in the absence of additive noise.
- In the case of non-binary data, it may happen that even for the same subcarrier some decisions are reliable, whereas other decisions are unreliable. For the example of QPSK, the inphase component of the received sample (after DFT) of a certain subcarrier may be close to the decision threshold, whereas the quadrature component of the same subcarrier may be more reliable.

For these reasons, in [21] we proposed to identify those bit decisions, which are close to the corresponding decision threshold. The proposed reduced-state symbol detector (RSSD) differs from the ML receiver in the fact that only hypotheses for the "weakest" bit decisions (i.e., bit decisions near the corresponding decision threshold) are evaluated. Since a memoryless nonlinearity is assumed, the computations can be done on an OFDM symbol basis. For more details see [21].

Numerical Results

The numerical results presented in this section are based on the following set-up: In the transmitter, an OFDM signal with $N = 128$, 256, or 512 subcarriers is generated. All subcarriers are QPSK modulated. A solid-state power amplifier according to [3] with $p = 2$ is used, where p controls the smoothness of the transition from the linear region to the saturation level. A predistorter according to (3.6) of order $L = 5$ with three non-zero coefficients f_1, f_3, and f_5 is applied optionally. Only the steady-state coefficient set $\mathbf{f}_\infty$ is considered. The channel model under consideration is an AWGN channel. The signal-to-noise ratio per information bit, E_b/N_0, shown in the following figures does not include the output back-off. At the receiver, the proposed RSSD with $H = 2$ or $H = 4$ hypotheses is applied. As a benchmark, the performance of the conventional OFDM receiver (which ignores nonlinear distortions) is shown as well. In all numerical results, 4x oversampling is performed in order to avoid aliasing.

In Fig. 3.1, the influence of the number of subcarriers with/without predistortion is studied. At the receiver, the proposed RSSD with $H = 2$ hypotheses is used. For $g_{NL}(.)$ the overall transmitter-side nonlinearity is assumed. As expected, the performance degrades with an increasing number of subcarriers, independently whether a predistorter is used or not. The most important result of Fig. 3.1 is the observation that the predistorter has a positive influence on the performance of the RSSD, particularly if the number of hypotheses (and hence the computational complexity) is small.

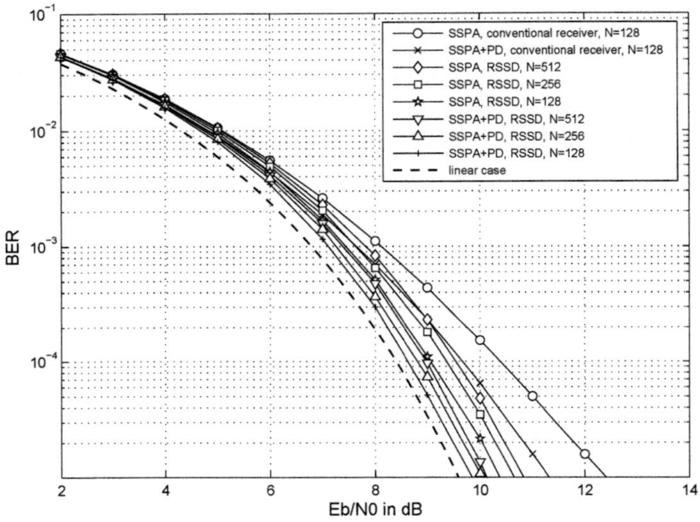

Figure 3.1: Raw BER performance of a QPSK/OFDM system with/without nonlinear distortion ($N = 128$, 256, and 512 subcarriers, $OBO = 3.16$ dB, reduced-state symbol detector with $H = 2$ hypotheses, with/without predistorter).

In Fig. 3.1, no modeling errors are considered in the RSSD. In order to study the influence of modeling errors, Fig. 3.2 displays the bit error performance for an RSSD with complete knowledge and partial knowledge of the transmitter-side nonlinearity, respectively. At the transmitter, no predistortion is applied. Complete knowledge means that an SSPA with known OBO is available in the RSSD. By partial knowledge we mean that a soft envelope limiter with known OBO is assumed in the RSSD. Two different output back-offs and two different number of hypotheses are considered. It can be noticed that the performance difference for complete and partial knowledge is negligible. In case of predistortion there would be even less degradation, since the modeling error would be smaller. This demonstrates the robustness of the proposed nonlinear detector.

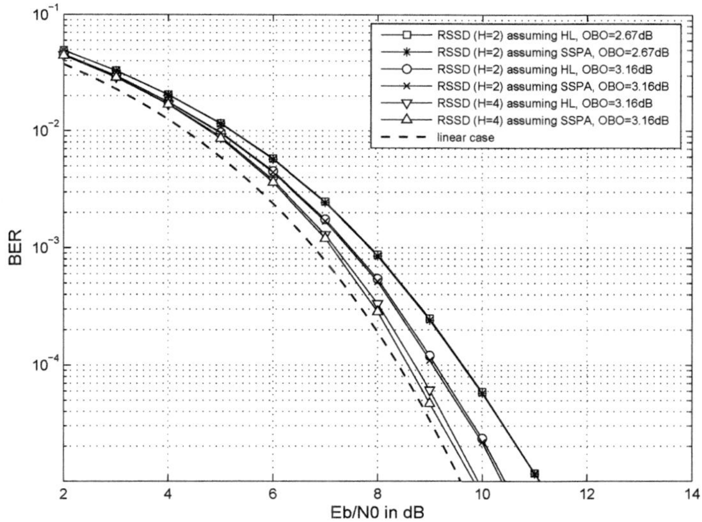

Figure 3.2: Raw BER performance of a QPSK/OFDM system with/without non-linear distortion (N = 128 subcarriers, OBO = 2.67 dB and 3.16 dB, no predistorter, reduced-state symbol detector with H = 2 and H = 4 hypotheses, complete and partial knowledge of SSPA).

3.1.3 Joint Data Detection and Channel Estimation

In most OFDM systems, coherent data detection is applied, which a few exceptions like Digital Audio Broadcasting (DAB). In order to support coherent detection, training symbols (pilot symbols) known at the receiver side are commonly used. In wireless OFDM systems, the training symbols may be either multiplexed within the data symbols in time and frequency domain, as firstly proposed in [22], or superimposed onto the data symbols, as suggested in [23]. For space limitations, we restrict ourself to the first case, since this technique became common in most state-of-the-art wireless OFDM systems. The main idea is to distribute the training symbols in time and frequency domain. Assuming that the fading process is bandlimited in time domain (since the Doppler power spectral density is typically bandlimited) and time-limited in frequency domain (since the delay power profile of a multipath channel is typically time-limited). Given initial estimates at the positions of the training symbols, the channel coefficients (channel gains) can be reconstructed at the data positions by means of interpolation or filtering [24]. A significant number of papers on related techniques have been published.

The main drawback of training-based channel estimation is the overhead: any additional training symbol has an impact on power efficiency and bandwidth efficiency. If channel coding is not taken into account for channel estimation, the minimum number of training symbols is dictated by the sampling theorem [22, 24]. The training overhead is often acceptable in SISO systems, but in MIMO systems the required training overhead grows proportionally with the number of transmit antennas [25, 26]. Hence, any means to reduce the number of training symbols is highly desirable particularly in MIMO systems.

Receivers that jointly estimate the channel coefficients and detect the information-carrying data symbols typically reduce the required training overhead [27, 28] and were adapted to OFDM in [29]. Unfortunately, the computational complexity of conventional joint data detection and channel estimation techniques is typically significant. Iterative receivers that utilize the turbo principle partly solve the problem by taking reliably detected data symbols as pseudo training symbols into account [30–32]. However, these iterative receivers require tentative channel estimates based on training symbols and are therefore unable to reduce the training overhead. More recently, soft-output data detection in conjunction with soft channel estimation based on iterative message passing over a general factor graph has been applied to single-carrier MIMO systems for the case of block-fading channels [33]. A substantial reduction of training overhead at modest computational complexity has been reported.

In this work, the graph-based soft data and channel estimation technique in [33] is extended and applied to MIMO-OFDM for time-varying frequency-selective channels. The key idea is to extend the underlying factor graph to two dimensions (2D) with only linear increase in complexity [34]. Towards this goal, code constraints are taking into account, i.e., the redundancy of the channel code is used to improve data detection and channel estimation. Initial channel estimates at training symbol locations are dispersed through the factor graph by a message exchange in time and frequency domain. We observe that in conjunction with channel coding the training overhead may be substantially reduced, even violating the sampling theorem, which plays an important role when code constraints are not taken into account. In the limit only one training symbol per transmit antenna is sufficient for proper convergence of the message passing algorithm.

Transmission Model

A MIMO-OFDM system with N subcarriers and K OFDM symbols per frame is considered. The equivalent discrete-time MIMO channel model with N_T transmit and N_R receive antennas is given by

$$\begin{aligned} y_n[k][l] &= \sum_{m=1}^{N_T} h_{n,m}[k][l]\, x_m[k][l] + w_n[k][l] \\ &= h_{n,m}[k][l]\, x_m[k][l] + \underbrace{\sum_{i=1,i\neq m}^{N_T} h_{n,i}[k][l]\, x_i[k][l]}_{\text{MAI}} + \underbrace{w_n[k][l]}_{\text{AWGN}} \end{aligned} \tag{3.11}$$

where $k \in \{0, 1, \ldots, K-1\}$ is the time index and $l \in \{0, 1, \ldots, L-1\}$ is the subcarrier index. The channel coefficient between the nth Rx antenna and the mth Tx antenna is denoted as $h_{n,m}[k][l]$. The data symbols transmitted by the mth transmit antenna are denoted as $x_m[k][l]$, and $w_n[k][l]$ is an additive white Gaussian noise sample with zero mean and variance σ_w^2. The channel coefficients are assumed to be wide-sense stationary with zero mean.

Graph-Based Receiver Structure

Factor graphs are graphical models that have been adopted to a variety of problems in digital communications [35, 36]. In combination with the sum-product algorithm for message exchange, powerful channel estimators have been derived for example in [33, 37, 38].

The objective of joint data detection and channel estimation is to estimate two variables, the data symbols $x_m[k][l]$ and the channel coefficients $h_{n,m}[k][l]$. Provided that encoded data symbols are sufficiently interleaved, adjacent data symbols will be independent in time and frequency domain. The corresponding factor graph for two transmit and two receive antennas is displayed in Fig. 3.3 with the unknown data symbols (symbol nodes) and channel coefficients (coefficient nodes) in circles and the received (known) symbols $y_n[k][l]$ (observation nodes) in rectangles. Without loss of generality the time and subcarrier index are omitted in Fig. 3.3, thus representing one subcarrier at one time index.

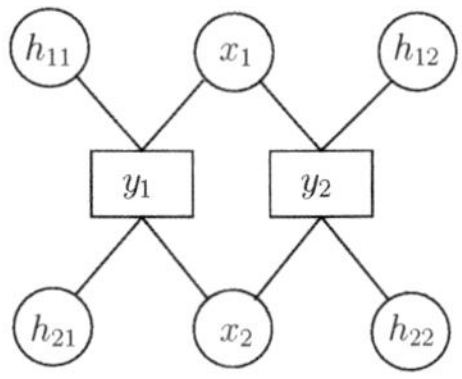

Figure 3.3: Structure of a factor graph with two Tx and two Rx antennas for one subcarrier at one time index ignoring channel coding.

If channel coding is applied in time and/or frequency domain, the corresponding connections of the data symbols can be either (i) directly implemented as a part of the factor graph or (i) separated from the factor graph. In case (i), Fig. 3.3 would also incorporate connections to code nodes, whereas in case (ii) the factor graph remains unchanged. A large variety of channel codes have been successfully implemented in factor graphs, cf. [39].

The key idea of the proposed 2D channel estimator is to connect the channel coefficients of neighboring time indices and subcarriers via so-called Δ-transfer nodes. Hence, messages can be exchanged in time and frequency domain throughout the graph. Figure 3.4 illustrates these connections for a time-varying frequency-selective fading channel. For simplicity, the symbol nodes and observation nodes are omitted, since their connections are not affected by changing channel conditions. Also, the indices n and m are dropped.

Δ-Transfer Nodes

Information about the channel coefficients and data symbols can be exchanged throughout the graph by means of the sum-product algorithm [35]. The Δ-transfer nodes introduced in the previous section describe the relation of neighboring channel coefficients, hence converting the message of a channel coefficient $h[k][l]$ to $h[k+1][l]$ and $h[k][l]$ to $h[k][l+1]$, respectively. The time and frequency indices will be omitted in the following in order to improve readability. A neighboring coefficient in

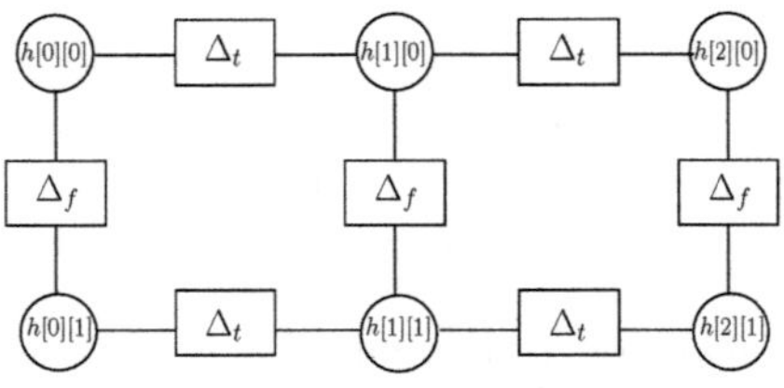

Figure 3.4: Simplified structure of a factor graph for 2D channel estimation for three time indices and two subcarriers.

either time or frequency domain is marked with a prime. Information exchange via a transfer node can be modeled as follows:

$$\mu_{h'} = \mu_h \qquad \sigma_{h'}^2 = \sigma_h^2 + \sigma_\Delta^2, \tag{3.12}$$

where Δ is defined as

$$\Delta \doteq h - h'. \tag{3.13}$$

Δ can be approximated by a Gaussian probability density function:

$$\Delta \sim \mathcal{N}\left(0, \sigma_\Delta^2\right). \tag{3.14}$$

The mean value μ_h represents the estimate of a channel coefficient and the variance σ_h^2 can be interpreted as its reliability information. The exchange of information about the channel coefficients is started at the training positions. The corresponding mean values are then distributed throughout the graph. The reliability of the estimated coefficients decreases with increasing distance to the training positions. The variance $\sigma_\Delta^2 \doteq \mathrm{E}\left\{|h-h'|^2\right\}$ can be pre-computed for a given Doppler bandwidth and a given maximum propagation delay, respectively.

Information Exchange at Coefficient Nodes

A coefficient node is connected with two Δ-transfer nodes in each domain and one observation node as illustrated in the left part of Fig. 3.5. Suppose the channel coefficient receives the messages $p_i \sim \mathcal{CN}\left(\mu_i, \sigma_i^2\right)$, $i = 1, \ldots, 5$. The messages leaving a coefficient node are generated as depicted in the right part of Fig. 3.5.

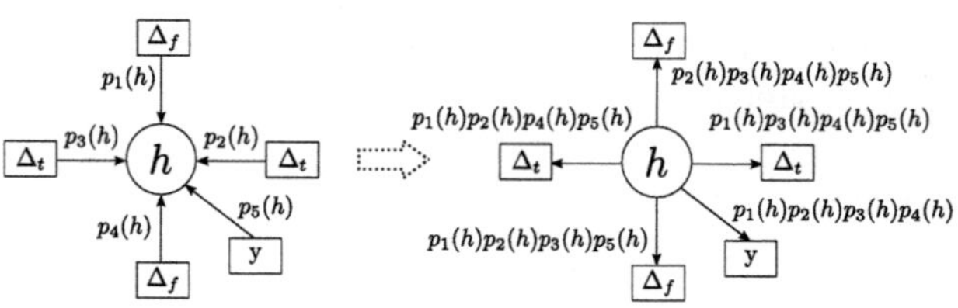

Figure 3.5: Information exchange at a channel coefficient node.

Information exchange at symbol nodes can be performed as in [33]. For further information, see [34].

Summary

The presented concept can nicely be extended in several directions. For example, one may take additional dimensions into account, such as correlated Tx antennas and/or correlated Rx antennas. Also, imperfections due to "dirty RF" can be considered. Furthermore, co-channel interference or adjacent channel interference can be compensated in this framework.

Acknowledgement

The author would like to thank H.D. Han and J. Araabab for valuable contributions to the first part of this work, and T. Wo, Ch. Knievel, Z. Shi, and G. Auer for significant contributions to the second part.

Bibliography

[1] J.A.C. Bingham, "Multicarrier modulation for data transmission: An idea whose time has come," *IEEE Communications Magazine*, vol. 28, pp. 5-14, May 1990.

[2] G. Fettweis, M. Loehning, D. Petrovic, M. Windisch, P. Zillmann, and W. Rave, "Dirty RF: a new paradigm," *Int. J. Wireless Information Networks (IJWIN)*, vol. 14, no. 2, pp. 133-148, June 2007.

[3] C. Rapp, "Effects of HPA-nonlinearity on a 4-PSK/OFDM-signal for a digital sound broadcasting system," in *Proc. 2nd European Conference on Satellite Communications (ECSC)*, Liége, Belgium, Oct. 1991, pp. 22-24.

[4] D.J.G. Mestdagh, P. Spruyt, and B. Biran, "Analysis of clipping effect in DMT-based ADSL systems," in *Proc. IEEE International Conference on Communications (ICC)*, New Orleans, Louisiana, May 1994, pp. 293-300.

[5] P. Banelli and S. Cacopardi, "Theoretical analysis and performance of OFDM signals in nonlinear AWGN channels," *IEEE Transactions on Communications*, vol. 48, no. 3, pp. 430-441, Mar. 2000.

[6] P. Banelli, G. Baruffa, and S. Cacopardi, "Effects of HPA non linearity on frequency multiplexed OFDM signals," *IEEE Transaction on Broadcasting*, vol. 47, no. 2, pp. 123-136, June 2001.

[7] C. van den Bos, M.H.L. Kouwenhoven, and W.A. Serdijin, "Effect of smooth nonlinear distortion on OFDM symbol error rate," *IEEE Transactions on Communications*, vol. 49, vol. 9, pp. 1510-1514, Sept. 2001.

[8] H. Ochiai and H. Imai, "Performance analysis of deliberately clipped OFDM signals," *IEEE Transactions on Communications*, vol. 50, no. 1, pp. 89-101, Jan. 2002.

[9] E. Costa and S. Pupolin, "M-QAM-OFDM system performance in the presence of a nonlinear amplifier and phase noise," *IEEE Transactions on Communications*, vol. 50, no. 3, pp. 462-472, Mar. 2002.

[10] K.R. Panta and J. Armstrong, "Effects of clipping on the error performance of OFDM in frequency selective fading channels," *IEEE Transactions on Wireless Communications*, vol. 3, no. 2, pp. 668-671, Mar. 2004.

[11] G. Karam and H. Sari, "Analysis of predistortion, equalization, and ISI cancellation techniques in digital radio systems with nonlinear transmit amplifiers," *IEEE Transactions on Communications*, vol. 37, no. 12, pp. 1245-1253, Dec. 1989.

[12] K. Wesolowski and J. Pochmara, "Efficient algorithm for adjustment of adaptive predistorter in OFDM transmitter," in *Proc. IEEE Vehicular Technology Conference (VTC)*, Boston, Massachusetts, Sept. 2000, pp. 2491-2496.

[13] H. Besbes and T. Le-Ngoc, "A fast adaptive predistorter for nonlinearly amplified M-QAM signals," in *Proc. IEEE Global Communications Conference (GLOBECOM)*, San Francisco, California, Nov.-Dec. 2000, pp. 108-112.

[14] Y. Guo and J.R. Cavallaro, "A novel adaptive pre-distorter using LS estimation of SSPA non-linearity in mobile OFDM systems," in *Proc. IEEE International Symposium on Circuits and Systems (ISCAS)*, Phoenix, Arizona, May 2002, pp. 453-456.

[15] A. Behravan and T. Eriksson, "Baseband compensation techniques for bandpass nonlinearities," in *Proc. IEEE Vehicular Technology Conference*, Orlando, Florida, Oct. 2003, pp. 279-283.

[16] S.H. Han and J.H. Lee, "An overview of peak-to-average power ratio reduction techniques for multicarrier transmission," *IEEE Wireless Communications*, vol. 12, no. 2, pp. 56-65, Apr. 2005.

[17] D. Declercq and G.B. Giannakis, "Recovering clipped OFDM symbols with Bayesian inference," in *Proc. IEEE Int. Conference on Acoustics, Speech and Signal Processing (ICASSP)*, Istanbul, Turkey, June 2000, pp. 157-160.

[18] L. Junqiang, M. Zhengxin, and C. Zhigang, "Compensation of nonlinear distortion using a new method combining predistortion with reconstruction in OFDM systems," in *Proc. Int. Conference on Communication Technology (ICCT)*, Beijing, China, Aug. 2000, pp. 769-772.

[19] J. Tellado, L.M.C Hoo, and J.M. Cioffi, "Maximum-likelihood detection of nonlinearly distorted multicarrier symbols by iterative decoding," *IEEE Transactions on Communications*, vol. 51, no. 2, pp. 218-228, Feb. 2003.

[20] P. Zillmann, "Soft detection and decoding of clipped and filtered COFDM signals," in *Proc. IEEE Vehicular Technology Conference (VTC '07-Spring)*, Dublin, Ireland, Apr. 2007.

[21] H.D. Han and P.A. Hoeher, "Predistortion and nonlinear detection for OFDM signals in the presence of nonlinear high power amplification," *European Transactions on Telecommunications (ETT)*, vol. 18, pp. 411-418, June 2007.

[22] P. Hoeher, "TCM on frequency-selective land-mobile fading channels," in *Proc. 5th Tirrenia Intern. Workshop on Digital Commun.*, Tirrenia, Italy, Sept. 1991, pp. 317-328.

[23] F. Tufvesson, M. Faulkner, P. Hoeher, and O. Edfors, "OFDM time and frequency synchronization by spread spectrum pilot technique," in *Proc. IEEE International Conference on Communications (ICC)*, Vancouver, Canada, June 1999, pp. 115-119.

[24] P. Hoeher, S. Kaiser, and P. Robertson, "Two-dimensional pilot-symbol aided channel estimation by Wiener filtering," in *Proc. IEEE International Conference on Acoustics, Speech, and Signal Processing (ICASSP)*, Munich, Germany, Apr. 1997, pp. 1845-1848.

[25] B. Hassibi and B.M. Hochwald, "How much training is needed in multiple-antenna wireless links?" *IEEE Transactions on Information Theory*, vol. 49, no. 4, pp. 951-963, Apr. 2003.

[26] I. Cosovic and G. Auer, "Capacity of MIMO-OFDM with pilot aided channel estimation," *EURASIP Journal on Wireless Communications and Networking*, vol. 2007, no. 5, Jan. 2007.

[27] N. Seshadri, "Joint data and channel estimation using blind trellis search techniques," *IEEE Transactions on Communications*, vol. 42, no. 2/3/4, pp. 1000-1011, Feb./Mar./Apr. 1994.

[28] E. Baccarelli and R. Cusani, "Combined channel estimation and data detection using soft-statistics for frequency-selective fast-fading digital links," *IEEE Transactions on Communications*, vol. 46, no. 4, pp. 424-427, Apr. 1998.

[29] T. Cui and C. Tellambura, "Joint channel estimation and data detection for OFDM systems via sphere decoding," in *Proc. IEEE Global Communications Conference (GLOBECOM)*, Nov. 2004, pp. 3656-3660.

[30] M.C. Valenti and B.D. Woerner, "Iterative channel estimation and decoding of pilot symbol assisted turbo codes over flat-fading channels," *IEEE Journal on Selected Areas in Communications*, vol. 19, no. 9, pp. 1697-1705, Sept. 2001.

[31] F. Sanzi, S. Jelting, and J. Speidel, "A comparative study of iterative channel estimators for mobile OFDM systems," *IEEE Transaction on Wireless Communications*, vol. 5, no. 2, pp. 849-859, Sept. 2003.

[32] G. Auer and J. Bonnet, "Threshold controlled iterative channel estimation for coded OFDM," in *Proc. IEEE Vehicular Technology Conference (VTC-Spring)*, Dublin, Ireland, Apr. 2007, pp. 1737-1741.

[33] T. Wo, C. Liu, and P.A. Hoeher, "Graph-based soft channel and data estimation for MIMO systems with asymmetric LDPC codes," in *Proc. IEEE International Conference on Communications (ICC)*, Beijing, China, May 2008, pp. 620-624.

[34] Ch. Knievel, Z. Shi, P.A. Hoeher, and G. Auer, "2D graph-based soft channel estimation for MIMO-OFDM," in *Proc. IEEE International Conference on Communications (ICC)*, Capetown, South Africa, May 2010.

[35] F.R. Kschischang, B.J. Frey, and H.-A. Loeliger, "Factor graphs and the sum-product algorithm," *IEEE Transactions on Information Theory*, vol. 47, no. 2, pp. 498-519, Feb. 2001.

[36] H.A. Loeliger, J. Dauwels, J. Hu, S. Korl, L. Ping, and F.R. Kschischang, "The factor graph approach to model-based signal processing," *Proceedings of the IEEE*, vol. 95, no. 6, pp. 1295-1322, June 2007.

[37] T. Wo, C. Liu, and P.A. Hoeher, "Graph-based iterative Gaussian detection with soft channel estimation for MIMO systems," in *Proc. ITG Conference on Source and Channel Coding (SCC)*, Ulm, Germany, Jan. 2008.

[38] T. Wo, J.Ch. Fricke, and P.A. Hoeher, "A graph-based iterative Gaussian detector for frequency-selective MIMO channels," in *Proc. IEEE Information Theory Workshop (ITW)*, Chengdu, China, Oct. 2006, pp. 581-585.

[39] H.-A. Loeliger, "An introduction to factor graphs," *IEEE Signal Processing Magazine*, vol. 21, no. 1, pp. 28-41, Jan. 2004.

3.2 Spreading

J. Lindner, University of Ulm, Germany

3.2.1 Introduction

Spreading is a well known technique which can cope with the frequency-selective behavior of common wireless transmission channels. At the transmit side no knowledge of the current channel is needed. For OFDM this means that the energy of the symbols to be transmitted is spread across subcarriers. If subcarriers are faded out by the channel, then the receiver can still recover the symbols on the remaining unaffected subcarriers, i.e., the frequency diversity of the channel can be exploited. In 1993 various system variants based on these ideas came up, see, e.g., [1] and also [2]. It is straightforward to apply the spreading concept to MIMO-OFDM to get access to both, spatial and frequency diversity, while keeping the transmission rate constant. For spreading there is no rate loss like in case of orthogonal space-time codes, see, e.g., [3]. The drawback is that spreading causes mutual interference between subcarriers and antenna signals and powerful detection methods must be used in the receiver [4]. This project dealt with finding proper spreading schemes for MIMO-OFDM allowing to achieve best BER performance while keeping the complexity in limits.

Spreading can be represented by a multiplication of the symbol vector to be transmitted with a spreading matrix $\underline{\underline{U}}$. Figure 3.6 shows this as part of a vector transmission model. Further blocks represent the MIMO-OFDM channel matrix $\underline{\underline{H}}$ and

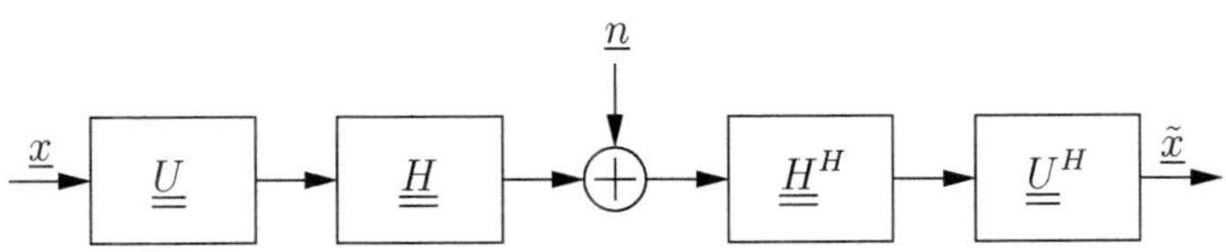

Figure 3.6: Block transmission model with spreading.

its matched filter matrix $\underline{\underline{H}}^H$ as well as the despreading matrix $\underline{\underline{U}}^H$. $\underline{n}$ is a sample vector of the assumed additive white Gaussian noise vector process. $\underline{\underline{H}}$ is an $(N\,n_R) \times (N\,n_T)$ matrix containing the transfer functions of channel impulse responses between all n_T transmit and n_R receive antennas, and N is the number of OFDM subcarriers [6]. The influence of the MIMO-OFDM channel including the matched filter matrix can be described by an equivalent channel matrix (on symbol basis)

$$\underline{\underline{R}}_{MO} = \frac{1}{n_R}\underline{\underline{H}}^H\underline{\underline{H}}. \tag{3.15}$$

Spreading and despreading can be included by defining

$$\underline{\underline{R}}_S = \underline{\underline{U}}^H\underline{\underline{R}}_{MO}\underline{\underline{U}}. \tag{3.16}$$

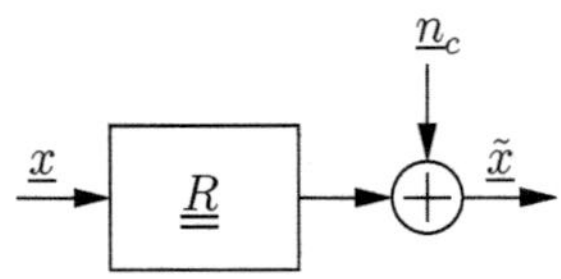

Figure 3.7: Transmission model using the equivalent channel matrix on symbol basis.

This leads to an equivalent matrix vector transmission model which is given by

$$\tilde{\underline{x}} = \underline{\underline{R}}\,\underline{x} + \underline{n}_c. \tag{3.17}$$

In the following $\underline{\underline{R}}$ belongs either to the unspread case ($\underline{\underline{R}}_{MO}$) or spread case ($\underline{\underline{R}}_S$), and $\underline{n}_c$ is the colored noise vector with correlation matrix $2N_0\underline{\underline{R}}$. Figure 3.7 shows in a qualitative way an example for $\underline{\underline{R}} = \underline{\underline{R}}_{MO}$ (i.e. for the unspread case). Spreading techniques do not eliminate the interference between symbol vector components, so in general $\underline{\underline{R}}$ is not a diagonal matrix. Thus vector equalization is needed.

3.2.2 MC-CDM and MC-CAFS

For MC-CDM the symbols are spread only in frequency direction. The spreading matrix can be described as

$$\underline{\underline{U}}_{MC-CDM} = \begin{pmatrix} \underline{\underline{S}} & \underline{\underline{0}} & \underline{\underline{0}} & \cdots & \underline{\underline{0}} \\ \underline{\underline{0}} & \underline{\underline{S}} & \underline{\underline{0}} & \cdots & \underline{\underline{0}} \\ \vdots & \ddots & \ddots & \ddots & \vdots \\ \underline{\underline{0}} & \cdots & \underline{\underline{0}} & \underline{\underline{S}} & \underline{\underline{0}} \\ \underline{\underline{0}} & \cdots & \underline{\underline{0}} & \underline{\underline{0}} & \underline{\underline{S}} \end{pmatrix}, \tag{3.18}$$

where $\underline{\underline{S}}$ is an $N \times N$ spreading matrix with $\underline{\underline{S}}^H\underline{\underline{S}} = \underline{\underline{I}}$, which is repeated n_T times on the main diagonal, i.e., each $\underline{\underline{S}}$ block corresponds to one transmit antenna. One example for an orthogonal spreading matrix $\underline{\underline{S}}$ is the normalized Walsh Hadamard matrix. As can be seen in (3.18), each symbol is spread only over the subcarriers of one transmit antenna. Therefore the maximum achievable diversity for MC-CDM is $L\,n_R$, where L denotes the number of channel taps. The scheme can easily be adapted to any number of transmit antennas.

Multi-carrier cyclic antenna frequency spreading (MC-CAFS) defines a family of spreading matrices that make use of the frequency as well as the spatial dimension offered by the MIMO channel [7]. The spreading matrices spread over all transmit antennas, and in addition, each symbol is spread over a set of subcarriers (frequencies) for each transmit antenna. The frequency sets are different for each antenna. MC-CAFS can achieve full diversity, which is $n_T\,n_R\,L$. The structure of the spreading matrices $\underline{\underline{U}}_{MC-CAFS}$ can be found in [6] [7]. To exploit full diversity while maintaining the orthogonality of the spreading matrix, the number of frequencies

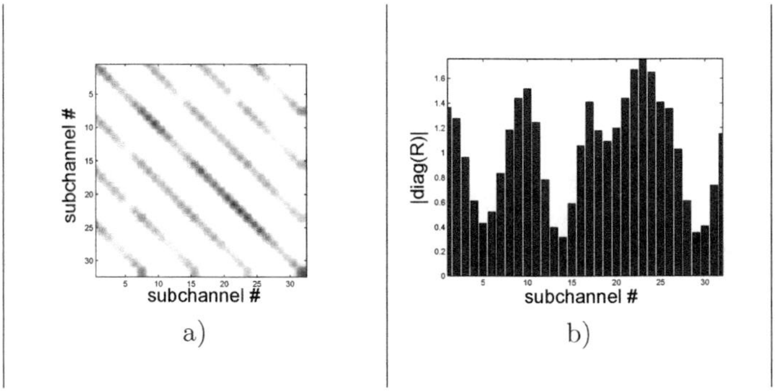

Figure 3.8: Equivalent channel matrix for unspread MIMO-OFDM. a) $\underline{\underline{R}}_{MO}$ for $n_T = n_R = 4$, $N = 8$ subcarriers, and $L = 4$ paths. (b) Absolute value of the diagonal elements of $\underline{\underline{R}}_{MO}$.

B, over which each symbol is spread, should fulfill the following conditions:

$$L \leq B \leq N/n_T. \tag{3.19}$$

Figures 3.8 to 3.10 demonstrate the effect of spreading on the equivalent channel matrix $\underline{\underline{R}}$ for a frequency-selective time-invariant channel with $L = 4$ taps. Figure 3.8 is for the unspread case. Although the probability of a deep fade is reduced by the multiple receive antennas and maximum ratio combining, the elements on the main diagonal have a large variance. With MC-CDM spreading (see Fig. 3.9) we can observe that the diagonal elements have the same value within each transmit antenna block, but vary from block to block. This is due to the fact that the symbol energy is spread over all frequencies of each antenna separately, but no spreading in antenna direction is applied. The new scheme, i.e., MC-CAFS, includes spreading in frequency and space direction. If the parameters of the spreading matrix are chosen properly, it is possible to achieve constant diagonal elements on the main diagonal of $\underline{\underline{R}}$ as can be seen in Fig. 3.10. This means, the matched filter bound (MFB) coincides with the AWGN performance.

3.2.3 Simulation Results

Figures 3.11 and 3.12 show a comparison of different spreading schemes for uncoded and coded transmissions, respectively. The channel was a block fading channel staying constant during one OFDM symbol block, but changes independently from block to block. For equalization, a soft Cholesky equalizer (SCE) [8] was used. The results for uncoded transmission in Fig. 3.11 show that exploiting additional diversity by spreading improves the performance substantially compared to unspread OFDM. MC-CAFS outperforms MC-CDM both for a channel length of $L = 2$ and $L = 4$ due to the additional transmit antenna diversity. Figure 3.12 shows the results for coded transmissions using a convolutional code with memory 2 and iterative equalization and decoding. The results for code rate 3/4 in Fig. 3.12 b) were obtained

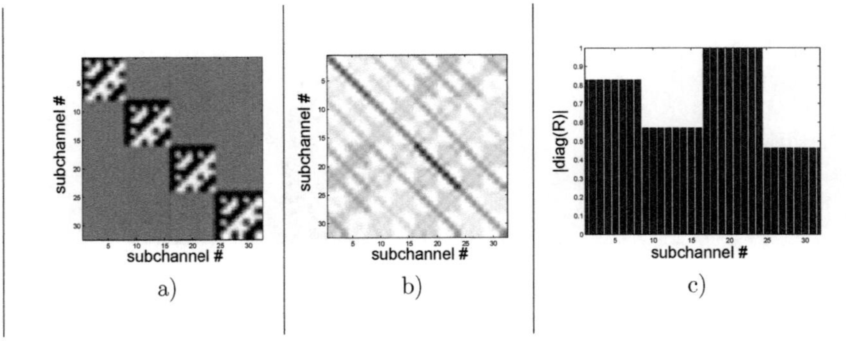

Figure 3.9: Equivalent channel matrix for MC-CDM: a) spreading matrix $\underline{\underline{U}}_{MC-CDM}$; b) $\underline{\underline{R}}_{MC-CDM}$ for $n_T = n_R = 4$, $N = 8$ subcarriers, and $L = 4$ paths; c) absolute value of the diagonal elements of $\underline{\underline{R}}_{MC-CDM}$.

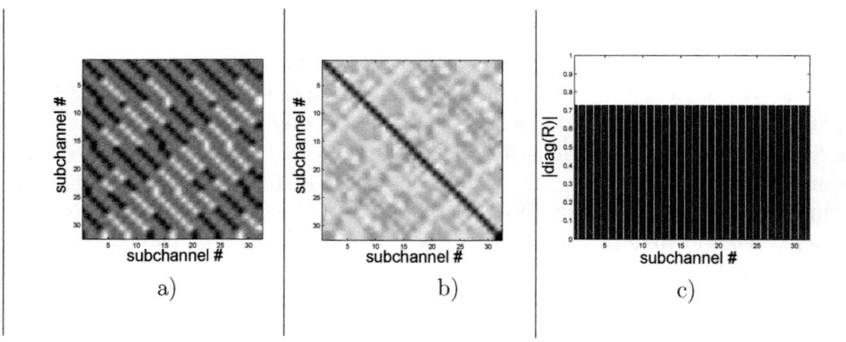

Figure 3.10: Equivalent channel matrix for MC-CAFS: a) spreading matrix $\underline{\underline{U}}_{MC-CAFS}$; b) $\underline{\underline{R}}_{MC-CAFS}$ for $n_T = n_R = 4$, $N = 8$ subcarriers, and $L = 4$ paths; c) absolute value of the diagonal elements of $\underline{\underline{R}}_{MC-CAFS}$.

by puncturing the rate 1/2 mother code from Fig. 3.12 a). The gain obtained through spreading increases as the code rate increases. Also the BER reduction of MC-CAFS compared to MC-CDM is larger for code rate 3/4. These results show that the channel code alone cannot exploit the maximum diversity even though the symbols of the codewords are interleaved over the whole codeword, covering many MIMO-OFDM symbols. More results and further details can be found in [6], where also the case of precoding in the transmitter is considered.

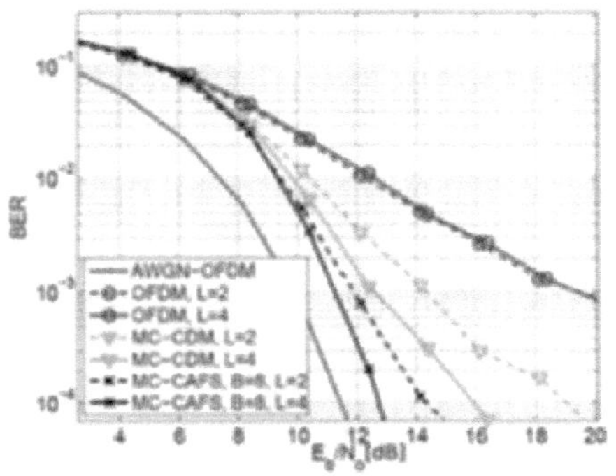

Figure 3.11: BER for MIMO-OFDM with spreading and equalization using SCE, 4PSK, $N = 32$, and $n_T = n_R = 4$.

3.2.4 Concluding Remarks

Spreading can help to exploit frequency as well as spatial diversity, but the spreading scheme must be designed properly. Multi-carrier cyclic antenna frequency spreading (MC-CAFS) can achieve excellent performance with higher transmission rates than with channel coding only (i.e. pure COFDM), but powerful detection schemes must be applied in the receiver, e.g., an SCE equalizer in a turbo loop with soft-in-soft-out decoding. Compared to orthogonal space-time codes, where the rate is at most 1, the rate of MC-CAFS is – independent of the current channel – the maximum rate, which is identical to the number n_T of transmit antennas. If the MIMO-OFDM channel does not have the potential for spatial multiplexing, zero or small eigenvalues of the channel matrix $\underline{\underline{R}}$ occur. As a result, the BER performance decreases, but the transmission rate remains always constant. This is comparable to MC-CDM (or MC-CDMA) if subchannels are faded out, because the subchannel gains of OFDM correspond directly to the eigenvalues in case of MIMO-OFDM. Spreading is a very general concept. This contribution is based on the work of Doris Yacoub [5–7]. More about this topic and its relation to general space-time or space-frequency coding and also dispersion codes can be found in [9]. The author thanks Matthias Wetz for preparing some parts of this text.

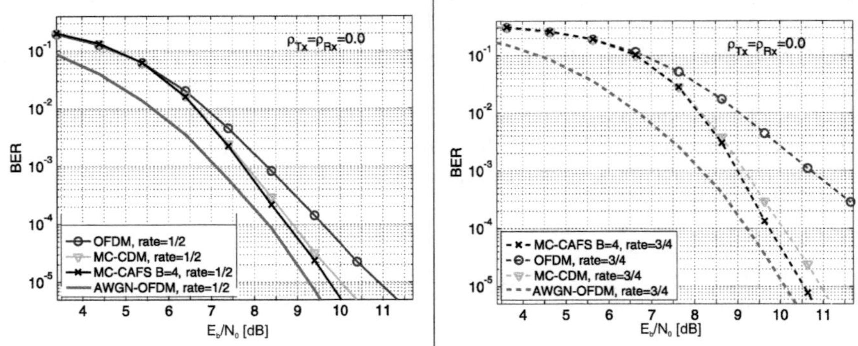

a) code rate 1/2 b) code rate 3/4

Figure 3.12: BER of MIMO-OFDM 4PSK with convolutional code, memory 2, $n_T = n_R = 4$, $N = 32$, and $L = 4$.

Bibliography

[1] K. Fazel. “Performance of CDMA/OFDM for Mobile Communication Systems,” *Proc. ICUPC ’93,* pp. 975-979, 1993.

[2] J. Lindner, “MC-CDMA in the context of general multiuser/multisubchannel transmission methods,” *ETT, European Transactions on Telecommunications,* vol. 10, no. 4, pp. 351-367, 1999.

[3] V. Tarokh, H. Jafarkhani, and A.R. Calderbank, “Space-time block codes from orthogonal designs,” *IEEE Trans. Inform. Theory,* vol. 45, no 5, 1999.

[4] D. Y. Yacoub, W. G. Teich, and J. Lindner, “Effect of Antenna Correlations on Interference and Performance of a Spread MIMO-OFDM System (MC-CAFS),” *11th International OFDM-Workshop 2006*, Hamburg / Germany, Aug. 2006.

[5] D. Y. Yacoub, W. G. Teich, and J. Lindner, “Precoding and Spreading for MIMO-OFDM in the Presence of Antenna Correlations,” *12th International OFDM-Workshop 2007*, Hamburg / Germany, Aug. 2007.

[6] D. Y. Yacoub, *Spreading and Coding for Wireless MIMO-OFDM Systems,* Dissertation, Institute for Information Technology, University of Ulm, June 2008.

[7] D. Y. Yacoub, W. G. Teich, and J. Lindner, “MC-Cyclic Antenna Frequency Spread: A Novel Space-Frequency Spreading for MIMO-OFDM,” *8th International Symposium on Communication Theory and Applications (ISCTA ’05),* Ambleside, Lake District / UK, July 2005.

[8] J. Egle, C. Sgraja, and J. Lindner, “Iterative soft Cholesky block decision feedback equalizer - a promising approach to combat interference,” *in Proc. IEEE*

Vehicular Technology Conference (VTC Spring), vol. 3, Rhodes / Greece, May 2001, pp. 1604-1608.

[9] C. Pietsch, *Coherent Space Time Block Codes from Sets of Subspaces*, Dissertation, Institute for Information Technology, University of Ulm, October 2008.

3.3 Iterative Diversity Reception for Coded OFDM Transmission Over Fading Channels

M. Matuszak, R. Urbansky, University of Kaiserslautern, Germany

3.3.1 Introduction

Time- and frequency-selective fading resulting from time-variant multipath propagation can be mitigated by forward error correction (FEC) channel coding in combination with time and frequency interleaving. In broadband systems coded orthogonal frequency-division multiplexing (COFDM) is a well-known implementation of this concept, which is applied, e.g., in digital terrestrial video broadcasting (DVB-T), Digital Audio Broadcasting (DAB), and Digital Radio Mondiale (DRM) [1] [4].

A bandwidth of 1.5 MHz for DAB or 8 MHz for DVB-T enables efficient frequency interleaving to mitigate fading. However, DRM is restricted to a bandwidth of up to 20 kHz for compatibility to existing services [3]. In addition, DRM transmission especially in short-wave bands is characterized by time varying ionospheric fading. Whereas fast fading is covered by time interleaving, long-term frequency-selective fading severely affects transmission, since a narrow bandwidth results in a high percentage of subcarriers with low signal-to-noise ratio (SNR), which may exceed FEC capabilities.

Diversity techniques also allow to mitigate fading [5]. COFDM systems, e.g., can inherently utilize delay diversity or path diversity in single frequency networks, provided the OFDM guard interval covers the maximum path or delay spread [1]. In addition, for narrow-band systems like DRM, antenna diversity, polarization diversity and especially frequency diversity may also be taken into account. Receiver concepts for frequency or antenna diversity usually apply combining techniques, like selection combining, equal gain combining or maximum ratio combining (MRC). In general, these methods combine the properly equalized and synchronized analog signals before FEC decoding.

We proposed a different approach: since diversity transmission of FEC encoded data can be regarded as a parallel concatenated coding scheme which allows for turbo decoding, we combine the received and appropriately equalized signals in an iterative decoding process, see Fig. 3.13 [3].

Propagating extrinsic information in terms of log-likelihood-ratio (LLR), the turbo diversity (TD) scheme delivers additional iteration gains compared to MRC [2]. This requires that the constituent component codes, usually punctured convolutional codes (CC), have to be chosen appropriately by applying extrinsic information transfer (EXIT) chart methods. Instead of CC, codes can be also applied, where again these methods have been used [7]. The project focuses mainly on LDPC codes as constituent codes, because they are known to approach the Shannon limit as close as Turbo codes (TC) [8] and efficient soft-input soft-output decoding algorithms are available.

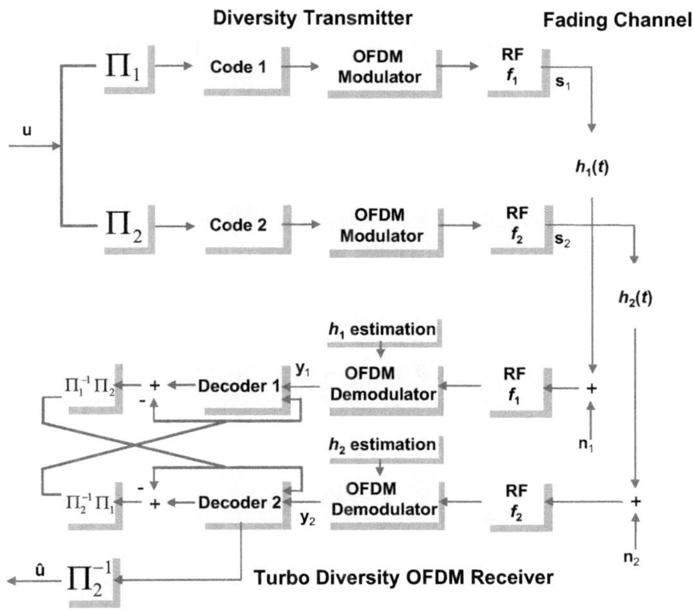

Figure 3.13: Iterative diversity receiver for OFDM systems.

3.3.2 Turbo Diversity

A soft-input soft-output (SISO) decoder which allows for propagating extrinsic information is the key element of the turbo decoding principle [6]. Therefore, an optimum maximum-a-posteriori symbol-by-symbol estimator like the Bahl-Cocke-Jelinek-Raviv (BCJR) algorithm for turbo codes or the sum-product algorithm [7] for LDPC codes have be applied, calculating for each bit c_k the LLR conditioned on the received sequence

$$\Lambda\left(c_k|\mathbf{r}\right) = \ln\left(P\left(c_k = 1|\mathbf{r}\right)/P\left(c_k = 0|\mathbf{r}\right)\right) \tag{3.20}$$

According to the turbo decoding principle, extrinsic information Λ_e is exchanged between the decoding stages, where for Λ_e any a-priori information has to be subtracted from the BCJR output reliability information, see Fig. 3.13.

The iteration process can be visualized by EXIT charts, where the mutual information (MI) of one decoder is plotted versus its a-priori input, i.e., the MI of the other decoder [9]. Using optimum decoders, the knowledge of the MI contained in its a-priori information is sufficient to derive the MI of Λ_e delivered by this decoder.

Figure 3.13 motivates the similarity of TC and TD. Consequently, EXIT chart methods considering the constituent LDPC codes allow for analyzing the iteration process of TD. In this case EXIT functions are derived from the parameters of LDPC codes, namely coefficients of degree distributions of variable nodes $\lambda(x)$ and of check nodes $\rho(x)$, denoted here shortly as (λ, ρ) [7]. Properties of TD can be examined using a scenario with two diversity paths, where the SNR of both paths differ by ΔSNR dB.

3.3.3 Optimization for Turbo Diversity

Proper LDPC code selection is the crucial issue of TD system design. Here we assume identical codes for both diversity branches. In the consecutive, different code optimization methods are considered.

The idea of the design method No. 1 is depicted in Fig. 3.14. The tunnel between the EXIT function of the first decoder $\mathcal{T}_{D1,n}$ and the inverse EXIT function $\mathcal{T}_{D2,n}^{-1}$ should be open. Taking into account the criterion of the minimal area gives rise to the definition of the first design method [2].

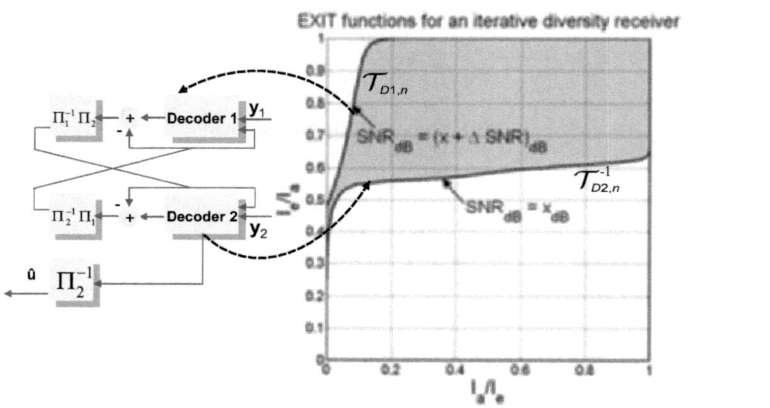

Figure 3.14: Idea of design method No. 1: area between EXIT functions of both decoders should be minimal.

Design method No. 2 makes use of the fact that for better SNR values the inverse EXIT function of the second decoder $\mathcal{T}_{D2,n}^{-1}$ calculated from (λ, ρ) "shifts" to the lower part of the EXIT chart. Hence, if it is feasible for a bad SNR value to shift an EXIT curve to the lower part of EXIT chart, then (λ, ρ) of the resulting EXIT function is expected to improve the performance of the iteration process. Besides, the requirement of an opened tunnel between $\mathcal{T}_{D1,n}$ and $\mathcal{T}_{D2,n}^{-1}$ will not be affected by this procedure.

The input LDPC code parameters for the proposed method No. 2 are (R, $d_{c_{min}}$, $d_{c_{max}}$) where R is the code rate, $d_{c_{min}}$ and $d_{c_{max}}$ are minimum and maximal acceptable check node degree in the Tanner graph of the LDPC code, respectively [7]. Using these parameters and exploiting relations to the code rate, other parameters ($d_{l_{max}}$, $\lambda_{2_{min}}$, $\lambda_{2_{max}}$, d_c^-, i, λ_2, λ_j, j) can be obtained, from which the last five are related to the code rate as follows [3]:

$$R = 1 - \frac{\frac{1}{d_c}}{\frac{\lambda_2}{2} + \frac{\lambda_i}{i} + \frac{\lambda_j}{j} + \frac{1-\lambda_2-\lambda_i-\lambda_j}{d_{l_{max}}}} \tag{3.21}$$

Method No. 2 proceeds in an iterative way and utilizes N five-dimensional parameter vectors $\mathbf{A} = (d_c, i, \lambda_2, \lambda_j, j)$ serving for determination of a specified number of MI values (I_e, I_a), and a search is made for the best vector which fulfills a certain condition at each iteration. This condition reads as follows: the sum of distances of MI values corresponding to a vector $\mathbf{A}$ for a low SNR value to the predefined straight line should be minimal. The line is placed in the lower part of the EXIT chart and is selected from a group of parallel lines at each iteration according to the procedure presented in the Fig. 3.15. The best vector at each iteration participates in the next iteration and the parameter vector selection takes places using methods of differential evolution [7]. As a result, LDPC code parameters are achieved for which an $\mathcal{T}_{D2,n}^{-1}$ for a certain SNR value is shifted to the preferred area (lower) of the EXIT chart as opposed to initial LDPC code parameters in the first iteration.

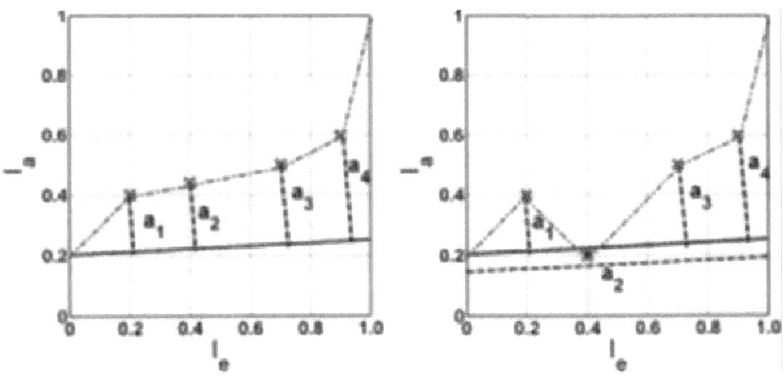

Figure 3.15: Design method No. 2: if $\sum_i a_i < 10^{-9}$ or there is no improvement in the following iteration, points (I_e, I_a) serve as solution. (a) If all points (I_e, I_a) lie above the blue solid line, the same line is used in the next iteration (b) If one of the points (I_e, I_a) lies below the blue solid line, the blue dashed line will be used in the next iteration.

3.3.4 Performance Evaluation

Turbo Diversity vs. MRC

Figure 3.16 demonstrates the performance gain of TD compared to MRC for an SNR difference of $\Delta SNR = 4$ dB and $\Delta SNR = 6$ dB. To assess TD in a COFDM framework we have focused on a system design considering the DRM broadcasting standard, mode B, spectrum occupancy 0 [10] and 16-QAM. For LDPC codes at BER $= 10^{-3}$, the performance gain after 25 iterations of a turbo loop and 10 iterations of the LDPC decoder compared to MRC with 250 LDPC decoding iterations is in the order of 1.5 dB and 3 dB, for $\Delta SNR = 4$ dB and $\Delta SNR = 6$ dB, respectively. In the case of turbo codes, where only 10 iterations of a turbo loop and 10 iterations of the inner decoder were considered in order to ensure a comparable computational effort, TD outperforms MRC by ca. 1.5 dB.

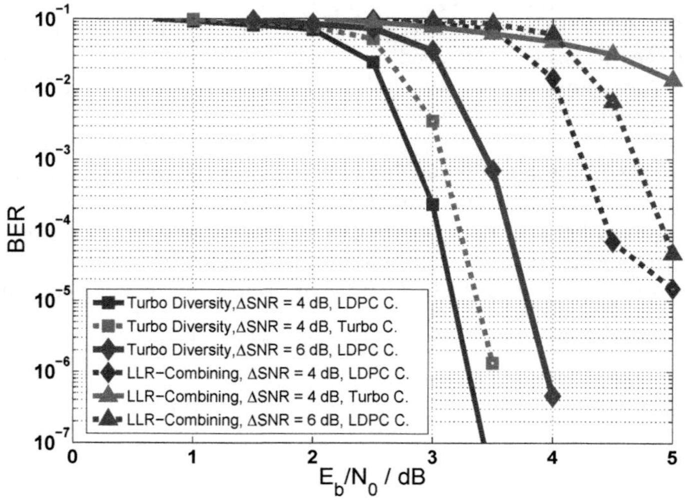

Figure 3.16: BER performance of MRC and TD using LDPC codes, turbo codes and RSCC (CC) for a DRM-based scenario.

LDPC Code Design

Wiener filter

The performance analysis was carried out for an OFDM scenario in which the channel profile followed "US Consortium" with path delays (1, 0.7, 1.5, 2.2 ms), path gains (1, 0.7, 0.5, 0.25) and normalized Doppler frequency $f_D = 0.05$. Channel estimation was implemented using the Wiener filter method [11] and LDPC codes of rate 3/4 and block length 9000 were used.

EXIT characteristics are depicted for both code design techniques in Fig. 3.17 a), where the corresponding LDPC code parameters are ($\lambda_2 = 0.198$, $\lambda_4 = 0.2441$, $\lambda_5 = 0.1421$, $\lambda_{29} = 0.093$, $\lambda_{30} = 0.151$, $\lambda_{34} = 0.1715$, $\rho_{19} = 0.1655$, $\rho_{20} = 0.8345$) and ($\lambda_2 = 0.441$, $\lambda_{30} = 0.5589$, $\rho_{16} = 0.2644$, $\rho_{17} = 0.7356$), respectively. The resulting EXIT functions of method No. 2, for low SNRs are wider opened than EXIT functions of method No. 1, what can be understood as better performance for a lower number of iterations. Despite of another design criterion for method No. 2, the area between resulting EXIT functions is smaller than for method No. 1, which is related to a lower distance to channel capacity.

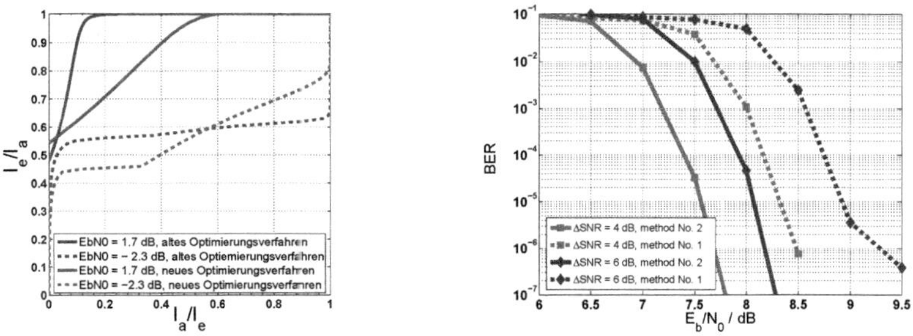

(a) (b)

Figure 3.17: Performance analysis of Turbo Diversity with optimized LDPC codes for different code design methods a) EXIT functions, b) BER performance of TD with difference between diversity paths $\Delta SNR = 4$ dB and $\Delta SNR = 6$ dB

Comparing the performance of these methods for the first and the second scenario (Fig. 3.17 b)), considering BER=10^{-3} it can be seen that for TD method No. 2 outperforms method No. 1. For LDPC codes with blocks of 9000 bits this gap is nearly 0.9 dB. Superiority of method No. 2 does not depend on the fading scenario but it should be mentioned that LDPC codes achieved by means of the method No. 2 turn to be worse for the system without TD (only one path) and on Additive White Gaussian Noise channel.

3.3.5 Summary

In this contribution we investigated Turbo Diversity as an iterative decoding concept for diversity systems which outperforms MRC in single carrier and OFDM systems. Additional gains are achieved by using optimized LDPC codes as constituent codes. Two different criteria were proposed for the LDPC code design adapted to TD with OFDM. The method using the design goal to position the inverse EXIT function in the preferred area outperforms the classical criterion aimed at minimal area between EXIT functions. Despite of the fact that the area between EXIT characteristics does not have first priority in our new method, this area turns out to be less than in the case of the classical method, thus fulfilling both criteria in a better way and reducing the distance to channel capacity. The new method is better suited for different fading scenarios and block lengths of LDPC codes, what proves its robustness. For practical applications of TD the new method to design constituent LDPC codes delivers a robust approach for different fading scenarios and block lengths of LDPC codes.

Bibliography

[1] R. van Nee and R. Prassad, *OFDM for Wireless Multimedia Communications*, Artech House Publishers, Norwood, MA., USA, 2000.

[2] M. Matuszak, W. Sauer-Greff and R. Urbansky, "Iterative Diversity Receiver Concept for Narrow-Band OFDM Systems," in *Proc. 13. International OFDM Workshop*, pp. 251-255, Hamburg, Germany, 2008.

[3] M. Matuszak, W. Sauer-Greff and R. Urbansky, "EXIT Chart Based LDPC Code Design for Iterative Diversity Receivers in OFDM Systems with Fading Channels," in *Proc. 14. International OFDM Workshop*, pp. 20-24, Hamburg, Germany, 2009.

[4] A. Dittrich, T. Schorr and R. Urbansky, "Diorama - A MATLAB Based Open Source Software Radio for Digital Radio Mondiale (DRM)," in *Proc. 10. International OFDM-Workshop*, pp. 391-395, Hamburg, Germany, 2005.

[5] A. Paulraj, "Diversity Techniques," in J.D. Gibson (Ed.) *The Mobile Communication Handbook*, pp. 166-176, CRC Press, Boca Raton, FL., USA, 1996.

[6] J. Hagenauer, "Das Turbo-Prinzip in Detektion und Decodierung," in *ITG-Fachberichte*, Vol. 146, pp. 131-136, 1998.

[7] T. Richardson, M. Shokrollahi and R. Urbanke, "Design of Capacity-Approaching Irregular Low-Density Parity-Check Codes," *IEEE Trans. on Inform. Theory*, Vol. 47, No. 2, pp. 619-637, Febr. 2001.

[8] L. Bahl, J. Cocke, F. Jelinek and J. Raviv, "Optimal Decoding of Linear Codes for Minimizing Symbol Error Rate," *IEEE Trans. on Inform. Theory*, Vol. 20, No. 2, pp. 284-287, March 1974.

[9] S. ten Brink, "Convergence Behavior of Iteratively Decoded Parallel Concatenated Codes," *IEEE Transactions on Communications*, Vol. 49, No. 10, pp. 1727-1737, Oct. 2001.

[10] European Telecommunications Standard Institute (ETSI), *Digital Radio Mondiale (DRM); System Specification,* ETSI ES 201 980 V2.1.1, Sophia Antipolis Cedex, France, 2004.

[11] C. Sgraja and J. Linder, "Estimation of Rapid Time-Variant Channels for OFDM using Wiener Filtering," in *Proc. IEEE International Conference on Communications (ICC)*, Vol. 4, pp. 2390-2395, AK., USA, May 2003.

3.4 MMSE-based Turbo Equalization Principles for Frequency Selective Fading Channels

M. Grossmann, R. Thomä, Ilmenau University of Technology, Germany

3.4.1 Introduction

Turbo equalization [1–11] is one of the most promising techniques to implement powerful equalizers, without requiring high computational complexity, for coded communication systems with frequency-selective fading channels. The complexity advantage of turbo equalizers is due to the separation of channel equalization and decoding into two basic processors, while the high performance is achieved by exchanging soft information between these two components in an iterative manner. The turbo equalization approach was originally proposed in [1], utilizing a maximum a posteriori probability (MAP) algorithm for iterative equalization in frequency-selective fading channels. However, because of its exponentially increasing complexity, the MAP-based equalizer is only practical for simple modulation formats, like binary phase shift keying (BPSK), and for channels with less multi-path components. In [2], the optimal MAP algorithm has been replaced by a low-cost alternative based on the soft cancellation (SC) and minimum mean-squared error (MMSE) principle. The SC-MMSE filtering approach in [2], originally proposed for detection of random coded code-division multiple-access (CDMA) signals, has been applied to channel equalization in [3], and to multiple-input multiple-output (MIMO) channel equalization in [4]. Recently, the SC-MMSE turbo concept has also been used to equalization of OFDM systems in time-varying channels [5].

In this contribution, we discuss three extensions of the basic SC-MMSE filtering concept for turbo equalization. In particular, we first propose a novel frequency domain (FD) turbo equalizer for MIMO-OFDM transmissions with insufficient guard interval. The SC-MMSE-based equalizer exploits the banded structure of the FD channel matrices, allowing the implementation of the equalizer with a complexity which is only linear in the number of sub-carriers. We then present a hybrid turbo equalizer, suitable for multi-user OFDM transmissions with spatially-correlated channels, that combines SC-MMSE filtering and MAP-detection. Finally, we propose a nonlinear MMSE-based turbo equalizer for single-carrier spatially-multiplexed MIMO transmissions with high-rate codes.

3.4.2 System Model

Consider the discrete-time baseband equivalent model of a cyclic-prefix (CP) assisted block transmission single-/multi-user system with M receive antennas, and U active users, each equipped with K transmit antennas in Fig. 3.18. The transmission scheme of the u-th user is based on bit interleaved coded modulation, where the information bit sequence is independently encoded by a binary encoder, and mapped to complex symbols according to the applied mapping scheme. The encoded symbol sequence is then grouped into several blocks, OFDM modulated, and finally transmitted over the frequency-selective multiple-access MIMO fading channel.

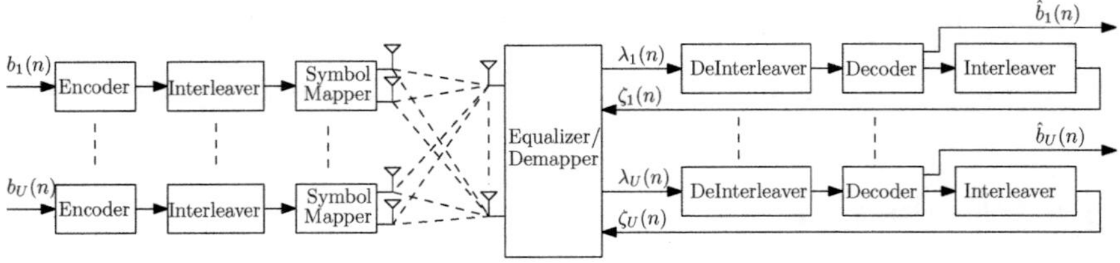

Figure 3.18: Structure for a coded single-/multi-user MIMO system with turbo equalization.

At the receiver side, iterative processing for joint equalization and decoding is performed. The receiver consists of an equalizer and several single-user channel decoders. Within the iterative processing, the *extrinsic* LLR sequences $\{\lambda_u(n)\}$ and $\{\zeta_u(n)\}$ of the coded bits $\{b_u(n)\}$ are exchanged between the equalizer and both decoders, each separated by the interleaver and deinterleaver in their iteration loop, following the turbo principle [2].

3.4.3 MMSE Turbo Equalization Principles

MMSE Turbo Equalization for MIMO OFDM Transmission with Insufficient Cyclic Prefix

In single-user MIMO-OFDM transmissions, the CP, located between neighboring OFDM symbols, should be longer than the expected length of the channel impulse response, to maintain orthogonality among sub-carriers. However, in channels with some far clusters, the maximum excess delay may exceed the length of the CP, resulting in a loss of sub-carriers' orthogonality. The related inter-block and inter-carrier interference severely degrades the performance of the standard MIMO-OFDM receiver [12]. Several approaches have been proposed to cope with this problem [13], [14]. Among all these equalization schemes for OFDM and MIMO-OFDM, one of the the most promising approaches is the iterative (turbo) SC-MMSE equalizer of [6]. The equalizer utilizes the soft feedback from channel decoding for the separation of the spatially multiplexed streams jointly with the cancellation of inter-block and inter-carrier interference. However, for transmissions with a large number of sub-carriers, the receiver has still a high complexity. In [7], an MMSE turbo equalizer that exploits the banded structure of the FD channel matrices has been proposed. The equalizer in [7] uses a sliding window to enforce this banded structure, resulting in a complexity which is only linear in the number of sub-carriers.

The performance of the proposed turbo receiver was evaluated using measurement data-based off-line simulations. The measurements were performed in a macro cell environment. The major specifications of the measurement campaign, the measurement device and setup are summarized in [15]. The main MIMO-OFDM simulation parameters follow the extended specification of the 5 GHz wireless LAN standard in [16]. A single-user MIMO-OFDM transmission ($U = 1$) with $K = M = 2$ antennas and 64 sub-carriers were assumed. Channel coding was performed with a

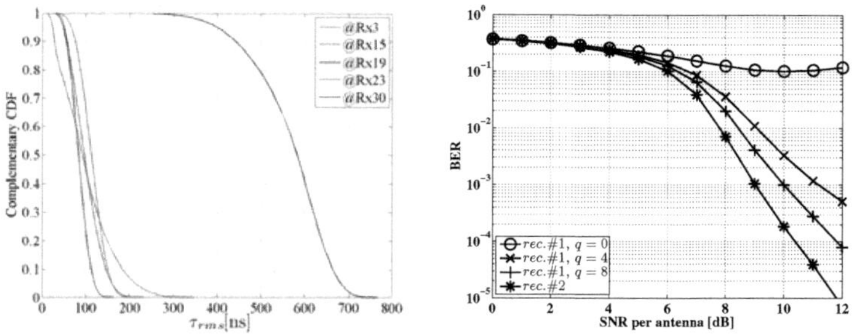

(a) (b)

Figure 3.19: (a) Complementary CDF of channel delay spread (τ_{rms}) for different positions of the measurement route. (b) BER performance versus average E_b/N_0 for the measured channel at position Rx19, 16-QAM transmission.

rate-1/2 convolutional code. The OFDM symbol duration was set to $T_S = 4\mu s$ and the length of the cyclic prefix to $T_c = 0.8\mu s$.

Figure 3.19 (a) shows the complementary cumulative probability density function (CDF) of the root-mean-squared channel delay spread for different positions of the measurement route. As observed from Fig. 3.19 (a), the delay spreads at position Rx19 are significantly large and exceed in some cases the length of the cyclic prefix of the MIMO-OFDM system, compared to other locations.

The BER performance of the proposed turbo receiver at location Rx19, referred to as 'rec. #1' for different values of the FD window length q is shown in Fig. 3.19 (b). As a reference the BER performance of the conventional MIMO-OFDM receiver ($q = 0$) and the non-banded SC-MMSE turbo receiver, referred to as 'rec. #2', are also shown. It can be seen that with $q = 8$, the performance of the banded SC-MMSE turbo equalizer is very close to the non-banded turbo equalizer. Thus, the proposed technique works with properly chosen q values even in environments with relatively large channel delay spreads.

Hybrid MMSE Turbo Equalization

The SC-MMSE turbo detector in [3], [6] achieves excellent performance in frequency-selective *spatially-uncorrelated* Rayleigh multiple-access fading channels. However, the performance of the detector drops significantly in scenarios where the users' channels are spatially correlated [9]. In [9], a hybrid turbo detection approach based on group-wise SC-MMSE filtering combined with MAP signal detection for OFDM multi-user systems has been proposed. The technique in [9], referred to as Hy SC-MMSE-MAP turbo detection, offers a design flexibility in terms of complexity in computation and robustness against spatial correlation of the users' channels.

The Hy SC-MMSE-MAP detector separates the U users into H disjoint groups $\{\mathcal{G}_1, \mathcal{G}_2, ..., \mathcal{G}_H\}$, such that each group contains G_h integers corresponding to user

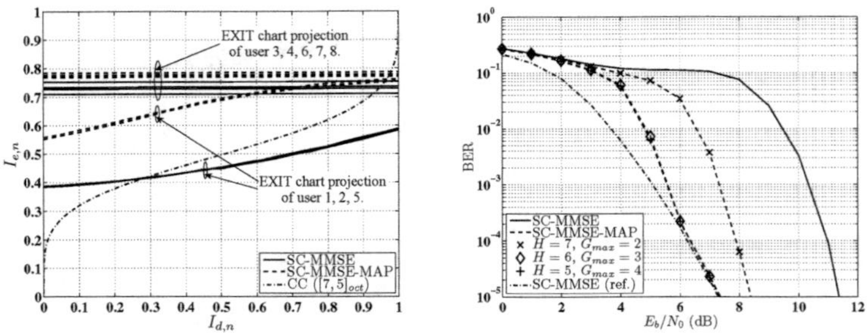

(a) (b)

Figure 3.20: (a) EXIT chart projection for each user of the conventional SC-MMSE and the proposed Hy SC-MMSE-MAP turbo detector with $(H; G_{max}) = (6; 3)$ for a random channel realization at 5 dB E_b/N_0. (b) BER performance of the SC-MMSE and Hy SC-MMSE-MAP turbo detector for BPSK transmission.

indexes. A trellis-based user grouping algorithm is used to allocate the user signals that have a high pairwise receive correlation into one group. For each group $\mathcal{G}_h$, soft cancellation of multiple-access components that originate from the remaining $H-1$ groups is performed by utilizing the extrinsic information fed back from channel decoding. For further suppression of residual interference components, group-wise SC-MMSE filtering is applied, jointly detecting the users' signals in one group. Based on a Gaussian approximation of the MMSE filter output signals, MAP (symbol) detection of the user signals in each group is then performed. The extrinsic LLRs on the coded transmit bits calculated for each user are then forwarded to the corresponding channel decoder. Note that the computational complexity for the MAP part of the detector is at an exponential order of the group size, which may dominate the required computational effort. On the contrary, for small group sizes, the most computationally complex part is due to the inversion of the MMSE filtering matrix. It should also be mentioned here that when each user is regarded as one group by itself, i.e., $H = U$, the Hy SC-MMSE-MAP turbo detector becomes equivalent to the SC-MMSE turbo detector of [3], [6].

The convergence property of the proposed receiver was analyzed by extrinsic information transfer (EXIT) charts [8]. An OFDM system in a Rayleigh fading environment having $U = 8$ active users, each having a single transmit antenna, and $M = 8$ uncorrelated receive antennas was considered, where three of the eight users' channels are significantly spatially correlated (correlation coefficient of $\rho = 0.95$) and the remaining users' channels are close to orthogonal ($\rho = 0$).

Figure 3.20 (a) illustrates the EXIT curves for each user of the conventional SC-MMSE and the Hy SC-MMSE-MAP detector with $(H; G_{max}) = (6; 3)$, where $G_{max} = \max G_h$. Note that the two-dimensional EXIT curves for each user were obtained from the multi-dimensional EXIT surfaces of the detector by the projection technique [17]. As observed in Fig. 3.20 (a), the Hy SC-MMSE-MAP detector improves significantly the convergence threshold for the three highly correlated user signals which stems from the group selection that allocates those users into one group. This indicates that the proposed detector can achieve better performance in the presence of high spatial channel correlation.

In addition to the EXIT analysis, simulations were carried out to evaluate the BER performance of the proposed detector. The effect of the group size on the BER performance of the Hy SC-MMSE-MAP detector for BPSK transmission with rate-1/2 convolutional codes is shown in Fig. 3.20 (b). For comparison, the performance of the conventional SC-MMSE detector in the spatially-correlated fading channel and in a spatially-uncorrelated fading channel are shown as well, and are referred to as SC-MMSE and SC-MMSE (ref), respectively. It is shown that Hy SC-MMSE-MAP can achieve the same performance as the conventional SC-MMSE detector in uncorrelated fading channels, even when strong correlation among the user signals exists.

Nonlinear MMSE Turbo Equalization using Probabilistic Data Association

In [11], a FD SC-MMSE turbo equalizer for spatial multiplexing single-carrier MIMO systems based on the framework of nonlinear MMSE (NMMSE) estimation was derived. It is shown that the computation of the NMMSE estimate of the coded transmitted symbols involves a sum of terms, which grows exponentially in the number of sub-carriers and transmit antennas. To reduce the complexity in computation, the probabilistic data association (PDA) filtering idea [18] is adopted, where the composite inter-symbol and multiple-access interference component is approximated by a multivariate Gaussian random process. The expression resulting from this Gaussian approximation can be iteratively solved following the PDA principle. As a result, the structure of the proposed turbo equalizer, denoted as PDA FD SC-MMSE turbo equalizer in the following, is similar to the FD SC-MMSE turbo equalizer of [10]. However, with the presented method, internal iterations within the equalizer following the PDA principle are used to improve the NMMSE estimates.

The BER performance of the proposed PDA FD SC-MMSE equalizer after 10 turbo iterations, five internal iterations in each turbo iteration, is shown in Fig. 3.21. For comparison, the performance of the conventional FD SC-MMSE equalizer is shown as well, and is referred to as FD SC-MMSE (ref). For the simulations, we considered a single-user $K = M = 2$ MIMO system utilizing constraint length three convolutional codes [19] with rates $r = 7/8$, $2/3$, and $1/2$. A spatially-uncorrelated Rayleigh fading environment with equal average power delay profile was assumed. The length of the CP was set to the maximum channel delay. As observed in Fig. 3.21, the PDA FD SC-MMSE equalizer outperforms the FD SC-MMSE equalizer, where the larger the rate r the larger the performance gain. Moreover, the FD SC-MMSE equalizer using the rate $r = 7/8$ code fails to converge for channels with less channel multi-path components for high E_b/N_0 values. In contrast, the additional internal iterations of the PDA FD SC-MMSE equalizer improve the convergence threshold, and hence, it can achieve better performance. Thus, the proposed equalizer significantly improves the convergence properties over the FD SC-MMSE

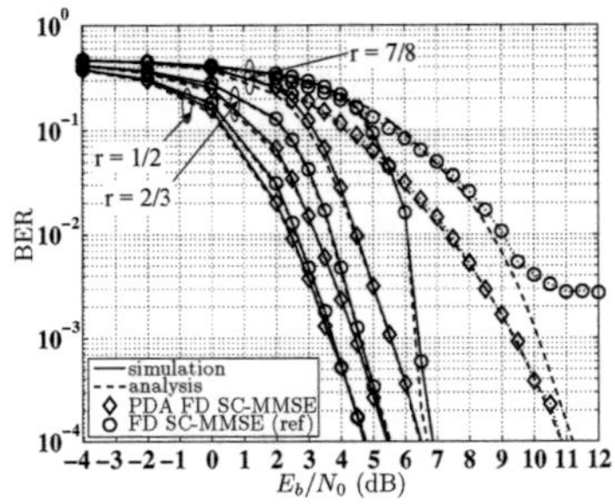

Figure 3.21: BER performance of the FD SC-MMSE and the proposed PDA FD SC-MMSE turbo equalizer utilizing convolutional codes with rates $r = 7/8$, $2/3$, and $1/2$ for BPSK transmission over spatially-uncorrelated Rayleigh fading channels with 32-path components (solid curves) and 2-path components (dotted curves).

technique, although they both have the same order of computational complexity.

3.4.4 Summary

We have discussed three possible extensions of the basic SC-MMSE filtering concept for turbo equalization. First, we have proposed a low-complexity SC-MMSE-based turbo equalizer for MIMO-OFDM transmissions with insufficient guard interval. The equalizer exploits the banded structure of the FD channel matrices, resulting in a complexity which is only linear in the number of sub-carriers. We also have proposed a hybrid turbo equalizer, Hy SC-MMSE-MAP, that combines SC-MMSE filtering and MAP-detection. Numerical results show that Hy SC-MMSE-MAP offers robustness against the users' channel correlation over the conventional SC-MMSE turbo detector in spatially-correlated multiple-access fading channels. The third turbo equalizer proposed, PDA FD SC-MMSE, is based on the PDA filtering concept in the framework of NMMSE estimation. It has been shown through simulations that PDA FD SC-MMSE significantly improves the convergence properties over the conventional FD SC-MMSE technique for transmission with high-rate codes.

Bibliography

[1] C. Douillard, et al., "Iterative correction of intersymbol interference: Turbo equalization," *European Trans. Telecomm.*, vol. 6, no. 5, pp. 507-511, Sept. 1995.

[2] X. Wang, and H. V. Poor, "Iterative (turbo) soft interference cancellation and decoding for coded CDMA," *IEEE Trans. Commun.*, vol. 47, no. 7, pp. 1046-1061, July 1999.

[3] M. Tüchler, and J. Hagenauer, "Turbo equalization: principles and new results," *IEEE Trans. Commun.*, vol. 50, pp. 754-767, May 2002.

[4] T. Abe, and T. Matsumoto, "Space-time turbo equalization in frequency-selective MIMO channels," *IEEE Trans. Veh. Techn.*, vol. 52, no. 3, pp. 469-475, May 2003.

[5] K. Fang, L. Rugini, and G. Leus, "Low-complexity block turbo equalization for OFDM systems with time-varying channels," *IEEE Trans. Signal Process.*, vol. 56, pp. 5555-5566, Nov. 2008.

[6] C. Schneider, M. Grossmann, and R. S. Thomä, "Measurement based performance evaluation of MIMO-OFDM with turbo-equalization," *in Proc. IEEE VTC2005 Spring*, Stockholm, Sweden, June 2005.

[7] M. Grossmann, C. Schneider, and R. Thomä, "Turbo equalization for MIMO-OFDM transmission with insufficient guard interval," *International Zurich Seminar on Communications*, Zurich, Switzerland, pp. 114-117 , Feb. 2006.

[8] S. ten Brink, "Convergence behavior of iteratively decoded parallel concatenated codes," *IEEE Trans. Commun.*, vol. 49, no. 10, pp. 1727-1737, Oct. 2001.

[9] M. Grossmann, and T. Matsumoto, "Hybrid turbo multiuser detection for OFDM transmission with spatially-correlated channels," *IEEE Commun. Letters*, vol. 11, pp. 420-422, May 2007.

[10] K. Kansanen, and T. Matsumoto, "An analytical method for MMSE MIMO turbo equalizer EXIT chart computation," *IEEE Trans. Wireless Commun.*, vol. 6, no. 1, pp. 59-63, Jan. 2007.

[11] M. Grossmann, and T. Matsumoto, "Nonlinear Frequency Domain MMSE Turbo Equalization using Probabilistic Data Association," *IEEE Commun. Letters*, vol. 12, pp. 295-297, April 2008.

[12] G.L. Stüber et al., "Broadband MIMO-OFDM wireless communications," *Proceedings of the IEEE*, vol. 92, no. 2, pp. 271-294, Feb 2004.

[13] D. Kim, and G.L. Stüber, "Residual ISI cancellation for OFDM with application to HDTV broadcasting," *IEEE J. Select. Areas Commun.*, vol. 16, pp. 1590-1599, Oct. 1998.

[14] G. Leus, and M. Moonen, "Per-Tone equalization for MIMO OFDM systems," *IEEE Trans. Signal Process.*, vol. 51, pp. 2965-2975, Nov. 2003.

[15] M. Landmann et al., "Polarisation Behaviour of Discrete Multipath and Diffuse Scattering in Urban Environments at 4.5GHz,"*EURASIP Journal on Wireless Communications and Networking, Special Issue on Space-Time Channel Modeling for Wireless Communications and Networking*, Article ID 57980, 16 pages, Jan 2007.

[16] IEEE Standard 802.11a-1999, "High-speed physical layer in the 5 GHz band," 1999.

[17] F. Brännström et al., "Convergence analysis and optimal scheduling for multiple concatenated codes," *IEEE Trans. Inform. Theory*, vol. 51, pp. 3354-3364, Sept. 2005.

[18] J. Luo et al., "Near Optimal Multiuser Detection in Synchronous CDMA using Probabilistic Data Association," *IEEE Comm. Letters*, vol. 5, no. 9, pp. 361-363, Sep. 2001.

[19] J. G. Proakis, "Digital Communications," *4th ed. New York: McGraw-Hill*, 2001.

3.5 Peak-to-Average Power Ratio Reduction in Multi-Antenna Scenarios

R. Fischer, C. Siegl, University of Erlangen-Nürnberg, Germany

3.5.1 Introduction and Overview on PAR Reduction Schemes

Regardless its considerable advantages, like, e.g., simple equalization, high adaptivity, etc., orthogonal frequency-divsion multiplexing (OFDM) has one significant problem: due to the superposition of a large number of individual components within the (inverse) discrete Fourier transform ((I)DFT), the complex amplitude OFDM transmit signal tends to be Gaussian distributed. Hence it exhibits a very large *peak-to-average power ratio (PAR)* [29].

Denoting the discrete-time transmit symbols (after the IDFT of length D) by a_k, $k = 0, \ldots, D-1$, the PAR is defined as[1] ($\mathrm{E}\{\cdot\}$ denotes expectation)

$$\mathrm{PAR} \overset{\mathrm{def}}{=} \frac{\max_{k=0,\ldots,D-1} |a_k|^2}{\mathrm{E}\{|a_k|^2\}} \tag{3.22}$$

As common, the probability that the PAR of an OFDM frame exceeds a given threshold $\mathrm{PAR_{th}}$ serves as performance measure. Via the *complementary cumulative distribution function (ccdf)* of the PAR $\Pr\{\mathrm{PAR} > \mathrm{PAR_{th}}\}$ ($\Pr\{\cdot\}$ denotes probability), clipping probabilities can be assessed. Noteworthy, for conventional OFDM, assuming Gaussian time-domain samples, the simple approximation

$$\Pr\{\mathrm{PAR} > \mathrm{PAR_{th}}\} = 1 - (1 - \mathrm{e}^{-\mathrm{PAR_{th}}})^D \approx D\,\mathrm{e}^{-\mathrm{PAR_{th}}} \tag{3.23}$$

holds [28].

Future OFDM transmission systems will use multiple antennas, especially to increase data rate (spatial multiplexing). In such systems, the PAR problem even gets worse, since here the PAR of all N_T transmit signals should be simultaneously as small as possible. Hence, the *worst-case PAR*

$$\mathrm{PAR} \overset{\mathrm{def}}{=} \max_{\mu=1,\ldots,N_\mathrm{T}} \mathrm{PAR}_\mu = \frac{\max_{\mu=1,\ldots,N_\mathrm{T},\, k=0,\ldots,D-1} |a_{\mu,k}|^2}{\mathrm{E}\{|a_{\mu,k}^2\}} \,. \tag{3.24}$$

is as reasonable parameter, where $a_{\mu,k}$ denotes the samples and PAR_μ the PAR at the μth transmit antenna. As $N_\mathrm{T} D$ instead of D time-domain samples are present the same ccdf of PAR as in (single-antenna) OFDM with $N_\mathrm{T} D$ carriers results, which is worse than that for D carriers.

The occurrence of large signal peaks substantially complicates implementation of

[1] We consider a standard discrete-time OFDM system model. Insertion of the cyclic prefix, pulse shaping, and modulation to radio frequency are not taken into account. However, using an oversampled DFT, the samples a_k very closely reflect the continuous-time transmit signal [45, 46, 58] and hence are sufficient to assess PAR reduction algorithms.

the analog radio frequency frontend at the transmitter side since amplifiers operating linearly over a wide dynamic range have to be employed. Undeliberate clipping of peaks by non-linear amplifiers will, on the one hand, cause signal distortion and hence lower the power efficiency of the transmission system. On the other hand, and even more critical, any non-linear operation on the continuous-time transmit signal will cause intolerable out-of-band radiation which, in the last resort, may lead to a violation of spectral masks imposed by standards bodies.

The obvious solution to this problem would be to operate the amplifiers with sufficiently large power back-offs. This, however, also leads to very power-inefficient transmission systems. A much more preferable answer to the PAR problem of OFDM is to apply some *algorithmic control* to the transmit signal. Basically, the idea of *PAR reduction schemes* is to only generate OFDM frames with low, or at least tolerable PAR.

Since the mid 1990s, a variety of PAR reduction algorithms has been developed. Even though they are all based on (i) introducing new degrees of freedom for restriction to or selection of suited candidate OFDM signals and (ii) implicitly or explicitly adding redundancy, very different approaches are present in literature. The most relevant (among others) PAR reduction principles are (for a more detailed overview see, e.g., [33]):

Redundant Signal Representations: Multiple transmit signals are created which represent the given data. From that the "best" representation is selected. Here, in particular *selected mapping (SLM)*, e.g., [2, 27, 28, 30, 47, 57] and *partial transmit sequences (PTS)*, e.g., [8, 24, 31, 48] have to be mentioned.

Tone Reservation: Here, some carriers are not used for data transmission but are selected via an algorithmic search, e.g., [41, 54].

Clipping (and Filtering): The transmit signal is passed through a non-linear, memoryless device. Combinations of clipping and filtering are also popular, e.g., [26, 43, 50, 51].

Constellation Expansion: The signal constellations in the carriers are warped; deviations from the regular QAM grid are allowed, especially for points at the perimeter of the constellation, e.g., [40].

Coding Techniques: Here, an algebraic code construction over the carriers is used, e.g., [34, 38, 52, 53] to exclude undesired frames.

(Trellis) Shaping Techniques: Employing a signal shaping algorithm operating over the frequency domain symbols, the signal envelope can be influenced, e.g., [35, 44, 49].

When additionally employing multi-antenna or multiple-input/multiple-output (MIMO) techniques, PAR reduction methods tailored to this situation should be utilized, instead of simply performing single-antenna PAR reduction in parallel, as done, e.g., in [27]. The fundamental idea behind PAR reduction in MIMO OFDM can be paraphrased with "*redistributing the peak power over the antennas*". The gain of MIMO PAR reduction is an increased slope of the ccdf curves, i.e., the probability of occurrence of large signal peaks can significantly lowered compared to single-antenna schemes. In this regard, MIMO approaches show a similar behavior as do MIMO techniques with respect to error rate: the slope of the respective performance curves is increased; some kind of *diversity gain* is achieved.

In this chapter, the main ideas and algorithms for MIMO PAR reduction will be reviewed and the achievable gains over single-antenna techniques will be briefly discussed.

3.5.2 PAR Reduction Schemes for MIMO Transmission

In this section, we give an overview over PAR reduction schemes for MIMO transmission. Thereby, we have to distinguish two basic scenarios: on the one hand, we consider *point-to-point MIMO transmission* achieving spectral multiplexing gain, where the signals can be processed jointly at transmitter and receiver. On the other hand, variants for the *point-to-multipoint situation*, i.e., the broadcast channel, where joint signal processing is only possible at the transmitter side are briefly discussed. In the *multipoint-to-point scenario* (multiple-access channel) no joint optimization of the transmit signals can be performed, hence this case is not amenable for MIMO approaches.

SLM- and PTS-Type PAR Reduction

Much work has been invested in generalizing the SLM principle [28] to the point-to-point MIMO setting, e.g., [2, 4, 5, 7, 9, 11, 12, 15, 18, 27, 42]. The extension of PTS to MIMO transmission is addressed in [8, 10].

Ordinary and simplified SLM

In *ordinary SLM* [27] SISO SLM is applied N_T times in parallel. For each of the parallel OFDM frames the best mapping out of U possible is individually selected. In *simplified SLM* a restriction to select the lth candidate over all antennas is imposed. In total, in both cases complexity per antenna is determined by the U IDFTs.

It can be shown that the asymptotic behavior of the ccdf is given $\mathrm{const} \cdot \mathrm{e}^{-\mathrm{PAR}_\mathrm{th} U}$, which is the same as in single-antenna SLM. Both schemes do not achieve any MIMO gain.

Directed SLM

The first scheme, *directed SLM*, tailored to the MIMO situation has been presented in [2,4]. Main idea is to invest complexity only where PAR reduction is really needed. Instead of performing U trials for each of the N_T transmitters, the fixed budget of $N_\mathrm{T}U$ IDFTs (same as in ordinary SLM) is used to successively improve the currently highest PAR over the antennas. Complexity is hence adaptively distributed over the antennas.

In the first step the PARs of the N_T initial (original) OFDM frames are calculated. Then, in each successive step, the OFDM frame with instantaneously highest PAR is considered. Calculating a next candidate, a reduction of PAR is tried. This procedure is continued until the same number of IDFT and PAR calculations have been carried out.

Using this technique, the ccdf of PAR has the asymptote const$\cdot e^{-\mathrm{PAR_{th}}\,(N_\mathrm{T}U-(N_\mathrm{T}-1))}$, hence compared to above the exponent is multiplied by a factor of almost N_T; "diversity gain" is achieved.

Instead of performing directed SLM over the spatial dimension (MIMO OFDM), it can also be applied to single-antenna schemes when constituting frames in temporal direction [7], or even over space and time, cf. [9]. The gains are basically proportional to the total size of the hyper frames, i.e., the product of number of antennas and temporal frame length.

MIMO SLM based on Reed–Solomon Codes or the Simplex Code

The successive candidate generation and assessment can be circumvented if, given N_T initial OFDM frames, a number of $N_\mathrm{T}U$ candidates is generated in parallel, from which the N_T best are selected for transmission.

Such schemes, based on Reed–Solomon codes or the Simplex code for creating the candidates have been developed in [11, 12, 15]. Thereby, the codes are arranged over the N_T OFDM frames rather than over the carriers—these frames constitute the systematic symbols, the additional candidates are the parity symbols in the encoding process. A combination of the principles of multiple signal representation with selection and the use of channel coding is present.

The Reed–Solomon code scheme achieves exactly the same performance (and the same "diversity gain") as directed SLM but all signal generation and assessment can be done in parallel. The Simplex code scheme is only slightly inferior.

Noteworthy, besides multi-antenna transmission, the Reed–Solomon code scheme can advantageously be applied in packet transmission, which, moreover, exhibits appealing similarities with incremental redundancy check schemes in automatic repeat request (ARQ) applications.

MIMO PTS

Similar to SLM, the concept of PTS, where the candidates are different linear combinations of so-called *partial transmit sequences*, can also be extended to the MIMO situation, see [8, 10]. The same ordinary, simplified, and directed concepts from SLM can be used in PTS as well. Moreover, spatial permutations of the partial sequences is another degree of freedom.

Comparing PTS and SLM based on the same required computational complexity, PTS offers somewhat better performance, as this method is able to assess more candidates with a lower number of IDFTs.

Successive Schemes

In some situations it is sufficient that the PAR stays below a tolerable limit, e.g., depending on the radio frontend some PAR value may still be acceptable. Here, complexity (and battery power) can be saved if candidate generation and assessment is done successively and stopped if the tolerable value is reached. Such *successive*

PAR reduction schemes have been proposed in [16, 19, 23] and already mentioned in [4].

Given the tolerable PAR, PAR_{tol}, the average number of assessed candidates can be calculated. Main result is that this number is simply given by the inverse of the cdf of PAR of the underlying original OFDM scheme, i.e., $\bar{n} = 1/(1 - \Pr\{\text{PAR} > \text{PAR}_{\text{tol}}\})$, which using the Gaussian approximation calculates to $\bar{n} = (1 - \mathrm{e}^{-\text{PAR}_{\text{tol}}})^{-D}$. Noteworthy, for PAR_{tol} equal to the so-called "*critical PAR value*" $\text{PAR}_{\text{tol}} = \log(D)$ (cf., e.g., [59]) and reasonably large number D of carriers, average complexity amounts to only $\bar{n} = \mathrm{e} = 2.71828\ldots$ (Euler's number). Hence, using successive PAR reduction approaches, on average only very low computational complexity is required to achieve significant gains.

PAR Reduction for Downlink Transmission

Compared to PAR reduction in point-to-point MIMO systems, schemes applicable in point-to-multipoint scenarios (broadcast channel) are a much more challenging task. Since no joint receiver side signal processing is possible, at the transmitter side only operation for candidate generation/modification can be applied which can individually be reversed in each of the receivers.

Among the above discussed schemes, only simplified SLM can be used for the broadcast channel. Consequently, adopted PAR reduction algorithms have been developed [5, 14]. The *selected sorting* technique uses the presence of the transmitter side pre-equalization, specifically, the application of (sorted) Tomlinson–Harashima precoding (THP) [32]. Modifying the order of the successive encoding in THP, which is present in each carrier individually and which operates over the spatial (user) dimension, the PAR of the OFDM frames at the antenna array can be influenced. Since the choices of the ordering interact and the ordering in each carrier affects all antenna signals, no directed or successive approach can be designed. In turn, in the broadcast case no diversity gain compared to single-antenna transmission is possible.

Other Optimization Criteria

Besides PAR reduction, other optimization aims for OFDM transmission may be of interest, e.g., average transmit power, cubic metric, or out-of-band power. In general, the aim in future OFDM systems will be to generate signal with certain desired properties. This deliberate generation of signals is commonly referred to as *signal shaping* [32]; PAR reduction is one particular instance thereof.

Concentrating on point-to-point MIMO systems, the extension of the above schemes to the more general aim of signal shaping is easily possible. Employing the SLM approach of candidate generation and selection, due to this selection step any criterion of optimality can be taken into account. Therefore, the signal parameter to be optimized has to be expressed mathematically, defining a selection metric.

Work on the optimization of other criteria than the PAR can be found in [3, 17], where out-of-band power reduction is aspired and power-amplifier-oriented metrics have been introduced. The even more general framework of *probabilistic signal shaping* in MIMO OFDM has been presented in [23].

Representation of Side Information

In most of the above discussed PAR reduction algorithms, the transmission of side information to characterize the selected candidate / performed signal modification is required. Only with this knowledge, at the receiver the initial data can be recovered from the received OFDM frame (by inverting the applied "mapping" used in SLM). The side information problem has been addressed, e.g., in [25, 36, 37, 39, 55, 56]. However, up to now, only very few approaches, e.g. [30], exists in literature where no explicit side information is necessary.

The embedding of explicit side information has been studied in [22], where it is shown that spending only a few carriers to represent the required redundancy is sufficient to extract it with very high reliability and without noticeable deteriorating effect on the error rate of the data bits. In [20,21] an implicit transmission of the side information via rotations has been proposed. According to some defined pattern, the QAM constellations in the carrier are left as they are or rotated by 45°. No increase in transmit power occurs, and the rotation pattern—directly representing the side information bits—can be detected highly reliable with almost no additional effort.

3.5.3 Numerical Results

Figure 3.22 shows the PAR reduction performance of directed SLM in terms of the ccdf. Various numbers N_{T} of transmit antennas are studied. The ccdfs of the original signal (single- and multi-antenna case) and the one of ordinary SLM (individual parallel application of single-antenna SLM) serve as reference. Directed SLM (which is as efficient as the Reed–Solomon code version) outperforms the individual parallel application of the respective PAR reduction scheme. The benefit of the directed/RS code approach is that its performance increases the more transmit antennas are used. In contrast to that, for the original signal or oSLM the PAR statistics gets worse with the number of transmit antennas. The benefit of directed SLM can also be observed by considering the asymptotic slopes (large $\mathrm{PAR}_{\mathrm{th}}$) of the ccdf curves. For single-antenna SLM, the slope of the ccdf of ordinary SLM is determined by the number U of a assessed signal candidates. Applying directed SLM offers an increase of the slope by (approximately) a factor of N_{T}, compared to ordinary SLM when assessing the same total number of signal candidates. This effect is similar to the diversity order (slope of bit error rate curves) when considering digital transmission over flat-fading MIMO channels.

3.5.4 Summary

Over the last years, the field of PAR reduction for OFDM has experienced a renaissance. In particular PAR reduction in MIMO schemes has gained enormous interest and a huge number of publications has appeared in journals and conference proceedings. However, transferring the MIMO concepts back to single-antenna transmission—most simple by establishing hyper frames over time—new interesting schemes also for conventional OFDM have been designed. In the future, a still rapid growth of this research field is anticipated. In particular, the deliberate generation

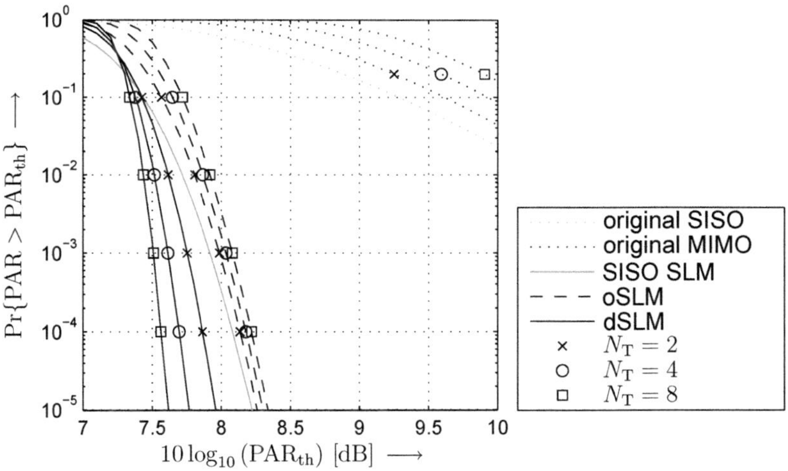

Figure 3.22: Ccdf of PAR of ordinary and directed SLM; $D = 512$, $M = 4$, $U = 16$.

of OFDM signals exhibiting some desirable properties via *signal shaping* to adapt the transmission systems even closely to the specific demands at hand is its very infancy.

Bibliography

Papers Emerged from the Project:

[1] R.F.H. Fischer, "Peak-to-Average Power Ratio (PAR) Reduction in OFDM based on Lattice Decoding," In *Proceedings of the 11th Int. OFDM Workshop*, pp. 71–75, Hamburg, Germany, August 2006.

[2] R.F.H. Fischer, M. Hoch, "Directed Selected Mapping for Peak-to-Average Power Ratio Reduction in MIMO OFDM," *IEE Electronics Letters*, Vol. 42, No. 22, pp. 1289–1290, October 2006.

[3] R.F.H. Fischer, C. Siegl, M. Hoch, "Out-of-Band Power Reduction in MIMO OFDM," in *Proceedings of International ITG/IEEE Workshop on Smart Antennas*, Vienna, Austria, February 2007.

[4] R.F.H. Fischer, M. Hoch, "Peak-to-Average Power Ratio Reduction in MIMO OFDM," in *Proceedings of the IEEE International Conference on Communications*, Glasgow, United Kingdom, June 2007.

[5] C. Siegl, R.F.H. Fischer, "Peak-to-Average Power Ratio Reduction in Multi-User OFDM," in *Proceedings of the International Symposium on Information Theory 2007*, Nice, France, June 2007.

[6] R.F.H. Fischer, "Widely-Linear Selected Mapping for Peak-to-Average Power Ratio Reduction in OFDM," *IEE Electronics Letters*, Vol. 43, No. 14, pp. 766–767, July 2007.

[7] C. Siegl, R.F.H. Fischer, "Directed Selected Mapping for Peak-to-Average Power Ratio Reduction in Single-Antenna OFDM," in *Proceedings of 12th Int. OFDM Workshop*, Hamburg, Germany, August 2007.

[8] C. Siegl, R.F.H. Fischer, "Partial Transmit Sequences for Peak-to-Average Power Ratio Reduction in Multi-Antenna OFDM," *EURASIP Journal on Wireless Communications and Networking, Special Issue on "Multicarrier Systems"*, Article ID 325829, 2008.

[9] C. Siegl, R.F.H. Fischer, "Joint Spatial and Temporal PAR Reduction in MIMO OFDM," in *International ITG Conference on Source and Channel Coding 2008*, Ulm, Germany, January 2008.

[10] C. Siegl, R.F.H. Fischer, "Comparison of Partial Transmit Sequences and Selected Mapping for Peak-to-Average Power Ratio Reduction in MIMO OFDM," in *International ITG/IEEE Workshop on Smart Antennas*, Darmstadt, Germany, February 2008.

[11] R.F.H. Fischer, C. Siegl, "Peak-to-Average Power Ratio Reduction in OFDM using Reed-Solomon Codes," in *International Zurich Seminar 2008*, pp. 40–43, Zurich, March 2008.

[12] R.F.H. Fischer, C. Siegl, "OFDM Peak-to-Average Power Ratio Reduction based on the Simplex Code," in *Proceedings of the International Symposium on Information Theory (ISIT 2008)*, Toronto, Canada, July 2008.

[13] C. Siegl, R.F.H. Fischer, "Tone Reservation for Peak-to-Average Power Ratio Reduction in OFDM under Different Optimization Constraints," in *Proceedings of 13th International OFDM Workshop*, Hamburg, Germany, August 2008.

[14] C. Siegl, R.F.H. Fischer, "Selected Sorting for PAR Reduction in OFDM Multi-User Broadcast Scenarios," in *Proceedings of International ITG Workshop on Smart Antennas (WSA 2009)*, Berlin, Germany, February 2009.

[15] R.F.H. Fischer, C. Siegl, "Reed-Solomon and Simplex Codes for Peak-to-Average Power Ratio Reduction in OFDM," *IEEE Transactions on Information Theory*, Vol. 55, No. 4, pp. 1519–1528, April 2009.

[16] R.F.H. Fischer, C. Siegl, "Successive PAR Reduction in (MIMO) OFDM," in *Proceedings of the IEEE International Conference on Communications (ICC 2009)*, Dresden, Germany, June 2009.

[17] C. Siegl, R.F.H. Fischer, "Out-of-Band Power Reduction using Selected Mapping with Power-Amplifier-Oriented Metrics," in *Proceedings of 14th International OFDM Workshop*, Hamburg, Germany, September 2009.

[18] R.F.H. Fischer, C. Siegl, "Peak-to-Average Power Ratio Reduction in Single- and Multi-Antenna OFDM via Directed Selected Mapping," *IEEE Transactions on Communications*, Vol. 11, No. 11, pp. 3205–3208, November 2009.

[19] C. Siegl, R.F.H. Fischer, "Asymptotic Performance Analysis and Successive Selected Mapping for PAR Reduction in OFDM," *IEEE Transactions on Signal Processing*, August 2009. Accepted for publication.

[20] C. Siegl, R.F.H. Fischer, "Selected Mapping with Implicit Transmission of Side Information using Discrete Phase Rotations," in *International ITG Conference on Source and Channel Coding*, Siegen, Germany, January 2010.

[21] C. Siegl, R.F.H. Fischer, "Transmission and Detection of Side Information for Selected Mapping in MIMO OFDM," in *Proceedings of International ITG Workshop on Smart Antennas (WSA 2010)*, Bremen, Germany, February 2010.

[22] C. Siegl, R.F.H. Fischer, "Selected Mapping with Explicit Transmission of Side Information," in *IEEE Wireless Communications and Networking Conference*, Sydney, Australia, April 2010.

[23] R.F.H. Fischer, C. Siegl, "Probabilistic Signal Shaping in MIMO OFDM," *Submitted for Publication*, Mai 2009.

Other References:

[24] A. Alavi, C. Tellambura, I. Fair, "PAPR Reduction of OFDM Signals Using Partial Transmit Sequence: An Optimal Approach Using Sphere Decoding," *IEEE Communications Letters*, pp. 982–984, Nov. 2005.

[25] E. Alsusa, L. Yang, "Redundancy-free and BER-Maintained Selective Mapping with Partial Phase-Randomising Sequences for Peak-to-Average Power Ratio Reduction in OFDM Systems," *IET Communications*, pp. 66-–74, January 2008.

[26] J. Armstrong, "Peak-to-Average Power Reduction for OFDM by Repeated Clipping and Frequency Domain Filtering," *Electronics Letters*, pp. 246–247, Feb. 2002.

[27] M.-S. Baek, M.-J. Kim, Y.-H. You, H.-K. Song, "Semi-Blind Channel Estimation and PAR Reduction for MIMO-OFDM System With Multiple Antennas," *IEEE Transactions on Broadcasting*, pp. 414–424, Dec. 2004.

[28] R. Bäuml, R.F.H. Fischer, J.B. Huber, "Reducing the Peak-to-Average Power Ratio of Multicarrier Modulation by Selected Mapping," *Electronics Letters*, pp. 2056–2057, Oct. 1996.

[29] J.A.C. Bingham, "Multicarrier Modulation for Data Transmission: An Idea Whose Time Has Come," *IEEE Communications Magazine*, pp. 5–14, May 1990.

[30] M. Breiling, S. Müller–Weinfurtner, J.B. Huber, "SLM Peak-Power Reduction without Explicit Side Information," *IEEE Communications Letters*, pp. 239–241, June 2001.

[31] L.J. Cimini, N.R. Sollenberger, "Peak-to-Average Power Ratio Reduciton of an OFDM signal using Partial Transmit Sequences," *IEEE Communications Letters*, pp. 86–88, Mar. 2000.

[32] R.F.H. Fischer, *Precoding and Signal Shaping for Digital Transmission*, John Wiley & Sons, Inc., New York, 2002.

[33] S.H. Han, J.H. Lee, "An Overview of Peak-to-Average Power Ratio Reduction Techniques for Multicarrier Transmission," *IEEE Wireless Communications*, pp. 56–65, April 2005.

[34] W. Henkel, "Analog Codes for Peak-to-Average Ratio Reduction," *ITG Conf. on Source and Channel Coding*, pp. 151–155, Munich, Jan. 2000.

[35] W. Henkel, B. Wagner, "Another Application for Trellis Shaping: PAR Reduction for DMT (OFDM)," *IEEE Transactions on Communications*, pp. 1471–1476, Sep. 2000.

[36] A.D.S. Jayalath, C. Tellambura, "A blind SLM receiver for PAR-reduced OFDM," in *IEEE Vehicular Technology Conference (VTC 2002-Fall)*, pp. 219–222, Sept. 2002.

[37] A.D.S. Jaylath, C. Tellambura, "SLM and PTS Peak-Power Reduction of OFDM Signals Without Side Information," *IEEE Transactions on Wireless Communications*, Vol. 4, No. 5, pp. 2006-–2013, Sept. 2005.

[38] A.E. Jones, T.A. Wilkinson, S.K. Barton, "Block Coding Scheme for Reduction of Peak to Mean Envelope Power Ratio of Multicarrier Transmission Scheme," *Electronics Letters*, pp. 2098–2099, Dec. 1994.

[39] B.K. Khoo, S.Y. Le Goff, C.C. Tsimendidis, B.S. Sharif, "OFDM PAPR Reduction Using Selected Mapping Without Side Information," in *IEEE International Conference on Communications (ICC 2007)*, Glasgow, United Kingdom, June 2007.

[40] B.S. Krongold, D.L. Jones, "PAR Reduction in OFDM via Active Constellation Extension," *IEEE Transactions Broadcasting*, pp. 258–268, Sep. 2003.

[41] B.S. Krongold, D.L. Jones, "An active-set approach for OFDM PAR reduction via tone reservation," *IEEE Transactions on Signal Processing*, pp. 495–509, Feb. 2004.

[42] Y.-L. Lee, Y.-H. You, W.-G. Jeon, J.-H. Paik, H.-K. Song, "Peak-to-Average Power Ratio in MIMO-OFDM Systems Using Selective Mapping," *IEEE Communications Letters*, pp. 575–577, Dec. 2003.

[43] X. Li, L.J. Cimini, "Effect of Clipping and Filtering on the Performance of OFDM," *IEEE Communications Letters*, pp. 131–133, May 1998.

[44] M. Litzenburger, W. Rupprecht, "Combined Trellis Shaping and Coding to Control the Envelope of a Bandlimited PSK-Signal," in *IEEE International Conference on Communications (ICC '94)*, pp. 630–634, May 1994.

[45] S. Litsyn, G. Wunder, "Generalized Bounds on the Crest-Factor Distribution of OFDM Signals With Application to Code Design," *IEEE Transactions on Information Theory*, pp. 992–1006, March 2006.

[46] S. Litsyn, *Peak Power Control in Multicarrier Communications*, Cambridge University Press, Cambridge, 2007.

[47] S.H. Müller, R.W. Bäuml, R.F.H. Fischer, J.B. Huber, "OFDM with Reduced Peak-to-Average Power Ratio by Multiple Signal Representation," *Annal of Telecommunications*, pp. 58–67, Feb. 1997.

[48] S.H. Müller, J.B. Huber, "OFDM with Reduced Peak-to-Average Power Ratio by Optimum Combination of Partial Transmit Sequences," *Electronics Letters*, Vol. 33, pp. 368–369, Feb. 1997.

[49] T.T. Nguyen, L. Lampe, "On Trellis Shaping for PAR Reduction in OFDM Systems," *IEEE Transactions on Communications*, pp. 1678–1682, Sep. 2007.

[50] R. O'Neill, L.B. Lopes, "Envelope Variations and Spectral Splatter in Clipped Multicarrier Signals," in *IEEE International Symposium on Personal, Indoor and Mobile Radio Communications (PIMRC '95)*, Toronto, Canada, pp. 71–75, Sep. 1995.

[51] M. Pauli, "*Zur Anwendung des Mehrträgerverfahrens OFDM mit reduzierter Außerbandstrahlung im Mobilfunk*," Ph.D. dissertation, University Hannover, 1999.

[52] K.-U. Schmidt, "On Cosets of the Generalized First-Order Reed-Muller Code with low PMEPR," *IEEE Transactions on Information Theory*, pp. 3220–3232, July 2006.

[53] K.-U. Schmidt, "Complementary Sets, Generalized Reed-Muller Codes, and Power Control for OFDM," *IEEE Transactions on Information Theory*, pp. 808–814, Feb. 2007.

[54] J. Tellado, *Peak to Average Power Reduction for Multicarrier Modulation*. Ph.D. dissertation, Stanford University, 2000.

[55] G.R. Tsouri, D. Wulich, "Efficient Coding of Side Information for Selected Mapping PAPR Reduction," in *Wireless Telecommunications Symposium (WTS 2008)*, pp. 286–290, April 2008.

[56] A. Vallavaraj, B.G. Stewart, D.K. Harrison, F.G. McIntosh, "Reducing the PAPR of OFDM Using a Simplified Scrambling SLM Technique with No Explicit Side Information," in *IEEE International Conference on Parallel and Distributed Systems (ICPADS '08)*, pp. 902–907, Dec. 2008.

[57] P. Van Eetvelt, G. Wade, M. Tomlinson, "Peak to average power reduction for OFDM schemes by selective scrambling," *Electronics Letters*, Vol. 32, pp. 1963–1964, Oct. 1996.

[58] G. Wunder, H. Boche, "Peak value estimation of bandlimited signals from their samples, noise enhancement, and a local characterization in the neighborhood of an extremum," *IEEE Transactions on Signal Processing*, pp. 771–780, March 2003.

[59] G. Wunder, H. Boche, "Upper Bounds on the Statistical Distribution of the Crest-Factor in OFDM Transmission," *IEEE Transactions on Information Theory*, pp. 488–494, Feb. 2003.

3.6 Single- vs. Multicarrier Transmission in MIMO and Multiuser Scenarios

R. Fischer, C. Stierstorfer, University of Erlangen-Nürnberg, Germany

3.6.1 Introduction

Since the invention of orthogonal frequency-division multiplexing (OFDM), the question has arisen whether multicarrier (MC) or singlecarrier modulation is superior or preferable over the competitor. In [25, 36] it has been shown, that in principle OFDM and singlecarrier pulse amplitude modulation (PAM) with decision-feedback equalization (DFE) can achieve the same transmission rate and both can utilize the capacity of the underlying channel (cf. also [22, 29]).

Since the fundamental insight (e.g., [28,30]) that by using antenna arrays in transmitter and receiver, hence creating a multiple-input/multiple-output (MIMO) channel, spectral efficiency and thus data rate can dramatically be increased, this question has been posed again: is MIMO OFDM or singlecarrier MIMO the better choice? Moreover, this comparison is not only of interest in point-to-point MIMO (multi-antenna) transmission, but as well for multipoint-to-point scenarios (multiple access problem, uplink) and for point-to-multipoint scenarios (broadcast channel, downlink), when the transmission paths exhibit intersymbol interference (ISI) and thus call for some kind of equalization.

In this chapter, we briefly compare single- and multicarrier MIMO transmission with respect to the achievable performance. Besides the study of information- and communication-theoretical bounds, the consequences on the code design when employing practical channel coding schemes are discussed, too.

3.6.2 Point-to-Point MIMO Transmission

In this section, we concentrate on point-to-point MIMO transmission, e.g., using antenna arrays in both transmitter and receiver, each with N_{T} antennas. When considering transmission over MIMO ISI channels, two sources of (self) interference occur: interference of the parallel data streams (spatial or multi-user interference (MUI)) and intersymbol (temporal) interference. The possible equalization schemes can be characterized by how they deal with these two types of interferences.

The first approach for transmission over MIMO ISI channels is to treat MUI and ISI jointly, i.e., to perform combined spatial/temporal equalization. An attractive scheme is DFE over space and time (MIMO DFE), see [27]. Here, by applying a feedforward filter, the end-to-end impulse response is shaped such that it exhibits spatial and temporal causality, i.e., symbols transmitted at some time instant should only interfere with receive symbols at subsequent time instances and symbols transmitted at the same time instance over the antennas should also only interfere with receive symbols for antennas with larger labels (spatial/temporal causality). Then, the symbols are processed in a zig-zag fashion over space and time, taking already detected symbols (via feedback) into account. Neglecting error propagation in the feedback loop, MIMO DFE decouples the MIMO ISI channel into a set of N_T parallel (near AWGN) channels with individual signal-to-noise ratios. Details on this MIMO DFE can be found in [1,27].

The competing approach is to treat MUI and ISI separately. In MIMO systems temporal equalization can be performed by applying an OFDM transmitter and OFDM receiver (using a (inverse) discrete Fourier transform ((I)DFT) of length D) to each transmit and receive antenna, respectively. Thereby, the MIMO ISI channel is decomposed into a set of D independent flat fading MIMO channels. Then, for each of this parallel flat fading MIMO channels (the tones), spatial equalization is performed individually, e.g., by (sorted) DFE (aka V-BLAST) [28] per carrier. In this situation, in total $N_\mathrm{T}D$ parallel, independent (almost AWGN) channels, each with its own signal-to-noise ratio, result.

Both strategies hence result in a set of parallel channels with varying reliability. In such cases, performance can be (significantly) improved by applying an optimized, nonuniform distribution of total rate and total power to the parallel channels. In particular, the loading algorithm given in [24] results in averaged channels; the SNR of each channel is then given by the geometric mean of the initial channels.

Main result of [5] is that only when using this type of loading, single- (joint spatial/temporal DFE) and multicarrier modulation (OFDM with DFE per carrier) perform the same. This result holds for both, uncoded error rate, as well as achievable sum capacity. If loading is not active, the SC scheme significantly outperforms the MC approach as, due to the combination in the feedforward filter, the SC schemes is able to utilize the full diversity offered by the channel. Without loading, OFDM exhibits only the diversity order one, that of flat fading channels. Employing rate and power loading is hence the key that both systems provide the same performance.

3.6.3 Up- and Downlink Scenarios in Multiuser Transmission

We now turn to the situation of multiuser communications, i.e., K individual users want to communicate with a central base station (using $N_\mathrm{T} \geq K$ antennas). Having the *uplink-downlink duality* [19,33–35] in mind, both cases can be treated similarly to a large extent. Compared to the problem discussed above, in the multiuser setting, not (only) the sum capacity of the error rate is of interest but the rate regions, i.e., the region of admissible rates of the users have to be considered.

First, transmission in the multipoint-to-point scenario (uplink), where a *multiple access channel (MAC)* is present, is studied. Here, the same receiver-side joint equalization schemes as in the point-to-point situation above can be used, as none of them requires joint transmitter-side operations. Consequently, the SC spatial/temporal DFE and the MC OFDM/DFE per carrier systems achieve the same sum rate.

However, when assuming perfect feedback in the DFE, the user detected last in each carrier of the MC scheme "sees" an interference-free (clean) channel. Hence, with respect to the rate of a particular user it is best to detect this user last in *each* carrier.[2] This user then has a rate as if there were no interference at all.

[2] In most situations a global sorting in the DFE part of the OFDM scheme is sufficient. Sorting per carrier offers a higher degree of freedom for rate distribution between the users but does

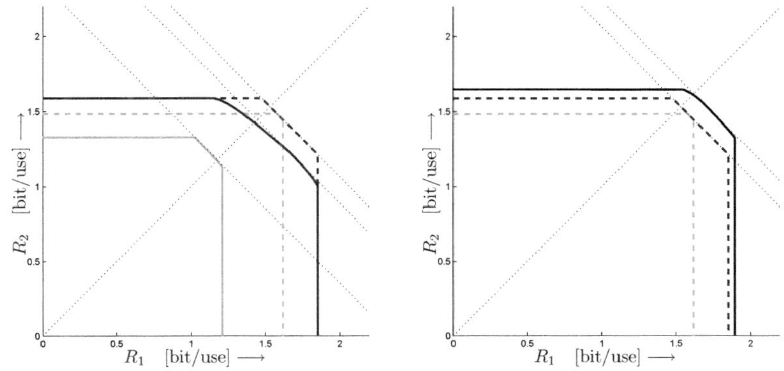

Figure 3.23: Uplink rate regions for exemplary 2×2 channel realization with four taps and equal gain power delay profile. Left: MIMO–DFE (light gray), MIMO–OFDM (dark gray); Right: MIMO–DFE (light gray), MIMO–OFDM (dark gray), MIMO–OFDM with water-filling (black); ZF approach (solid), MMSE approach (dashed).

Maximum-ratio combining can be used, leading to the maximum SNR among all receiving strategies [31] and hence to the maximum rate; no further increase is possible. In particular the feedforward processing in the SC scheme results in lower SNRs and rates. Hence, the bounds on the maximum individual rates of the users are lower in the SC scheme compared to the MC scheme (ignoring the loss due to the guard interval). Consequently, the rate regions of MC schemes are larger than those of SC schemes. The decoupling of temporal and spatial equalization is proved to be advantageous over a joint equalization by spatial/temporal DFE, both in performance and flexibility.

In Fig. 3.23, the achievable rate regions for an exemplary channel realization are depicted for both single- and multi-carrier transmission. Zero-forcing approaches are compared to the respective minimum mean-squared error (MMSE) variants. The advantages of a power allocation according to the water-filling principles can be also recognized.

Second, we turn to the point-to-multipoint scenarios (downlink), i.e., a *broadcast channel (BC)* is present. Basically, the transmission system for the downlink (joint transmitter side processing) can be obtained from the uplink system (joint receiver side processing) by flipping the entire structure. The DFE (either joint spatial/temporal or per carrier) is now replaced by Tomlinson-Harashima precoding, the simplest implementation of the optimum philosophy of *Costa precoding* [23]. When applying the uplink-downlink duality the encoding ordering in THP is the reversed decoding ordering of DFE, cf. [33] and the channel impulse responses as well as the temporal processing have to be time-reversed, cf. [8].

Moreover, in contrast to the uplink scenario with a *per user* power constraint, in downlink transmission only a *total* power constraint is active. Hence, the distribution

not offer additional gains in enlarging the rate region.

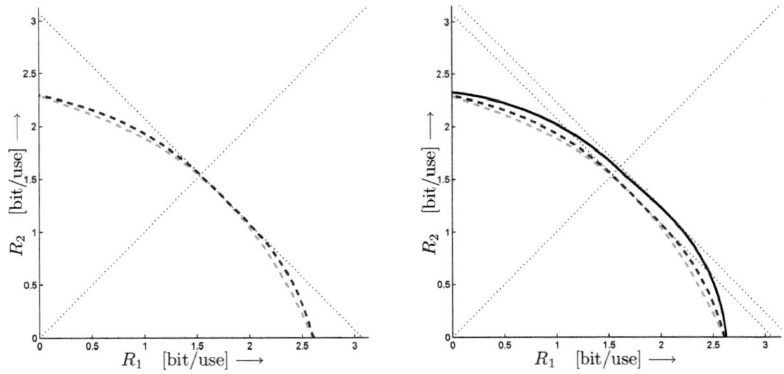

Figure 3.24: Downlink rate regions for exemplary 2×2 channel realization with four taps and equal gain power delay profile. Left: MIMO–DFE (light gray), MIMO–OFDM (dark gray); Right: MIMO–DFE (light gray), MIMO–OFDM (dark gray), MIMO–OFDM with water-filling (black); MMSE approach (dashed).

of the sum power on the users is an additional degree of freedom, and the rate region of the downlink scenario is given by the convex hull over the union of the uplink rate regions for all admissible power distributions [34]. The dominate face (rate tuples of maximum sum rate) of the uplink rate region immediately gives the dominate face of the downlink rate region; hence in the downlink case SC and MC approaches support the same sum rate. However, as the uplink rate region is larger for the MC setting, the downlink rate region is also larger for the MC system (OFDM with DFE per carrier) compared to the SC scheme (joint spatial/temporal DFE). Noteworthy, both approaches are clearly superior to frequency-division multiplexing in disjoint frequency bands (orthogonal transmission of the users).

In Fig. 3.24 the respective rate regions are shown for downlink transmission.

3.6.4 Aspects of Channel Coding

The application of channel coding schemes is the most prominent way to increase power efficiency of a digital transmission systems. When designing coding schemes for (point-to-point) MIMO OFDM, the particular characteristics of this approach have to be taken into account.

A stated above, the combination of OFDM and some spatial equalization scheme per carrier (e.g., DFE) leads to a set of $N_T D$ parallel, independent, approximately AWGN channels, each with its own signal-to-noise ratio. When coding over these channels, a situation similar to that when transmitting over a flat-fading channel is present. From information theory it is known, that in principle a single code, averaging over the fading states is sufficient to approach channel capacity, e.g., [17, 20]. However, using practical codes, the situation changes.

Additionally, it is well known that in OFDM transmission performance of uncoded transmission can be significantly increased by employing *rate and power loading* utilizing so-called *loading algorithms*, e.g., [12, 24]. However, loading is based on (at least partial) channel state information (CSI) at the transmitter and hence is not always applicable. Moreover, loading and channel coding interact and depend on each other.

It has been shown [3] that bit-interleaved coded modulation (BICM) [21] is a well-suited solution to improve reliability for highly bandwidth efficient transmission as long as no rate and power loading is used. Furthermore, large signal constellations should be avoided, i.e., the rate of the employed channel code has to be selected carefully and taking the required signal constellation for achieving a certain spectral efficiency into account (this holds for BICM in general and is not restricted to MIMO OFDM).

As soon as rate and power loading can be applied, BICM cannot be recommended. Here, the relevant performance measure is the parallel decoding capacity. Unfortunately, this quantity cannot be (significantly) improved by loading; some power loading strategies, well-suited for uncoded transmission, even lead to a decrease in capacity and hence are useless in coded transmission. In contrast to BICM, trellis-coded modulation (TCM) [32]—due to the uncoded bits—does only provide gains in combination with rate and power loading. However, neither conventional BICM nor TCM can achieve convincing performance in MIMO OFDM. Uncoded but loaded transmission offers already good performance a the price of requiring partial TX CSI.

When transmitter-side CSI is available, BICM can be improved by adapting the employed bit interleaver [9, 10]. Taking the actual SNRs of the parallel channels and the used mapping into consideration, a set of equivalent bit levels can be established; the entire modulation scheme is broken down into its fundamental building blocks. Based on the capacities of these equivalent binary channels, the interleaver is then designed such that within the relevant decoding window an optimized (equalized) metric arrangement is present. Such an interleaver design shows gains over both, conventional interleaving and BICM with rate and power loading. Concerning CSI, only the interleaver sequence has to be exchanged between transmitter and receiver and the additional complexity is almost negligible. Moreover, the capacities of these equivalent binary channels can also be utilized to perform bit-loading in uncoded transmission very efficiently, see [12, 13]. Finally, a bit-interleaver design for coded multicarrier transmission using rate and power loading is also possible using the tool of equivalent binary channels [14].

3.6.5 Summary

In summary it can be stated that MIMO OFDM is an attractive transmission scheme for future communication systems. In point-to-point applications, singlecarrier modulation with joint spatial/temporal DFE performs as good as MIMO OFDM with suited rate and power loading (uncoded transmission; the losses due to the guard interval and error propagation in the DFE are ignored). Both approaches offer the same sum rate. The situation changes when multipoint-to-point or point-to-multipoint scenarios are considered. Here, the rate region (which is the relevant assessment factor) is always larger for the multicarrier scheme. Moreover, the OFDM scheme inherently offers much more flexibility and adaptivity.

However, the advantages of OFDM are paid with the price of increased demands on the channel coding schemes. Without coding or loading, OFDM exhibits very poor performance. Rate and power loading requires CSI at the transmitter. Coding over parallel, independent channel with strongly varying quality is also a challenging task; the popular approach of bit-interleaved coded modulation gets worse for very large signal constellations, which, however, are required in MIMO OFDM. Again, having transmitter-side CSI available, using an adapted interleaver, solves the problem of coding for MIMO OFDM to a large extent.

Bibliography

Papers Emerged from the Project:

[1] R.F.H. Fischer, C. Stierstorfer, J.B. Huber, "Point-to-Point Transmission over MIMO Channels with Intersymbol Interference: Comparison of Single- and Multicarrier Modulation," in *Proceedings of the 9th International OFDM Workshop*, pp. 1–5, Dresden, Germany, September 2004.

[2] C. Stierstorfer, R.F.H. Fischer, "Comparison of Coding Strategies for Multi-Antenna Multi-Carrier Transmission," in *Proceedings of the 10th International OFDM Workshop*, Hamburg, Germany, August/September 2005.

[3] R.F.H. Fischer, C. Stierstorfer, "Comparison of Code Design Requirements for Singlecarrier and Multicarrier Transmission over Frequency-Selective MIMO Channels," in *Proceedings IEEE International Symposium on Information Theory (ISIT 2005)*, Adelaide, Australia, September 2005.

[4] C. Stierstorfer, R.F.H. Fischer, "Rate Regions for Multipoint-to-Point-Transmission over Channels with Intersymbol Interference," in *Proceedings of International ITG/IEEE Workshop on Smart Antennas*, Castle Reisensburg, Germany, March 2006.

[5] R.F.H. Fischer, C. Stierstorfer, "Single- and Multicarrier Transmission in Multipoint-to-Point Scenarios with Intersymbol Interference: A Comparison," in *Proceedings of 4th International Symposium on Turbo Codes in connection with the 6th International ITG-Conference on Source and Channel Coding*, München, Germany, April 2006.

[6] C. Stierstorfer, R.F.H. Fischer, "Rate Regions for Multicarrier Point-to-Multipoint Transmission over Channels with Intersymbol Interference," in *Proceedings of the 11th International OFDM Workshop*, pp. 293–297, Hamburg, Germany, August 2006.

[7] C. Stierstorfer, R.F.H. Fischer, "(Gray) Mappings for Bit-Interleaved Coded Modulation," in *Proceedings of 65th IEEE Vehicular Technology Conference (VTC 2007 Spring)*, Dublin, Ireland, April 2007.

[8] R.F.H. Fischer, C. Stierstorfer, "Single- and Multicarrier Transmission in Up- and Downlink Scenarios with Intersymbol Interference: A Comparison," *European Transactions on Telecommunications (ETT)*, Vol. 18, pp. 467–477, August 2007.

[9] C. Stierstorfer, R.F.H. Fischer, "Intralevel Interleaving for BICM in OFDM Scenarios," in *Proceedings of 12th International OFDM Workshop*, Hamburg, Germany, August 2007.

[10] C. Stierstorfer, R.F.H. Fischer, "Adaptive Interleaving for Bit-Interleaved Coded Modulation," in *Proceedings of 7th International ITG Conference on Source and Channel Coding (SCC)*, Ulm, Germany, January 2008.

[11] C. Stierstorfer, R.F.H. Fischer, "Mappings for BICM in UWB Scenarios," in *Proceedings of 7th International ITG Conference on Source and Channel Coding (SCC)*, Ulm, Germany, January 2008.

[12] C. Stierstorfer, R.F.H. Fischer, "Rate Loading in OFDM Based on Bit Level Capacities," in *Proceedings of the International Symposium on Information Theory (ISIT 2008)*, Toronto, Canada, July 2008.

[13] C. Stierstorfer, R.F.H. Fischer, "Aspects of Rate and Power Loading in (Coded-)OFDM," in *Proceedings of 13th International OFDM Workshop*, Hamburg, Germany, August 2008.

[14] C. Stierstorfer, R.F.H. Fischer, "Asymptotically Optimal Mappings for BICM with M-PAM and M^2-QAM," *IET Electronics Letters*, Vol. 45, No. 3, pp. 173–174, January 2009.

[15] C. Stierstorfer, R.F.H. Fischer, "Loading-Aware Bit-Interleaver Design," in *Proceedings of 14th International OFDM Workshop*, Hamburg, Germany, September 2009.

[16] C. Stierstorfer, *A Bit-Level-Based Approach to Coded Multicarrier Transmission*," Dissertation, Univerität Erlangen–Nürnberg, July 2009.

Other References:

[17] E. Biglieri, J. Proakis, S. Shamai (Shitz), "Fading Channels: Information-Theoretic and Communications Aspects," *IEEE Transactions on Information Theory*, pp. 2619–2692. June 1998.

[18] J.A.C. Bingham, "Multicarrier Modulation for Data Transmission: An Idea Whose Time Has Come," *IEEE Communications Magazine*, pp. 5–14, May 1990.

[19] H. Boche, M. Schubert, "A General Duality Theory for Uplink and Downlink Beamforming," in *Proceedings of IEEE Vehicular Technology Conference (VTC 2002 Fall)*, 2002.

[20] M. Bossert, A. Donder, V. Zyablov, "OFDM–Übertragung über Mobilfunkkanäle: Bemerkungen zur Kanalkapazität," in *2. OFDM–Fachgespräch*, Braunschweig, Sep. 1997.

[21] G. Caire, G. Taricco, E. Biglieri, "Bit–Interleaved Coded Modulation," *IEEE Transactions on Information Theory*, pp. 927-–946, May 1998.

[22] J.M. Cioffi, G.P. Dudevoir, M.V. Eyuboğlu, G.D. Forney, "MMSE Decision-Feedback Equalizers and Coding — Part I: Equalization Results, Part II: Coding Results," *IEEE Transactions on Communications*, pp. 2582–2604, Oct. 1995.

[23] M.H.M. Costa, "Writing on dirty paper," *IEEE Transactions on Information Theory*, pp. 439–441, May 1983.

[24] R.F.H. Fischer, J.B. Huber, "A New Loading Algorithm for Discrete Multitone Transmission," in *IEEE Global Telecommunications Conference (GLOBECOM '96)*, London, U.K., pp. 724–728, Nov. 1996.

[25] R.F.H. Fischer, J.B. Huber, "On The Equivalence of Single- and Multicarrier Modulation: A New View," in *IEEE International Symposium on Information Theory (ISIT '97)*, p. 197, Ulm, June-July 1997.

[26] R.F.H. Fischer, *Precoding and Signal Shaping for Digital Transmission*, John Wiley & Sons, Inc., New York, 2002.

[27] R.F.H. Fischer, J.B. Huber, C. Windpassinger, "Signal Processing in Decision-Feedback Equalization of Intersymbol-Interference and Multiple-Input / Multiple-Output Channels: A Unified View," *ELSEVIER Signal Processing*, Vol. 83, No. 8, pp. 1633–1642, Aug. 2003.

[28] G.J. Foschini, "Layered Space-Time Architecture for Wireless Communication in a Fading Environment When Using Multiple Antennas," *Bell Laboratories Technical Journal*, pp. 41–59, Autumn 1996.

[29] R. Price, "Nonlinear Feedback Equalized PAM versus Capacity for Noisy Filter Channels," *Proc. ICC'72*, pp. 22.12–22.17, 1972.

[30] E. Telatar, "Capacity of multi-antenna Gaussian channels," *European Transactions on Telecommunications*, pp. 585–596, Nov. 1999.

[31] D. Tse, P. Viswanath, *Fundamentals of Wireless Communication*, Cambridge Univ. Press, Cambridge, UK, 2005.

[32] G. Ungerböck, "Channel Coding with Multilevel/Phase Signals," *IEEE Transactions on Information Theory*, pp. 55—67, Jan. 1982.

[33] P. Viswanath, D. Tse, "Sum Capacity of the Vector Gaussian Broadcast Channel and Uplink-Downlink Duality," *IEEE Transactions on Information Theory*, pp. 1912–1921, Aug. 2003.

[34] S. Vishwanath, N. Jindal, A. Goldsmith, "Duality, Achievable Rates, and Sum-Rate Capacity of Gaussian MIMO Broadcast Channels," *IEEE Transactions on Information Theory*, pp. 2658–2668, Oct. 2003.

[35] W. Yu, J.M. Cioffi, "Sum Capacity of Gaussian Vector Broadcast Channels," *IEEE Transactions on Information Theory*, pp. 1875–1892, Sep. 2004.

[36] N.A. Zervos, I. Kalet, "Optimized Decision Feedback Equalization Versus Optimized Orthogonal Frequency Division Multiplexing for High–Speed Data Transmission Over the Local Cable Network," *Proc. ICC'89*, Boston, pp. 1080–1085, Sep. 1989.

3.7 Successive Bit Loading Concept

H. Rohling, C. Fellenberg, Hamburg University of Techology, Germany

3.7.1 Introduction

The Orthogonal Frequency-Division Multiplexing (OFDM) transmission technique is mainly applied in multi-path propagation and frequency-selective radio channels, where each subcarrier has an individual transfer factor H_n. Therefore, the subcarriers can have different bit error probabilities if the same modulation scheme is applied to all subcarriers.

It is assumed in this section that the subcarrier specific channel transfer factors H_n are perfectly known at the transmitter side for adaptive modulation purposes. The radio channel transfer factors H_n are continuously measured by the channel estimation procedure. The noise random variables N_n are expected to be additive white Gaussian noise with variance σ^2. If all subcarriers are transmitted with the same normalized transmit power, the subcarrier-specific signal-to-noise ratio (SNR_n) values are calculated in decibel as follows:

$$\mathrm{SNR}_n = 10 \cdot \log_{10} \frac{|H_n|^2}{\sigma^2} \tag{3.25}$$

In this frequency-selective channel, an adaptive modulation technique based on a bit loading procedure can be applied to improve the system performance. Subcarriers with a large SNR_n can carry more bits than others. That subcarrier which has the highest SNR_n value (Fig. 3.25) will carry the largest number of bits at the end of the loading procedure.

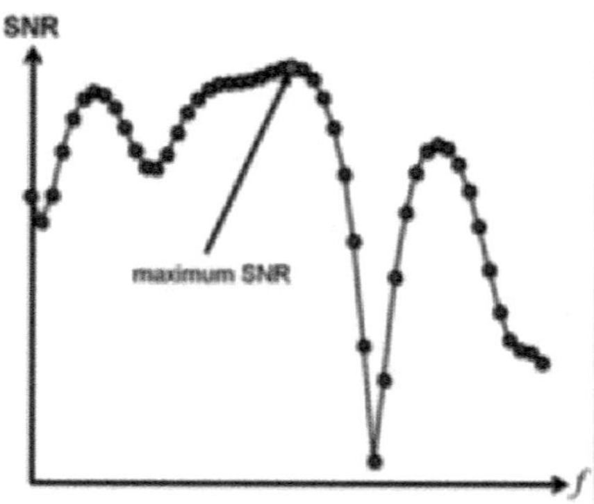

Figure 3.25: SNR values for each subcarrier in a frequency-selective radio channel

One of the first proposals for bit loading algorithms in OFDM-based transmission systems was presented in [1], which achieves either maximum capacity or minimum transmit power. Although this algorithm yields good performance, it is not optimal in the sense of minimizing the BER. Furthermore, the computation complexity of the proposed algorithm is very high. Therefore, in recent years several other bit loading algorithms have been proposed which have lower complexity, e.g., [3–9]. Most of them are based on a final rate prediction for each subcarrier, followed by an iterative adjustment to the radio channel conditions.

Alternatively, the proposed bit loading scheme is going to load the bits successively [1, 7] on a bit by bit basis, while the computation complexity is very low. It is important to note that this loading procedure directly achieves the minimum BER.

The discussed bit loading algorithm can be considered as a bottom-up procedure with a stepwise and successive bit-by-bit loading scheme. The loading algorithm was inspired by and can be compared with the well-known Huffman source coding scheme [2]. In this case, the source coding scheme also has a successive structure, which is based on the fact that symbols with the lowest probability will be assigned to the longest codeword. After each coding step, the successive structure can be applied in an unchanged form to the remaining set of source symbols.

In comparison to the Huffman source coding algorithm, the proposed bit loading procedure is based on the fact that the subcarrier with the largest SNR_n value will carry the largest number of bits relative to all other subcarriers. Furthermore, after each loaded bit, the loading procedure can be applied to the next step in an unchanged form but for the remaining bits and the remaining SNR values. Selecting the subcarrier with the lowest BER is equivalent to a maximum search of that subcarrier with the largest remaining SNR. Therefore, the meaning of *remaining SNR* on each subcarrier is very important for the considered loading scheme. It will be defined and discussed in section 3.7.3.

3.7.2 System Model

An OFDM-based transmission system with N subcarriers is considered. The radio channel is assumed to be frequency-selective and is modeled by a Wide Sense Stationary Uncorrelated Scattering (WSSUS) stochastic process. In case of a single subcarrier, Fig. 3.26 shows the BER curves for several and different well-known modulation schemes.

It is important to note that the BER curves are nearly parallel to each other. Therefore, it is assumed in this paper and for the successive loading algorithm that the SNR differences between adjacent BER curves are approximately constant independently of the considered BER. The SNR differences ($\Delta\mathrm{SNR}(b_n)$) between the adjacent BER curves are summarized in Table 3.1. The variable b_n describes the number of bits which have been loaded so far onto subcarrier n.

Table 3.1: SNR differences ($\Delta\mathrm{SNR}(b_n)$) between adjacent BER curves

b_n	modulation schemes	$\Delta\mathrm{SNR}(b_n)$
1	BPSK -> QPSK	3 dB
2	QPSK -> 8QAM	4.6 dB
3	8QAM -> 16QAM	2.1 dB
4	16QAM -> 32QAM	3.2 dB
5	32QAM -> 64QAM	2.8 dB

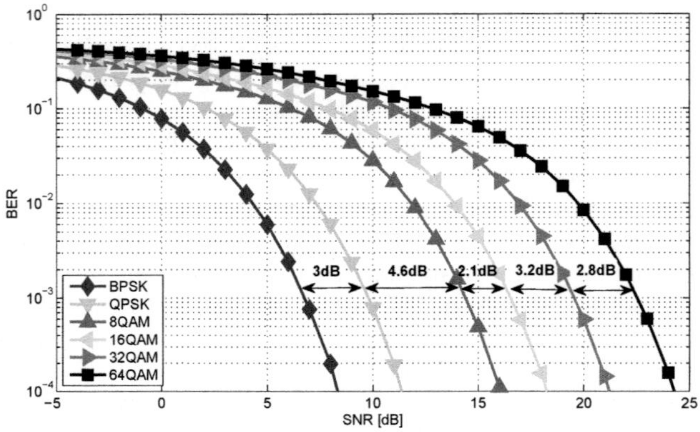

Figure 3.26: BER curves for a single subcarrier and different modulation schemes in an uncoded transmission system

For each OFDM symbol a fixed number of bits B_{target} is assigned which is the sum of all bits b_n per subcarrier n.

$$B_{\text{target}} = \sum_{n=1}^{N} b_n \tag{3.26}$$

The proposed bit loading algorithm minimizes the resulting BER, has a successive structure and an extremely low computational complexity, respectively.

3.7.3 Bit Loading Algorithm

The total number of bits B_{target} and the subcarrier-specific SNR values are the input parameters of the loading algorithm, Fig. 3.27. Additionally, the results of Table 3.1 and the SNR differences ($\Delta\text{SNR}(b_n)$) are used in the following loading procedure.

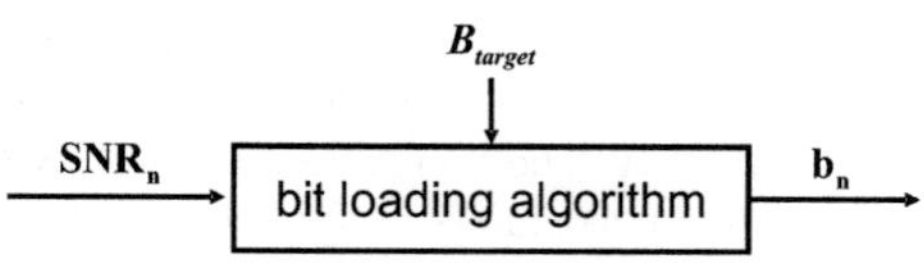

Figure 3.27: Overview of bit loading algorithm

The loading procedure is characterized by a successive bit-by-bit loading. The current status of the loading scheme is described by the number of bits b_n which have been loaded onto the different subcarrier n so far.

The general objective of the successive loading procedure is that in each step, the next bit should be loaded onto that subcarrier which has the lowest BER. In the first step and for the first loaded bit, the task to select the subcarrier with the lowest BER is equivalent with the task to find the subcarrier with the largest SNR.

However, when the first bit is loaded, the SNR on that subcarrier will be modified. The remaining SNR_n for subcarrier n and loaded bit number b_n is denoted by

$$\mathrm{SNR}_{\mathrm{rem}}(n, b_n). \tag{3.27}$$

Due to the assumed parallel behavior of the BER curves, this remaining SNR can be calculated as follows:

$$\begin{aligned}
\mathrm{SNR}_{\mathrm{rem}}(n,1) &= \mathrm{SNR}_{\mathrm{rem}}(n,0) - 3\,\mathrm{dB}\\
\mathrm{SNR}_{\mathrm{rem}}(n,2) &= \mathrm{SNR}_{\mathrm{rem}}(n,1) - 4.6\,\mathrm{dB}\\
\mathrm{SNR}_{\mathrm{rem}}(n,3) &= \mathrm{SNR}_{\mathrm{rem}}(n,2) - 2.1\,\mathrm{dB}\\
\mathrm{SNR}_{\mathrm{rem}}(n,4) &= \mathrm{SNR}_{\mathrm{rem}}(n,3) - 3.2\,\mathrm{dB}\\
\mathrm{SNR}_{\mathrm{rem}}(n,5) &= \mathrm{SNR}_{\mathrm{rem}}(n,4) - 2.8\,\mathrm{dB}\\
\mathrm{SNR}_{\mathrm{rem}}(n,6) &= -\infty
\end{aligned} \tag{3.28}$$

With this definition and understanding of remaining SNR values, again the subcarrier with the lowest BER is equivalent to the subcarrier with the largest remaining SNR. Therefore, in each loading step the sequence $\mathrm{SNR}_{\mathrm{rem}}(n, b_n)$, $n = 1, \cdots, N$ is considered and the largest value is selected. In this case, the loaded bit has the lowest BER due to the following relation:

$$\begin{aligned}
\mathrm{BER}\left[\mathrm{QPSK}(\mathrm{SNR}_{\mathrm{rem}}(n,1))\right] &= \mathrm{BER}\left[\mathrm{BPSK}(\mathrm{SNR}_{\mathrm{rem}}(n,0) - 3\,\mathrm{dB}\right]\\
\mathrm{BER}\left[\mathrm{8QAM}(\mathrm{SNR}_{\mathrm{rem}}(n,2))\right] &= \mathrm{BER}\left[\mathrm{BPSK}(\mathrm{SNR}_{\mathrm{rem}}(n,0) - 7.6\,\mathrm{dB})\right]\\
\mathrm{BER}\left[\mathrm{16QAM}(\mathrm{SNR}_{\mathrm{rem}}(n,3))\right] &= \mathrm{BER}\left[\mathrm{BPSK}(\mathrm{SNR}_{\mathrm{rem}}(n,0) - 9.7\,\mathrm{dB})\right]\\
\mathrm{BER}\left[\mathrm{32QAM}(\mathrm{SNR}_{\mathrm{rem}}(n,4))\right] &= \mathrm{BER}\left[\mathrm{BPSK}(\mathrm{SNR}_{\mathrm{rem}}(n,0) - 12.9\,\mathrm{dB})\right]\\
\mathrm{BER}\left[\mathrm{64QAM}(\mathrm{SNR}_{\mathrm{rem}}(n,5))\right] &= \mathrm{BER}\left[\mathrm{BPSK}(\mathrm{SNR}_{\mathrm{rem}}(n,0) - 15.7\,\mathrm{dB})\right]
\end{aligned}$$

This set of equations proves the equivalence between minimum BER and maximum remaining SNR search. Therefore, independently of how many bits have already been loaded onto subcarrier n, the remaining SNR value, $\mathrm{SNR}_{\mathrm{rem}}(n, b_n)$, for subcarrier n can be directly compared with all other remaining SNR values in the next maximum search.

In a mathematical description, the bit loading procedure can be designed and explained as follows: The next bit will always be loaded onto that subcarrier which has the largest $\mathrm{SNR}_{\mathrm{rem}}(n, b_n)$. After the loading decision and subcarrier selection, the SNR value of this subcarrier will be modified by the SNR differences as shown in Table 3.1 to get the new remaining SNR value for $b_n = 1, 2, \ldots, 5$.

$$\mathrm{SNR}_{\mathrm{rem}}(n, b_n) = \mathrm{SNR}_{\mathrm{rem}}(n, b_n - 1) - \Delta\mathrm{SNR}(b_n) \tag{3.29}$$

The successive bit loading scheme has the following recursive structure:

1. **Initialize $b_n = 0$ and $\mathbf{SNR_{rem}}(n, 0) = \mathbf{SNR}_n$ for all $n = 1, \ldots, N$**

2. **Select that subcarrier which has the largest remaining SNR value $\mathbf{SNR_{rem}}(i, b_i)$**

3. **Load the next bit onto the selected subcarrier with index i: $b_i = b_i + 1$**

4. **Calculate the remaining SNR value $\mathbf{SNR_{rem}}(i, b_i)$**

$$\mathrm{SNR_{rem}}(i, b_i) = \mathrm{SNR_{rem}}(i, b_i - 1) - \begin{cases} 3.0\,\mathrm{dB}, \text{ if } b_i = 1 \\ 4.6\,\mathrm{dB}, \text{ if } b_i = 2 \\ 2.1\,\mathrm{dB}, \text{ if } b_i = 3 \\ 3.2\,\mathrm{dB}, \text{ if } b_i = 4 \\ 2.8\,\mathrm{dB}, \text{ if } b_i = 5 \end{cases}$$

5. **Repeat step 2 to 4 until $\sum b_n = B_{\mathbf{target}}$**

This loading scheme has a pure bottom-up and successive structure. The BER is minimized and the bit loading procedure results in the number of bits b_n loaded onto subcarrier n. The transmit power is assumed to be identical for each subcarrier.

3.7.4 Results

An OFDM-based transmission system with $N = 256$ subcarriers is considered. The radio channel is assumed to be frequency-selective and will be simulated by a WSSUS model. Three different loading algorithms are analyzed and compared with a non-adaptive modulation procedure. In addition to the proposed loading algorithm, the algorithm described in [3] (maximum capacity) and the algorithm described in [4] (minimum SER) are considered in the simulation.

The resulting BER performance for this bit loading algorithm and for a bandwidth efficiency of 2 bit/s/Hz in an uncoded system is shown in Fig. 3.28. The transmit power is assumed to be uniformly distributed over all loaded subcarriers.

In comparison to the non-adaptive modulation scheme, all adaptive bit loading algorithms achieve tremendous gains. The proposed bit loading procedure has an advantage of 0.2 dB in SNR compared with the loading scheme described in [4]. The resulting computation complexity of this new bit loading procedure is very low since only the calculation of SNR differences and some sorting procedures have to be performed.

Although the described bit loading procedure is optimum in the sense of minimum BER, the subcarriers will have different bit error probabilities at the end of the loading procedure due to the subcarrier-specific remaining SNR values, $\mathrm{SNR_{rem}}(n, b_n - 1)$.

$$\mathrm{BER}_n = \mathrm{BER}[\mathrm{BPSK}(\mathrm{SNR_{rem}}(n, b_n - 1))] \tag{3.30}$$

However, the proposed bit loading algorithm can be extended by an additional power loading scheme. The objective of the power loading scheme is to have the same BER on all subcarriers.

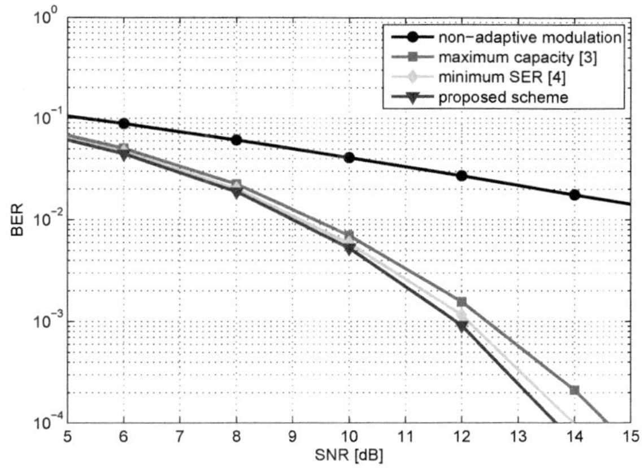

Figure 3.28: BER performance curves for different bit loading schemes in an uncoded system; bandwidth efficiency = 2 bit/s/Hz

The subcarrier-specific remaining SNR values, $\mathrm{SNR}_{\mathrm{rem}}(n, b_n - 1)$, are considered for the additional power loading procedure. First, the maximum remaining SNR is calculated.

$$\mathrm{SNR}_{\mathrm{rem,max}} = \max_i \left(\mathrm{SNR}_{\mathrm{rem}}(i, b_i - 1)\right) \tag{3.31}$$

The transmit power of each subcarrier is boosted by a scaling factor a_n until all subcarriers have the same BER. The scaling factor a_n in decibel is determined as follows:

$$a_n = \mathrm{SNR}_{\mathrm{rem,max}} - \mathrm{SNR}_{\mathrm{rem}}(n, b_n - 1). \tag{3.32}$$

This power loading procedure is illustrated in Fig. 3.29 and 3.30 as an example with 10 subcarriers, different remaining $\mathrm{SNR}_{\mathrm{rem}}(n, b_n - 1)$ values, and resulting BER figures, respectively.

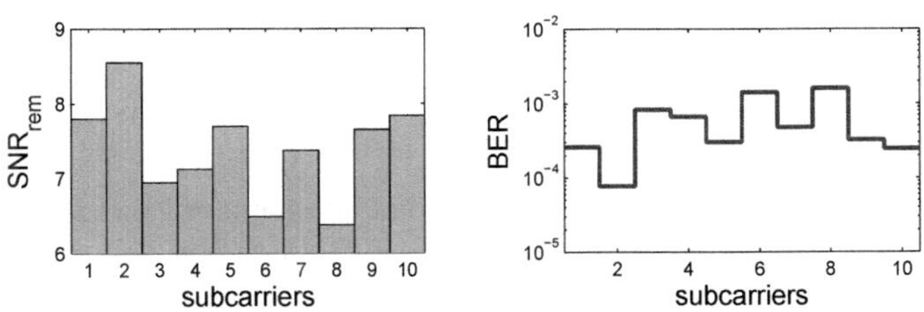

Figure 3.29: Remaining $\mathrm{SNR}_{\mathrm{rem}}(n, b_n - 1)$ and resulting BER for a uniform transmit power distribution

Figure 3.29 shows the resulting BER for each subcarrier if a uniform transmit power is assumed. The BER can be directly calculated by the subcarrier-specific remaining SNR value.

After the bit loading procedure, the transmit power is boosted by the scaling factor a_n to get the same BER for each subcarrier. The resulting BER values are shown in Fig. 3.30.

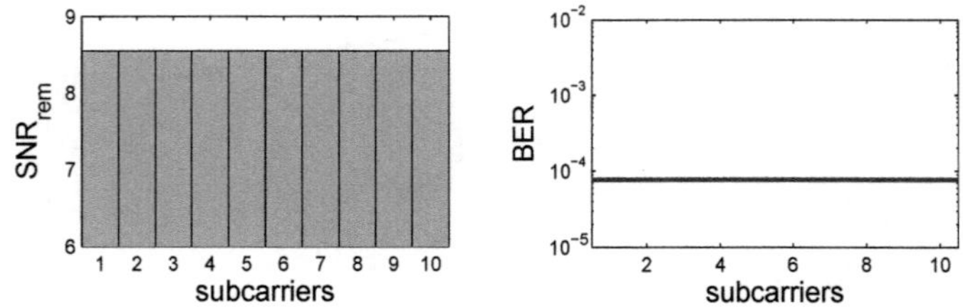

Figure 3.30: Subcarrier specific $\mathrm{SNR}_{\mathrm{rem}}(n, b_n - 1)$ with power loading based on the scaling factor a_n procedure

In Fig. 3.31, the results of a joint bit and power loading scheme are depicted. A clear difference in BER performance between the loading algorithms for a bandwidth efficiency of 2 bit/s/Hz can be observed. The proposed loading procedure has a 0.3 dB better performance compared to the loading scheme described in [4].

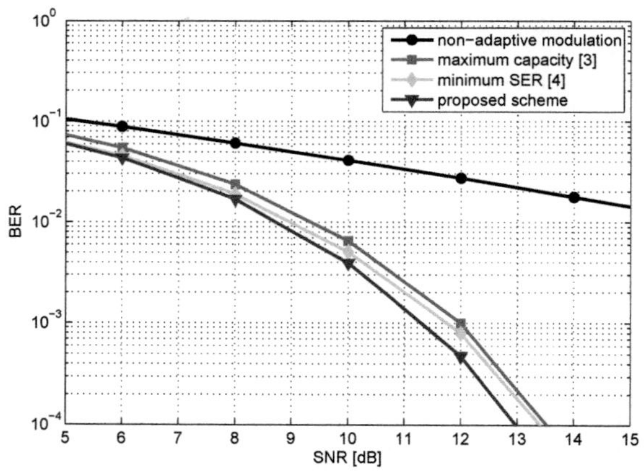

Figure 3.31: BER performance curves for the joint bit and power loading algorithms in an uncoded system, bandwidth efficiency = 2 bit/s/Hz

3.7.5 Summary

A new successive bit loading algorithm has been proposed which directly minimizes the resulting BER. Furthermore, this scheme has a very low computational complexity. An important point is the calculation of the remaining SNR. Independently of how many bits have been already loaded onto the different subcarriers, the values of remaining SNR can be directly compared with each other and can be used for a maximum search.

Therefore, it is possible to assign the next bit to that subcarrier which has the largest remaining SNR. In this case, the resulting BER is minimized and the proposed loading algorithm shows a difference of 0.2 dB in the BER performance compared with [4]. This is a small difference in system performance, but the computational complexity of this loading scheme is very low. In each step of the bit loading procedure, a single difference must be calculated and a maximum search is necessary.

This bit loading scheme can be extended by an additional power loading procedure based on the introduced logarithmic scaling factor a_n. In this way, the BER performance is improved by 0.3 dB.

Bibliography

[1] D. Hughes-Hartogs, "Ensemble modem structure for imperfect transmission media", *United States Patent* no. 4,833,796, May 1989

[2] D. A. Huffman, "A method for the construction of minimum-redundancy codes", *Proceedings IRE*, pp. 1098-1101, September 1952

[3] P. S. Chow, J. M. Cioffi and J. A. C. Bingham, "A Practical Discrete Multitone Transceiver Loading Algorithm for Data Transmission over Spectrally Shaped Channels", *IEEE Transactions on Communications*, vol. 43, no. 2/3/4, pp. 773-775, Feb./Mar./Apr., 1995

[4] R. F. H. Fischer and J. B. Huber, "A New Loading Algorithm for Discrete Multitone Transmission", *Proceedings of Global Telecommunications Conference*, pp. 724-728, London, UK, Novemeber 1996

[5] J. Campello, "Practical Bit Loading for DMT", *IEEE International Conference on Communications*, pp. 801-805, Vancouver, Canada, 1999

[6] W. Henkel and K. Hassan, "OFDM (DMT) Bit and Power Loading for Unequal Error Protection", *11th International OFDM-Workshop*, pp. 36-40, Hamburg, Germany, 2006

[7] D. A. Bui, C. Fellenberg, H. Rohling, "Successive bit loading scheme with low computation complexity", *13th International OFDM-Workshop*, pp. 115-118, Hamburg, Germany, 2008

[8] H. E. Levin, "A Complete and Optimal Data Allocation Method for Practical Discrete Multitone Systems", *Global Telecommunications Conference 2001*, pp. 369-374, San Antonio, USA, 2001

[9] R. Grünheid, E. Bolinth and H. Rohling, "A Blockwise Loading Algorithm for the Adaptive Modulation Technique in OFDM Systems", *Proc. IEEE VTC Fall 2001*, Atlantic City, October 2001

3.8 Adaptive Transmission Techniques

A. Czylwik, University of Duisburg-Essen, Germany

3.8.1 Introduction

The concept of adaptive modulation and coding is very well established for twisted pair communications. The corresponding method is discrete multitone transmission, see, e.g., [1]. In DMT, different modulation schemes with different bandwidth efficiencies are used in order to most efficiently transmit via the frequency-selective twisted pair channel. The process of distribution of information on the different subcarriers is called bit-loading, where the optimum bitloading method is described in [2]. The information bits are distributed in a way that the bit error probabilities are as small as possible and almost equal for each subcarrier.

In principle, this concept can also be applied for transmission via multipath radio channels. Besides the aspect that DMT transmission is a baseband concept, the only fundamental difference is the time variance of the radio channel. Therefore, it has been proposed to use adaptive modulation techniques also for radio communications [3].

Because of small-scale fading, the capacity of time-variant radio channels varies also with time. For strong multipath propagation it can be assumed that each subcarrier shows approximately Rayleigh fading statistics. The correlation between fading of subchannels decreases with increasing frequency separation. Thus, for a broadband channel these fluctuations cancel out partly so that the fluctuations of the overall broadband channel are reduced. It has been shown in [4], [5] that the remaining fluctuations of the channel capacity depend on the relation between available bandwidth and coherence bandwidth of the channel. For a system bandwidth significantly larger than the coherence bandwidth of the channel, the remaining fluctuations are proportional to the square root of the system bandwidth.

In case of time-variant channels with corresponding time-variant capacity, two different adaptation concepts can be applied. One possibility is to keep the transmission quality (bit error probability) constant and let the data rate fluctuate. The other possibility is to keep the data rate constant and let the transmission quality fluctuate. Since most transmission services require a piecewise constant data rate corresponding to a given transmission frame, only the case of constant data rate and fluctuating bit error probability is considered in the following. An overview about adaptive multicarrier modulation can be found in [6].

3.8.2 Adaptive Modulation and Coding

For single input single output (SISO) systems, by adaptive modulation a substantial gain compared with fixed modulation can be obtained. Simulation results show that performance gains of up to 12 dB can be achieved. If adaptive modulation is utilized, additional adaptive coding does not yield significant improvements, since by adaptive modulation the bit error probabilities for all subcarriers are approximately equalized. Therefore, forward error correction coding needs not to be adapted to the slightly different performances of subchannels.

3.8.3 Adaptive MIMO Transmission

Using multiple antennas on both sides of the radio link is an efficient method to increase the achievable data rate. A combination of OFDM and MIMO transmission has the advantage that MIMO techniques can be directly applied on frequency-flat subchannels. No simultaneous equalization and MIMO processing is necessary.

A first step towards adaptive MIMO transmission can be done by selecting a specific MIMO transmission method and adapting the modulation scheme. In [7] eigenmode transmission has been investigated as a fixed MIMO scheme. Since eigenmode transmission divides the MIMO channel into independent parallel channels, the eigenmodes can be used to create a second dimension of transmission channels in addition to the dimension of the frequency domain subchannels of OFDM. Therefore, bit-loading bit loading can be done with respect to both dimensions: OFDM subchannels and MIMO eigenmodes. It has been shown that for realistic channel conditions, the strongest eigenmodes exhibit the largest channel capacity, so that, for example, it is sufficient to distribute the information bits on the two strongest eigenmodes and neglect the others. An example for adaptive eigenmode transmission via a 4×4 MIMO picocell channel is shown in Fig. 3.32. It can be observed that the performance cannot be improved when distributing bits versus more than the two strongest eigenmodes.

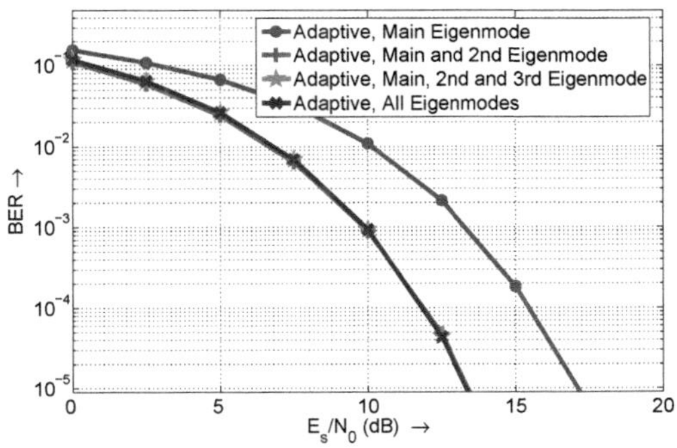

Figure 3.32: Bit error ratio (BER) performance with adaptive bit allocation over eigenmodes and subcarriers for a 4×4 MIMO picocell channel.

A more elaborate method of adapting the MIMO method is to combine different MIMO transmission methods by selecting that method which shows the best performance in the given transmission scenario. In general, there are three fundamental methods to use the MIMO concept in mobile radio communications.

- If the transmission channel shows poor quality (e.g. in case if transmitter and receiver are far away from each other), diversity concepts can be used to reduce error probabilities while transmitting with small data rates.

- If the channel is in good condition (including sufficient scattering), spatial multiplexing may be used to increase data rates.

- Finally, if the channel is in a quite good condition, but there is severe co-channel interference, multiple antennas may be used for beamforming in order to suppress the interference.

A criterion has to be elaborated which tells the system when to switch between the different MIMO transmission modes. A simple and very practical method for choosing the MIMO method is the *exponential effective SNR mapping* (EESM) approach [8] which has been originally developed for non-adaptive SISO-OFDM transmission. The EESM approach maps the N subcarrier SNRs γ_k at the receiver in case of a frequency-selective channel to an effective SNR γ_{eff}:

$$\gamma_{\text{eff}} = -\beta \ln \left(\frac{1}{N} \sum_{k=1}^{N} \mathrm{e}^{-\frac{\gamma_k}{\beta}} \right) \tag{3.33}$$

The effective SNR γ_{eff} is calculated in such a way that the bit error ratio (BER) of the OFDM system with the frequency-selective channel equals the BER of a frequency-flat AWGN channel with the SNR γ_{eff}. Therefore, the EESM method maps the performance of a frequency-selective channel to single value which is used as an interface between link and system level simulations. The EESM method can be extended to adaptively bit-loaded OFDM transmission systems with convolutional coding and soft decision decoding [9]. Furthermore, it can also be efficiently adapted to different MIMO schemes: In [10] the EESM approach is extended to space-frequency coding as well as spatial multiplexing.

The performance of adaptive MIMO transmission with bit interleaved convolutionally coded modulation has been analyzed in [11] and [12]. In Figs. 3.33 and 3.34 different adaptive 2×2 MIMO-OFDM techniques are compared with the average normalized channel capacity C/B:

- Spatial multiplexing (SM) with VBLAST detection and per antenna rate control (PARC) and fixed modulation on all subcarriers.

- Transmit diversity realized by rate 1 space-frequency coding (SFC) with fixed modulation on all subcarriers.

- Adaptive MIMO transmission (switching between SM and SFC) with fixed modulation on all subcarriers.

- Adaptive MIMO transmission (switching between SM and SFC) with optimized bit-loading.

The MIMO schemes are adapted for each OFDM block (symbol) - this means that for all subcarriers the same MIMO scheme is used. Obviously, for a multipath-rich channel and high SNR values (see Fig. 3.33) spatial multiplexing results in a higher information rate. In case of a channel with a small amount of propagation paths and large spatial correlation, space-frequency coding clearly outperforms spatial multiplexing. Adaptive MIMO transmission shows a performance at least as good as SM or SFC. It can be observed that – independent of the channel – bit-loading results in an additional performance gain.

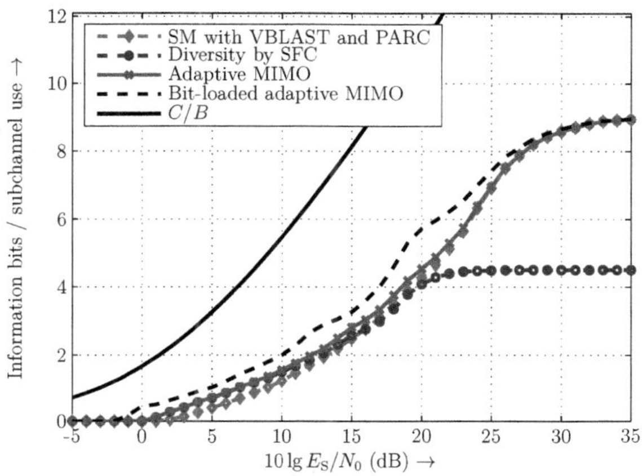

Figure 3.33: Comparison of adaptive MIMO transmission methods with fixed MIMO transmission. [12]. Channel type I: multipath-rich MIMO channel with low spatial correlation.

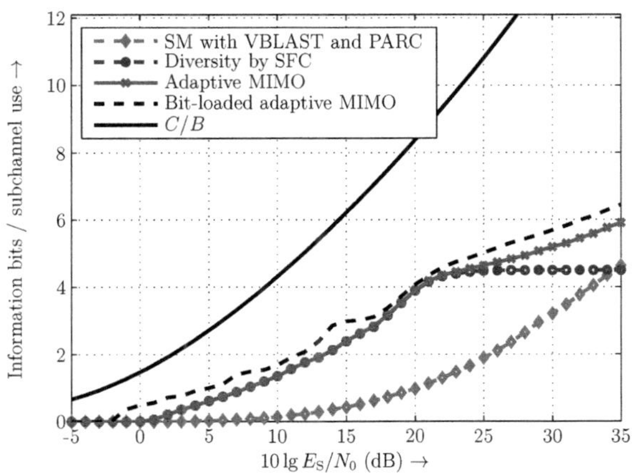

Figure 3.34: Comparison of adaptive MIMO transmission methods with fixed MIMO transmission. [12]. Channel type II: MIMO channel with a small amount of multipath contributions and large spatial correlation.

3.8.4 Signaling of the Bit Allocation Table

For a system with adaptive modulation, the use of different modulation schemes for the individual subcarriers is summarized in a so-called *bit allocation table* (BAT). The information of the BAT has to be synchronized between transmitter and receiver. A simple straight-forward solution of this synchronization is described in the following for a time division duplex system (TDD): The transmitter selects the modulation schemes based on its channel measurements carried out in the reverse link. Together with the payload data the transmitter transmits also the BAT via a signaling channel so that the receiver can use the same BAT. In order to reduce the signaling overhead, adjacent subcarriers can be grouped together so that the same modulation scheme is used for each group of subcarriers. Assuming that the total number of available data subcarriers is N_{DSC}, by grouping pairs/triplets of subcarriers the number of subcarrier groups is $N_{DSC}/2$ and $N_{DSC}/3$, respectively.

On the other hand, if subcarriers are grouped, the adaptation of modulation schemes becomes less flexible so that a degradation of the bit error probability results (see Fig. 3.35) [13].

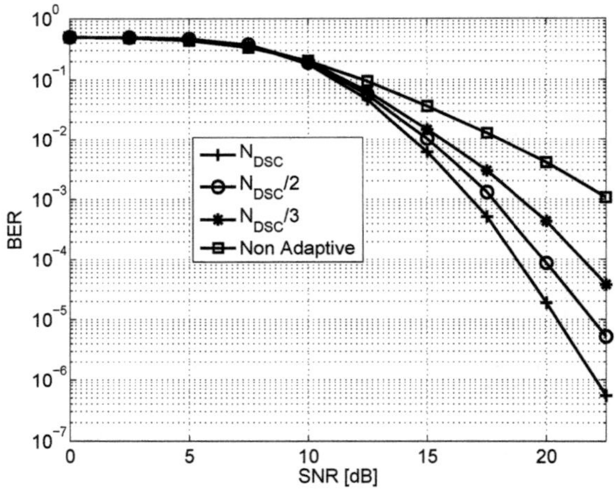

Figure 3.35: BER for different subband group sizes.

For a system with 64 subcarriers, $N_{DSC} = 48$ data subcarriers and five modulation schemes (no modulation, BPSK, QSPK, 16-QAM and 64-QAM) the required amount of signaling is investigated. The channel is modeled as a Rayleigh fading channel with Jakes' Doppler spectrum and AWGN. It is assumed that blocks of 10 OFDM data symbols are transmitted. The number of signaling bits depends on the coding scheme and is shown in Fig. 3.36:

- In a trivial coding method, 3 bits are needed to select one out of 5 modulation schemes ($\Rightarrow 3 \times 48 = 144$ bit/frame – curve 1).

- Combining pairs of subcarriers and encoding the corresponding $5^2 = 25$ alternatives jointly by 5 bits ($\Rightarrow 5 \times 24 = 120$ bit/frame – curve 2).

- For triplets of subcarriers, the lower bound of signaling overhead when combining multiple subcarriers is almost reached: Corresponding $5^3 = 125$ alternatives can be encoded jointly by 7 signaling bits (curve 3). ($\Rightarrow 7 \times 16 = 112$ bit/frame – curve 3).

- If Huffman coding is used, the signaling overhead can be further reduced. With Huffman coding the average codeword length becomes random with an average of 1.811 bit per subcarrier and transmission frame. ($\Rightarrow 1.811 \times 48 = 86.9$ bit/frame – curve 4).

- A further reduction of the signaling overhead is achieved by using state-dependent Huffman encoding, which takes advantage from the correlation of BATs between subsequent transmission frames. In average, the signaling overhead is reduced to 1.068 bit per subcarrier and transmission frame. ($\Rightarrow 1.068 \times 48 = 51.3$ bit/frame – curve 5).

- Using state-dependent Huffman coding not only with respect to time but also with respect to frequency yields an average signaling overhead of 0.56 bit per subcarrier and transmission frame. ($\Rightarrow 0.56 \times 48 = 26.9$ bit/frame – curve 6).

- When grouping pairs of subcarriers and using the same modulation scheme for these pairs, the signaling overhead is approximately halved. The lower bound corresponding to curve 3 is shown in curve 7.

- Grouping pairs of subcarriers and using state-dependent Huffman coding which utilizes the correlation of modulation schemes in subsequent transmission frames, results in curve 8.

- Grouping pairs of subcarriers and using state-dependent Huffman coding to jointly take into account the correlation in time and frequency domain, yields curve 9.

These simple source encoding approaches of the signaling information shows that the amount of signaling information can be reduced dramatically when the correlation of the signaling information is utilized.

3.8.5 Automatic Modulation Classification

Instead of transmitting the BAT by signaling channels in a more or less efficiently encoded way, an alternative approach is to blindly estimate the BAT at the receiver [14]. Blind estimation of the BAT completely avoids any signaling overhead and is therefore very attractive for a system with adaptive modulation.

For several decades, automatic modulation classification algorithms have been mainly investigated for military applications with a focus to distinguish between different modulation types – however, in this application field the classification of

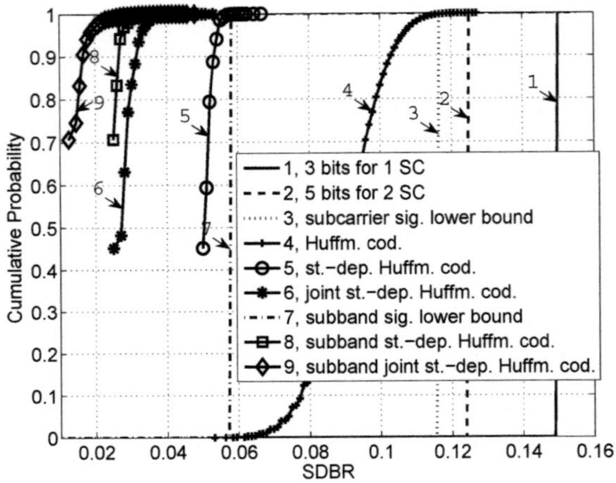

Figure 3.36: Cumulative distribution function of the signaling-to-data-bit ratio (SDBR).

different digital QAM schemes has not been considered. There are only a few contributions which discussed the application in adaptive OFDM systems [15–17].

Three different types of information which are available at the receiver, can be used for the BAT estimation process:

- The received complex symbols are created by scaling and rotating one of the known QAM constellation diagrams. From observing the received noisy complex symbols, the constellation diagram can be estimated.
- For a TDD-based system, the approximate reciprocity between the uplink and the downlink channel can be exploited. In case of ideal reciprocity and no channel estimation error, transmitter and receiver could carry out the same bit loading algorithm so that they independently find the same BAT.
- A given service is usually frame-oriented so that the number of data bits per transmission frame is fixed and known at the receiver.

In [14, 18, 19] optimized classification algorithms have been studied intensively. Maximizing the likelihood function of the received symbols with respect to all hypotheses of possible modulation schemes results in a classifier which does not take into account the approximate reciprocity of the channel. An improved classifier based on a maximum-a-posteriori approach is derived in [19]. It takes into account the strong correlation of channel transfer functions in the uplink and the downlink. Furthermore, an approximate maximum a-posteriori classifier with significantly less computational effort is proposed in [18]. In Fig. 3.37 corresponding simulation results show the probability of incorrect classifications. In addition Fig. 3.38 shows the resulting packet error probability (PER) if blind modulation classification is used. Obviously, using the approximate maximum-a-posteriori (AMAP) approach shows only a small degradation compared with adaptive modulation and perfect BAT knowledge. If the length of the transmission frame is larger, a further reduced degradation can be observed.

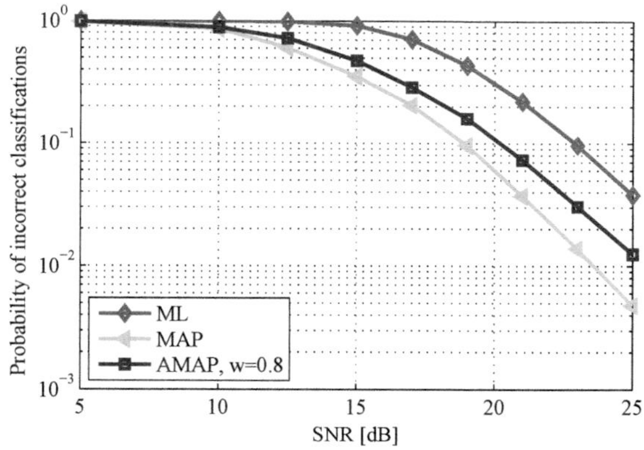

Figure 3.37: Probability of incorrect classifications of the BAT for different automatic modulation classification algorithms: ML = maximum likelihood approach, MAP = maximum-a-posteriori approach, AMAP = approximated maximum-a-posteriori approach. Parameters: FFT size $N = 64$, guard interval $N_g = 16$, bandwidth $B = 20$ MHz, bandwidth efficiencies: 0, 1, 2, 4, 6 bit/symbol, average bandwidth efficiency: 4 bit/symbol, pairs of subcarriers are grouped, frame length: $K = 10$ OFDM symbols $\Rightarrow$ modulation classifications are based on 20 received symbols, channel model: IEEE 802.11a indoor with delay spread $\tau_{ds} = 100$ ns and Doppler frequency $f_{dop} = 10$ Hz.

3.8.6 Summary

The capacity of an OFDM mobile radio link can be significantly increased using adaptive transmission techniques. For SISO systems, adaptive modulation can be utilized in order to adapt the transmission scheme to the radio channel as good as possible. In case of systems with multiple antennas, in addition to the modulation scheme also the MIMO scheme can be adapted to the radio channel.

The main drawback of adaptive modulation is that requires synchronization between transmitter and receiver with respect to the modulation scheme. If signaling is used for synchronization, the amount of signaling can be significantly reduced by using simple source coding concepts. As a very interesting alternative, synchronization can be realized by blind estimation of modulation schemes at the receiver, so that no signaling at all is necessary.

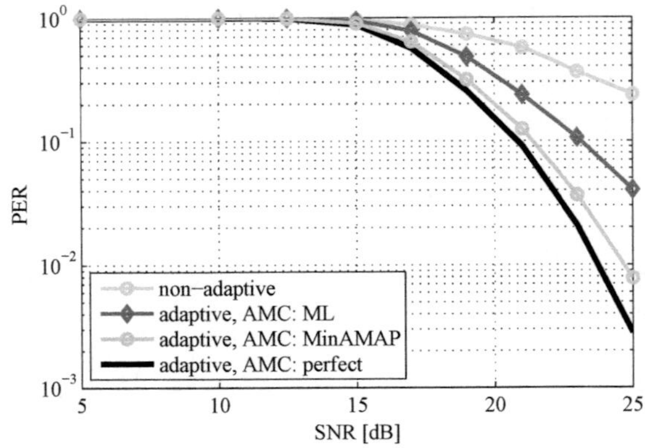

Figure 3.38: Packet error ratio (PER) for non-adaptive and adaptive OFDM systems using different automatic modulation classification algorithms (parameters see Fig. 3.37).

Bibliography

[1] J. S. Chow, J. C. Tu, and J. M. Cioffi, "A discrete multitone transceiver system for HDSL applications," *IEEE J. on Selected Areas in Communications* 9 (1991), pp. 895–908.

[2] D. Hughes-Hartogs, "Ensemble modem structure for imperfect transmission media," *U. S. Patent* 4,679,227 (1987).

[3] A. Czylwik, "Adaptive OFDM for wideband radio channels," in *Proceedings of the IEEE Global Telecommunications Conference (GLOBECOM '96), London*, pp. 713–718 (1996).

[4] A. Czylwik, "Temporal fluctuations of channel capacity in wideband radio channels," in *Proceedings of the IEEE International Symposium on Information Theory '97, Ulm*, pp. 468 (1997).

[5] A. Czylwik, "Fluctuations of the capacity of ultra-wideband radio channels," in *General Assembly of the International Union of Radio Science (URSI), New Delhi, India* (2005).

[6] T. Keller and L. Hanzo, "Adaptive multicarrier modulation: a convenient framework for time-frequency processing in wireless communications," *Proceedings of the IEEE* 88 (2000), pp. 611–640.

[7] A. Camargo and A. Czylwik, "Adaptive spatial multiplexing in MIMO-OFDM systems exploiting eigenmode statistics," in *Proceedings of the 10th International OFDM Workshop 2005, Hamburg*, pp. 313–317 (2005).

[8] Ericsson, *System-level evaluation of OFDM – further considerations.* 3GPP TSG-RAN WG 1 35, R1-031303, Nov. 17-21, 2003.

[9] A. Camargo and A. Czylwik, "PER prediction for bit-loaded BICM-OFDM with hard decision Viterbi decoding," in *Proceedings of the 11th International OFDM Workshop 2006, Hamburg*, pp. 26–30 (2006).

[10] A. Camargo, D. Yao, P. Paunov, and A. Czylwik, "Adaptive MIMO scheme for BICM OFDM by link adaptation and per-antenna rate control," in *Proceedings of the 12th International OFDM Workshop 2007, Hamburg*, pp. 286–290 (2007).

[11] A. Camargo, D. Yao, and A. Czylwik, "Bandwidth efficiency of practical MIMO-OFDM systems with adaptive MIMO schemes," in *Proceedings of the IEEE International Communications Conference (ICC'09), Dresden* (June 2009).

[12] A. Camargo, *Adaptive modulation, channel coding and MIMO schemes for practical OFDM systems (PhD thesis at the University Duisburg-Essen).* Shaker-Verlag, Aachen 2009.

[13] Y. Chen, L. Häring, and A. Czylwik, "Reduction of AM-induced signalling overhead in WLAN-based OFDM systems statistics," in *Proceedings of the 14th International OFDM Workshop 2009, Hamburg* (2009).

[14] L. Häring, A. Czylwik, and Y. Chen, "Automatic modulation classification in application to wireless OFDM systems with adaptive modulation in TDD mode," In *Proceedings of the 13th International OFDM Workshop 2008, Hamburg* (Aug. 2008).

[15] Q.-S. Huang, Q.-C. Peng, and H.-Z. Shao, "Blind modulation classification algorithm for adaptive OFDM systems," *IEICE Trans. Commun.* E.90-B (Febr. 2007), pp. 296–301.

[16] S. Edinger, M. Gaida, and N. J. Fliege, "Classification of QAM signals for multicarrier systems," in *Proceedings of the EUSIPCO, Poznan*, pp. 227–230 (Sept. 2007).

[17] M. Lampe, *Adaptive techniques for modulation and channel coding in OFDM communication systems (PhD thesis at the Technical University Hamburg-Harburg).* Apr. 2004.

[18] L. Häring, Y. Chen, and A. Czylwik, "Efficient modulation classification for adaptive wireless OFDM systems in TDD mode," in *Proceedings of the Wireless Communications and Networking Conference, Sydney* (Apr. 2010).

[19] L. Häring, Y. Chen, and A. Czylwik, "Automatic modulation classification methods for wireless OFDM systems in TDD mode," *IEEE Trans. on Communications* (accepted for publication 2010).

4 System Level Aspects for Single Cell Scenarios

4.1 Efficient Analysis of OFDM Channels

N. Grip, Luleå University of Technology, Sweden
G. E. Pfander, Jacobs University Bremen, Germany

4.1.1 Introduction

Narrowband finite lifelength systems such as wireless communications can be well modeled by smooth and compactly supported spreading functions. We show how to exploit this fact to derive a fast algorithm for computing the matrix representation of such operators with respect to well time-frequency localized Gabor bases (such as pulse shaped OFDM bases). Hereby we use a minimum of approximations, simplifications, and assumptions on the channel.

The derived algorithm and software can be used, for example, for comparing how different system settings and pulse shapes affect the diagonalization properties of an OFDM system acting on a given channel.

4.1.2 The Channel Matrix G

A *Gabor* (or *Weyl-Heisenberg*) system with *window* g and *lattice constants* a and b is the sequence $(g_{q,r})_{q,r\in\mathbb{Z}}$ of translated and modulated functions

$$g_{q,r} \stackrel{\text{def}}{=} T_{ra}M_{qb}g \stackrel{\text{def}}{=} \mathrm{e}^{\mathrm{i}2\pi qb(x-ra)}g(x-ra).$$

For OFDM communications applications, information is stored in the coefficients of the transmitted signal $s = \sum_{q,r\in\mathbb{Z}} c_{q,r}g_{q,r}$. In order to guarantee that the coefficients can be recovered from s in a numerically stable way, s and its coefficients should be equivalent in the sense that for some nonzero and finite A, B independent of s, $A\|s\|^2 \leq \|c\|^2 \leq B\|s\|^2$ with $\|c\|^2 \stackrel{\text{def}}{=} \sum_{q,r}|c_{q,r}|^2$ and $\|s\|^2 \stackrel{\text{def}}{=} \int_{\mathbb{R}}|s(t)|^2\,\mathrm{d}t$. This means that the sequence of functions $(g_{q,r})_{q,r\in\mathbb{Z}}$ is a *Riesz basis* for the function space $L^2(\mathbb{R})$ of square integrable functions. This guarantees the existence of a *dual basis* $(\widetilde{g}_{q,r})$ that also is a Gabor basis. Such bases are also called *biorthogonal*, or, in the special case $\widetilde{g} = g$ (or equivalently $A = B = 1$ [4]), *orthonormal.*

In communications applications, s is sent through a channel with linear channel operator H. With * notation for complex conjugate, the receiver typically tries to reconstruct the transmitted coefficients $c_{q,r} = \langle s, \widetilde{g}_{q,r}\rangle \stackrel{\text{def}}{=} \int_{\mathbb{R}} s(t)\widetilde{g}^*_{q,r}(t)$ from the

H. Rohling (ed.), *OFDM: Concepts for Future Communication Systems*,
Signals and Communication Technology, DOI: 10.1007/978-3-642-17496-4_4,

received signal Hs using some (possibly other) Gabor Riesz basis $(\gamma_{q',r'})$. A standard Riesz basis series expansion [4, 6] with this basis gives

$$Hs = \sum_{q',r'\in\mathbb{Z}} \langle Hs, \gamma_{q',r'}\rangle \widetilde{\gamma}_{q',r'} = \sum_{q',r'\in\mathbb{Z}} \left\langle H \sum_{q,r\in\mathbb{Z}} c_{q,r} g_{q,r} , \gamma_{q',r'} \right\rangle \widetilde{\gamma}_{q',r'}$$

$$= \sum_{q',r'\in\mathbb{Z}} \left(\sum_{q,r\in\mathbb{Z}} c_{q,r} \langle Hg_{q,r}, \gamma_{q',r'}\rangle \right) \widetilde{\gamma}_{q',r'}, = \sum_{q',r'\in\mathbb{Z}} (Gc)_{q',r'} \widetilde{\gamma}_{q',r'},$$

where G is the coefficient mapping $(c_{q,r})_{q,r} \mapsto \left(\sum_{q,r\in\mathbb{Z}} c_{q,r} \langle Hg_{q,r}, \gamma_{q',r'}\rangle\right)_{q',r'}$ with biinfinite matrix representation (the *channel matrix*)

$$\mathrm{G}_{q',r';q,r} = \langle Hg_{q,r}, \gamma_{q',r'}\rangle ,$$

and with indices (q', r') and (q, r) for rows and columns respectively. The matrix elements are usually called *intercarrier interference* (ICI) for $p = p'$ and $q \neq q'$. Similarly, the matrix elements are called *intersymbol interference* (ISI) when $p \neq p'$. Recovering the transmitted coefficients corresponds to inverting G, which is unreasonably time-consuming unless g and γ can be chosen so that G is diagonal or at least has fast off-diagonal decay.

We call H *time-invariant* if it commutes with the time-shift operator $T_{t_0} f(t) = f(t - t_0)$ for any t_0, that is, if $T_{t_0} H = HT_{t_0}$. Linear and time-invariant H are convolution operators, for which it is well-known that the family of complex exponentials $\mathrm{e}^{i2\pi\xi t}$ are "eigenfunctions" in the sense that for the restriction of such functions to an interval $[0, L]$, that is, $s(\cdot) = \mathrm{e}^{i2\pi\langle\xi,\cdot\rangle}\chi_{[0,L]}(\cdot)$, there is some complex scalar λ_ξ such that if h lives on $[0, L_h]$, then $Hs = \lambda_\xi s$ in the interval $[L_h, L]$. Thus G can easily be diagonalized by using Gabor windows $g = \chi_{[0,L]}$, $\gamma = \chi_{[L_h,L]}$ and lattice constants such that the resulting Gabor systems $(g_{k,l})$ and $(\gamma_{k,l})$ are biorthogonal bases [6]. This trick is used in wireline communications, where the smaller support of γ is obtained by removing a guard interval (often called *cyclic prefix*) from g. See, for example, [4, Section 2.3] for more details and further references.

In *wireless* communications, due to reflections on different structures in the environment, the transmitted signal reaches the receiver via a possibly infinite number of different wave propagation paths. Because of the highly time varying nature of this setup of paths and the corresponding channel operator, we can at most hope for approximate diagonalization of the channel operator. In fact, two different time-varying operators do in general not commute, so both cannot be diagonalized with the same choice of bases. Thus, diagonalization is usually only possible in the following sense: Typically, $(Hg_{q,r})$ is a finite and linearly independent sequence, and thus a Riesz basis with some dual basis $\left(\widetilde{Hg_{q,r}}\right)$, so for true diagonalization of G, we would have to set $\gamma_{q',r'} = \widetilde{Hg_{q,r}}$, but then $\gamma_{q',r'}$ would typically not be a Gabor basis or have any other simple structure that enables efficient computation of all $\gamma_{q',r'}$ and all the diagonal elements $\langle Hg_{q,r}, \gamma_{q',r'}\rangle$. Hence, for computational complexity to meet practical restrictions we have to settle for "almost dual" Gabor bases $(g_{q,r})$ and $(\gamma_{q',r'})$, such as the Gabor bases proposed in [7]. We are primarily interested in bases that are good candidates for providing low intersymbol and interchannel interference (ISI and ICI). As proposed in [7], we expect excellent joint time-frequency concentration of g and γ to be the most important requirement for achieving that goal.

For such g and γ we propose a fast algorithm for computing G in Section 4.1.4, based on a channel operator model described in Section 4.1.3. Our model is deterministic, so a typical example use is in coverage predictions for radio network planning [1, Section 3.1.3]. The algorithm computes the ISI and ICI dependence on, for example, pulse shaping and threshold choices from input data. It depends on describing a particular channel, that we assume to be known, for example, from measurements or computed from ray tracing, finite element or finite difference methods (described with more references in [1]). Moreover, the performance of a communication system is usually evaluated by means of extensive Monte-Carlo simulations [1], which also might be a potential future application where fast algorithms are required.

4.1.3 Common Channel Operator Models

The channel operator H maps an input signal s to a weighted superposition of time and frequency shifts of s:

$$Hs(\cdot) = \int_{K\times[A,\infty)} S_H(\nu,t)\, \mathrm{e}^{i2\pi\nu(t-t_0)} s(\,\cdot\, - t)\,\mathrm{d}(\nu,t)\,, \qquad K \text{ compact.}$$

This standard model is usually formulated for so-called *Hilbert–Schmidt operators* with the *spreading function* S_H in the space L^2 of square integrable functions (e.g., in [8,9]) or for S_H in some subspace of the tempered distributions S' (e.g., in [10,12]). The weakest such assumption is that $S_H \in S'$, which restricts the input signal s to be a Schwartz class function.

Alternatively, one can assume s to be in the Wiener amalgam space $W(A, l^1) = S_0$ (also named the Feichtinger algebra), which consists of all continuous $f\colon \mathbb{R} \to \mathbb{C}$ for which (with $\|g\|_1 \overset{\text{def}}{=} \int_{\mathbb{R}} |g(x)|\,\mathrm{d}x$ and $\hat{}$ denoting Fourier transform)

$$\sum_{n\in\mathbb{Z}} \|(f(\cdot)\psi(\cdot - n))^{\wedge}\|_1 < \infty$$

for some compactly supported[1] ψ having integrable Fourier transform and satisfying $\sum_{n\in\mathbb{Z}} \psi(x-n) = 1$. We write S_0' for the space of linear bounded functionals on S_0. S_0 is also a so-called modulation space, described at more depth and with notation $S_0 = M^{1,1} = M^1$ and $S_0' = M^{\infty,\infty} = M^\infty$ in [3,6].

Since the space $S_0'(\mathbb{R}\times\mathbb{R})$ includes Dirac delta distributions, this model includes important idealized borderline cases such as the following:

Line-of-sight path transmission: $S_H = a\delta_{\nu_0,t_0}$, a Dirac distribution at (ν_0, t_0) representing a time- and Doppler-shift with attenuation a.

Time-invariant systems: $S_H(\nu,t) = h(t)\delta_0(\nu)$.

Moreover, S_0' excludes derivatives of Dirac distributions, corresponding to complex-valued Hs with no physical meaning [11, Sec. 3.1.1]. Further, S_0 is the smallest

[1] A function is said to have *compact support* if it vanishes outside some finite length interval.

Banach space of test functions with some useful properties like invariance under time-frequency shifts [6, p. 253], thus allowing for time-frequency analysis on its dual S_0' which is, in that particular sense, the largest possible Banach space of tempered distributions that is useful for time-frequency analysis. One more motivation for considering spreading functions in S_0' is that Hilbert–Schmidt operators are compact, hence, they exclude both invertible operators (including line-of-sight channels) and small perturbations of invertible operators, which are useful in the theory of radar identification and in some mobile communication applications. For results using a Banach space setup, see for example [9, 12].

Nevertheless, for narrowband finite lifelength channels such as those typical for radio communications, all analysis can be restricted to the time window and frequency band of interest. We show in [5] that the full system behavior within this time-frequency window can be modeled with an infinitely many times differentiable spreading function $S_H(\nu, t)$ that vanishes for frequencies ν outside some finite interval and which has subexponential decay as a function of t. That a function f has *subexponential decay* means that for $0 < \varepsilon < 1$ there is some $C_\varepsilon > 0$ such that

$$|f(x)| \leq C_\varepsilon \mathrm{e}^{-|x|^{1-\varepsilon}} \qquad \text{for all } x \in \mathbb{R}.$$

Hence we can with negligible errors also do a smooth cutoff to a compactly supported and infinitely many times differentiable spreading function. A big advantage of this Hilbert–Schmidt model is that Fourier analysis can be applied without the need of deviating into distribution theory.

4.1.4 Computing the Channel Matrix G

For $\epsilon > 0$ we define the *ϵ-essential support* of a bounded continuous function $f\colon \mathbb{R} \to \mathbb{C}$ to be the closure of the set $\{x : |f(x)| \geq \epsilon \cdot \max_x |f(x)|\}$. For communications applications with Q carrier frequencies, at least Q samples of every received symbol are needed in the receiver. Thus a hasty and naive approach to computing the matrix elements could start with a $Q \times Q$ matrix representation of H for computing the samples of $Hg_{q,r}$. If up to R neighboring transmission symbols have overlapping ϵ-essential support, then we need to compute $(RQ)^2$ matrix elements $\langle Hg_{q,r}, \gamma_{q',r'} \rangle$, which, with this approach, would require $R^2 \cdot \mathcal{O}(Q^5)$ arithmetic operations with Q typically being at least of the size 256–1024 in radio communications, and with $R = 4$ for $\epsilon = 10^{-6}$ and the optimally well-localized Gaussian windows that we have used for the applications described in [5]. This is a quite demanding task, so therefore more efficient formulas and algorithms were derived in [5] for the Hilbert–Schmidt channel models described in last section. With notation $I_{C,B} \stackrel{\text{def}}{=} \left[C - \frac{B}{2}, C + \frac{B}{2}\right]$, the resulting model is based on the following assumptions about compact supports and index sets for the involved functions:

$$\begin{aligned}
\operatorname{supp}\widehat{g} &\subseteq I_{\Omega_c,\Omega}, \qquad T_g \stackrel{\text{def}}{=} \tfrac{1}{\Omega}, \qquad T_\gamma \stackrel{\text{def}}{=} \tfrac{1}{\Omega+\omega},\\
\operatorname{supp} S_H &\subseteq I_{\omega_c,\omega} \times I_{C,L}, \qquad \operatorname{supp}\widehat{Hg} \subseteq \operatorname{supp}\widehat{\gamma} \subseteq I_{\Omega_c+\omega_c,\Omega+\omega},\\
\mathcal{K},\mathcal{M} &\subset \mathbb{Z}, \qquad |\mathcal{K}| < \infty, \qquad |\mathcal{M}| < \infty \qquad \text{and}\\
g(mT_g) &= \gamma(kT_\gamma) = (Hg)(kT_\gamma) = 0 \quad \text{for} \quad k \in \mathbb{Z}\setminus\mathcal{K} \text{ and } m \in \mathbb{Z}\setminus\mathcal{M}.
\end{aligned}$$

The analysis takes place in an interval $I_{C_0+t_0,L_0}$ containing the support of all perturbed basis functions $Hg_{q,r}$. We refer to [5] for details, but in short, the algorithm is based on a smooth truncation of $\widehat{S_H(\nu,\cdot)}$ to a band of width $1/T''$ containing the full transmission frequency band, in which $S_H(\nu,\cdot)$ can be fully represented by sample values $S_{n,p}$, from which the spreading function S_H^q experienced by the functions $(g_{q,r})_r$ can be computed:

$$\widehat{S_H^q(\cdot,t)}(t_0) = \omega_0 T'' \chi_{I_{C_0,L_0}}(t-t_0) \sum_{p\in\mathcal{P}} \mathrm{e}^{\mathrm{i}2\pi\Omega_{c,q}(t-pT'')} \operatorname{sinc}_\Omega(t-pT'') \times \\ \times \sum_{n\in\mathcal{N}} S_{n,p} \mathrm{e}^{\mathrm{i}2\pi\frac{t-t_0-pT''}{L_0}}, \tag{4.2}$$

with $\Omega_{c,q}$ being the centerpoint of the support of $\widehat{g_{q,r}}$ and $\operatorname{sinc}_\Omega(x) \stackrel{\text{def}}{=} \frac{\sin(\pi\Omega x)}{\pi x}$ extended continuously to $\mathbb{R}$. Using (4.2), we can compute the samples $(Hg_{q,r})(kT_\gamma) = T_g \sum_{m\in\mathbb{Z}} f(mT_g)\,(S_H^q(\cdot,kT_\gamma - mT_g))\widehat{\ }(-mT_g)$ and finally the matrix element $\langle Hg_{q,r},\gamma_{q',r'}\rangle$ using the formula

$$\langle u,v\rangle_{L^2(\mathbb{R})} = T \sum_{k\in\mathcal{I}_u} u(kT)\overline{v_{\text{bpf}}(kT)}$$

for functions with supports

$$\operatorname{supp}\widehat{u} \subseteq I_{C_u,B}, \quad \operatorname{supp}\widehat{v} \subseteq I_{C_v,B}, \quad I_{C_{uv},B_{uv}} \stackrel{\text{def}}{=} I_{C_u,B} \cap I_{C_v,B} \neq \emptyset, \qquad T = \tfrac{1}{B}$$

and with v_{bpf} being defined by its Fourier transform $\widehat{v_{\text{bpf}}}(\xi) \stackrel{\text{def}}{=} \widehat{v}(\xi)\chi_{I_{C_{uv},B_{uv}}}(\xi)$.

As explained in [5], this way the full matrix G can be computed in $R^2\cdot\mathcal{O}(M^2\cdot Q^2)$ arithmetic operations with $M \stackrel{\text{def}}{=} |\mathcal{M}|$, which can be compared to the $R^2\cdot\mathcal{O}(Q^5)$ operations of the more naive and straightforward matrix computation approach described above.

Bibliography

[1] E. Bonek, H. Asplund, C. Brennan, C. Bergljung, P. Cullen, D. Didascalou, P. C.F. Eggers, J. Fernandes, C. Grangeat, R. Heddergott, P. Karlsson, R. Kattenbach, M. B. Knudsen, P. E. Mogensen, A. F. Molisch, B. Olsson, J. Pamp, G. F. Pedersen, I. Pedersen, M. Steinbauer, M. Weckerle, and T. Zwick, *Antennas and Propagation*, chapter 3, pages 77–306, John Wiley & Sons, 2001.

[2] B. Delyon and A. Juditsky, "On minimax wavelet estimators," *Appl. Comput. Harmon. Anal.*, 3(3):215–228, 1996.

[3] H. G. Feichtinger and G. Zimmermann, "A Banach space of test functions for Gabor analysis," in Hans G. Feichtinger and Thomas Strohmer, editors, *Gabor Analysis and Algorithms*, chapter 3, pages 123–170. Birkhäuser, Boston, MA, USA, 1998. WWW: `http://www.uni-hohenheim.de/~gzim/Publications/bsotffga.pdf`.

[4] N. Grip, *Wavelet and Gabor Frames and Bases: Approximation, Sampling and Applications*, Doctoral thesis 2002:49, Luleå University of Technology, SE-971 87 Luleå, 2002, WWW: `http://pure.ltu.se/ws/fbspretrieve/1334581`.

[5] N. Grip and G. Pfander, "A discrete model for the efficient analysis of time-varying narrowband communication channels," *Multidim. Syst. Sign. Process.*, 19(1):3–40, March 2008. WWW: `http://pure.ltu.se/ws/fbspretrieve/1329566`.

[6] K. Gröchenig, *Foundations of Time-Frequency Analysis*, Birkhäuser, 2000.

[7] G. Matz, D. Schafhuber, K. Gröchenig, M. Hartmann, and F. Hlawatsch, "Analysis, optimization, and implementation of low-interference wireless multicarrier systems," *IEEE Trans. Wireless Comm.*, 6(5):1921–1931, May 2007. WWW: `http://ibb.gsf.de/homepage/karlheinz.groechenig/preprints/matz_twc05.pdf`.

[8] G. Matz and F. Hlawatsch, "Time-frequency transfer function calculus (symbolic calculus) of linear time-varying systems (linear operators) based on a generalized underspread theory," *J. Math. Phys.*, 39(8):4041–4070, August 1998, (Special issue on Wavelet and Time-Frequency Analysis.)

[9] G. E. Pfander and D. F. Walnut, "Measurement of time-variant linear channels," *IEEE Trans. Inform. Theory*, 52(11):4808–4820, November 2006, WWW: 5 `http://www.math.jacobs-university.de/pfander/pubs/timevariant.pdf`.

[10] G. E. Pfander and D. F. Walnut, "Operator identification and Feichtinger's algebra," *Sampl. Theory Signal Image Process*, 5(2):183–200, May 2006, WWW: `http://www.math.jacobs-university.de/pfander/pubs/operatoridentfei.pdf`.

[11] S. Rickard, *Time-frequency and time-scale representations of doubly spread channels*, Ph.D. dissertation, Princeton University, November 2003. WWW: `http://sparse.ucd.ie/publications/rickard03time-frequency.pdf`.

[12] T. Strohmer, "Pseudodifferential operators and Banach algebras in mobile communications," *Appl. Comput. Harmon. Anal.*, 20(2):237–249, March 2006, WWW: `http:///www.math.ucdavis.edu/~strohmer/papers/2005/pseudodiff.pdf`.

4.2 Generic Description of a MIMO-OFDM-Radio-Transmission-Link

R. Amling, V. Kühn, University of Rostock, Germany

4.2.1 Introduction

As more and more mobile devices are supporting multiple air interfaces it is necessary to choose the best radio access system for a requested service. Therefore, several quality parameters are needed to accomplish an automatic selection of the optimal access network. In order to prevent mobile devices from complex calculations a generic model is regarded to be useful. This model should allow the prediction of important parameters like error rate, data rate and latency as reliably as possible based on a usually imperfect channel estimation and related system parameters.

The focus of this project is the analysis of a multiple-input multiple-output (MIMO) link in combination with orthogonal frequency-division multiplexing (OFDM). Channel coding, interleaving and further system parameters have been taken from the LTE-specifications. As the first half of the project time has elapsed, this summary presents only the results of a coded single-input single-output (SISO) OFDM system.

4.2.2 System Model

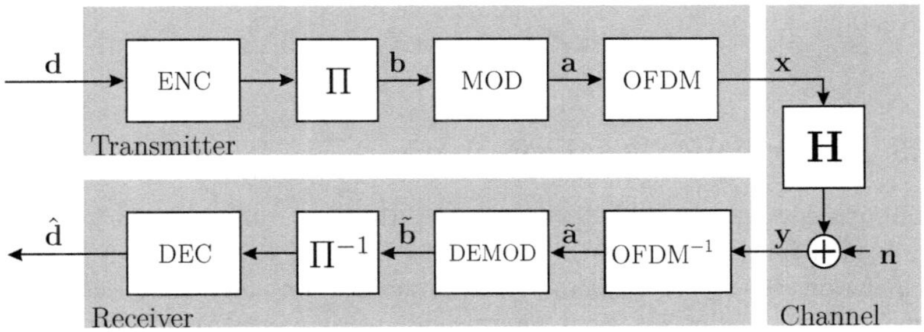

Figure 4.1: System model of MIMO-OFDM link

A typical OFDM link as depicted in Fig. 4.1 is considered. The binary information sequence $\mathbf{d}\ \epsilon\{0,1\}^{N_d\times 1}$ is encoded using one of the following three forward error correction schemes [1–3].

Code 1: convolutional code, constraint length $L_c = 3$, code rate $R_c = \frac{1}{2}$, generators $G = [7;5]_8$

Code 2: convolutional code, constraint length $L_c = 9$, code rate $R_c = \frac{1}{2}$, generators $G = [561;715]_8$

Code 3: parallel turbo code, constraint length $L_c = 4$, code rate $R_c = \frac{1}{3}$, generators $G = [13; 15]_8$

After encoding, a random bit interleaver is employed. The encoded and interleaved binary sequence $\mathbf{b} \in \{0,1\}^{N_b \times 1}$ is mapped onto M-QAM symbols $\mathbf{a}$ with constellation sizes $M = \{4, 16, 64\}$, using binary reflected Gray mapping [4, 5].

According to the LTE specifications [6], the OFDM symbol can consist of $N_c = 256$ subcarriers and a guard interval whose length N_G is 20% of the core OFDM symbol. The corresponding sequence $\mathbf{x}$ is transmitted over a block fading channel whose impulse response has a length $N_h \in \{2, 10, 20\}$. Its coefficients are i.i.d. complex Gaussian distributed random variables with zero mean and variance $1/N_h$. The additive noise is assumed to be white and Gaussian with $n \sim \mathcal{N}(0, \sigma_N^2)$.

A major advantage of OFDM is the very efficient equalization of the received signal $\mathbf{y}$ due to frequency nonselective conditions on each subcarrier. The demodulation block provides LLRs $\tilde{\mathbf{b}}$ of each code bit which are de-interleaved and decoded by a conventional Viterbi algorithm. As no iterative turbo detection is performed, this approach is termed bit-interleaved coded modulation with parallel decoding (BICM-PD) [5]. The estimated information word is denoted by $\hat{\mathbf{d}} \in \{0,1\}^{N_d \times 1}$.

For channel estimation, a typical OFDM pilot symbol based approach is applied. In the simulations, $N_P = 2$ OFDM pilot symbols with unit power are inserted in front of each OFDM frame. At the receiver, the estimation of the channel transfer function is improved by a noise reduction approach exploiting the fact that the channel impulse response does not exceed the guard interval [7, 8]. This leads to the estimated channel coefficient $\hat{H}_k = H_k + \Delta H_k$ on subcarrier k with a variance of

$$\sigma_H^2 = \frac{\sigma_N^2}{N_P} \cdot \frac{N_G}{N_c} \,. \tag{4.3}$$

4.2.3 Performance Analysis

In order to develop a feasible generic model, the achievable bit error rate (BER) for a fixed channel and a given signal to noise ratio (SNR) have been determined by simulations for different channel models, modulation and coding schemes. The channel capacity was chosen as an intermediate parameter representing the channel transfer function and the SNR by a single value. As will be shown later, this mapping is not bijective but allows a tight prediction of the BER.

For subcarrier k and a perfectly known channel coefficient H_k at the receiver, the link capacity for a discrete input alphabet $\mathbb{X}$ and a continuous output set $\mathbb{Y}$ is

$$C_k = I(X; Y \mid H_k) = \sum_{x \in \mathbb{X}} \Pr\{x\} \int_{\mathbb{Y}} p(y \mid x, H_k) \cdot \log_2 \frac{p(y \mid x, H_k)}{p(y \mid H_k)} \mathrm{d}y \,. \tag{4.4}$$

Since BICM-PD is employed, the capacity C_k in (4.4) cannot be achieved. Instead, the bit-level capacities $C_{k,\mu}^{\mathrm{bl}}$ for bit-level μ, $\mu = 0, 1, \ldots, m-1$, with $m = \log_2 M$ as introduced in [5] have to be determined. This can be accomplished by applying the chain rule of mutual information

$$\begin{aligned} C_k &= I(b_0, b_1, \ldots, b_{m-1}; Y \mid H_k) \\ &= I(b_0; Y \mid H_k) + I(b_1; Y \mid b_0, H_k) + \ldots + \\ &\quad + I(b_{m-1}; Y \mid b_0, b_1, \ldots, b_{m-2}, H_k) \,. \end{aligned} \tag{4.5}$$

Neglecting the constraint of known bit-levels leads to a reduction of the mutual information in (4.4) and the capacity for parallel decoding becomes

$$C_k^{\mathrm{pd}} = I(b_0; Y \mid H_k) + I(b_1; Y \mid H_k) + \ldots + I(b_{m-1}; Y \mid H_k) = \sum_{\mu=0}^{m-1} C_{k,\mu}^{\mathrm{bl}} < C_k \,. \tag{4.6}$$

Finally, the average parallel-decoding capacity for an OFDM symbol with N_c subcarriers and m bit-levels becomes

$$C^{\mathrm{pd}} = \frac{1}{N_c} \sum_{k=1}^{N_c} C_k^{\mathrm{pd}} = \frac{1}{N_c} \sum_{k=1}^{N_c} \sum_{\mu=0}^{m-1} C_{k,\mu}^{\mathrm{bl}} \,. \tag{4.7}$$

In order to infer from the bit-level capacities to the BER, simulations have been performed for different channel realizations, codes and modulation schemes. As an example, Fig. 4.2 illustrates the obtained results for a 16-QAM and all considered channel impulse response lengths with uniform power delay profile.

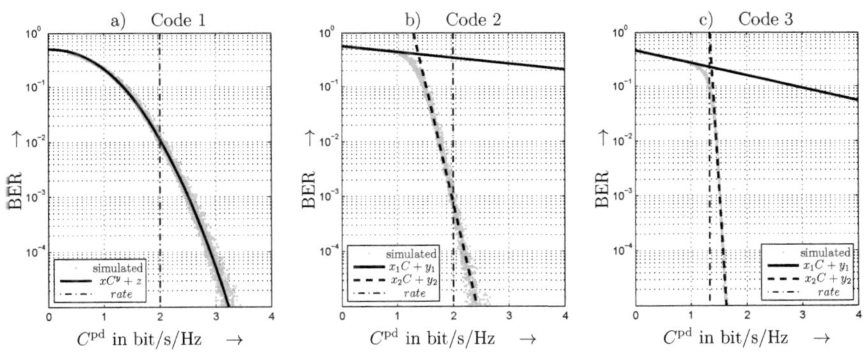

Figure 4.2: Simulation results (gray), generated models (black) and for different FEC codes and 16-QAM

In Fig. 4.2a), the error rates for the simple convolution code show an exponential slope with an increasing variance at higher capacities. For the convolution code with $L_c = 9$ and the turbo code in the diagrams b) and c), a waterfall-like behavior can be observed. Moreover, the turbo code's BER can be predicted quite accurately as the variations are very small even at high capacities. The largest capacities required for a reference BER of 10^{-5} are 3.4 bit/s/Hz, 2.6 bit/s/Hz and 1.7 bit/s/Hz for code 1, 2, and 3, respectively. Comparing these values with the spectral efficiencies $R = m \cdot R_c$ indicated by the dashed vertical lines, gaps $\Delta R(\mathrm{BER}) = C^{\mathrm{pd}}(\mathrm{BER}) - R$ can be defined. They take the values 1.4 bit/s/Hz, 0.6 bit/s/Hz and 0.3 bit/s/Hz, respectively. As expected, the turbo code's efficiency R approaches the capacity most closely. Further simulations have been performed for different modulation schemes and channels which cannot be shown in this survey. The modulation schemes 4-QAM and 64-QAM show a similar behavior as 16-QAM. The 4-QAM reaches a BER of 10^{-5} at 1.8 bit/s/Hz with code 1, 1.4 bit/s/Hz with code 2 and 1 bit/s/Hz with code 3 and has smaller variations at low error rates. For 64-QAM, the BER variations increase at higher capacities especially for the weak code 1. A BER of 10^{-5} can be reached at capacities from 4 bit/s/Hz to 5 bit/s/Hz.

Figure 4.3 illustrates that the relationship between bit error rate and bit-level capacity does not depend on the length of the channel impulse response. For different lengths N_h, the same quantitative behavior can be observed.

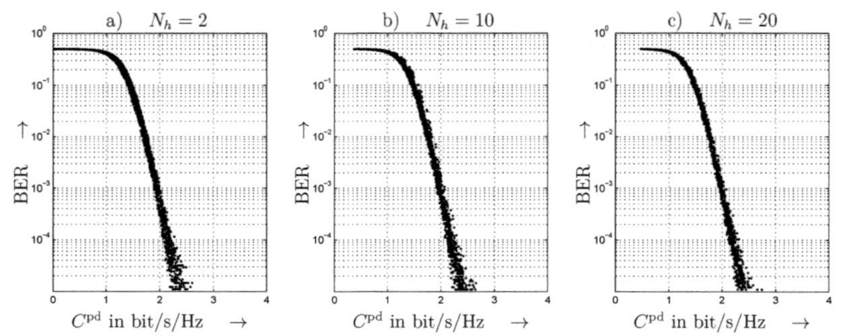

Figure 4.3: Simulation results comparison for 16-QAM, Code 2 and different channel impulse response lengths N_h

4.2.4 Generic Model

The aim of the project is to describe the dependency between the capacity C^{pd} and the BER by a generic model. This is accomplished by applying curve fitting algorithms to the results shown in Fig. 4.2. They provide results with a confidence level of 95%. For the code 1 with memory 2, an exponential function of the form

$$\log_{10}(\ BER(C)\ ) = x \cdot C^y + z \tag{4.8}$$

emerged as the best description. The function is depicted in Fig. 4.2a). The parameters x, y and z are provided in Table 4.1 for 16-QAM modulation. For the other two codes, the curves were separated into two regions. The first region covers low capacities and high error rates, the second is the one of interest and includes the waterfall region. In both of them, the logarithm of the BER can be approximated by a straight line. We obtain

$$\log_{10}(\ BER_1(C)\ ) = x_1 C + y_1 \quad \text{for} \quad 0 < C < C_P \tag{4.9a}$$

$$\log_{10}(\ BER_2(C)\ ) = x_2 C + y_2 \quad \text{for} \quad C_P < C < \log_2(M)\ , \tag{4.9b}$$

where C_P denotes the capacity where both lines intersect. Figures 4.2b) and c) illustrate the results, the corresponding model parameters are summarized in Table 4.1. It also contains the gaps between the spectral efficiency R and the required capacity C^{pd} predicted by the model.

Table 4.1: Generic model parameters for 16-QAM

	Code 1	Code 2	Code 3
generic model $\log_{10}(BER(C))$	$x \cdot C^y + z$	$x_{1,2} \cdot C + y_{1,2}$	$x_{1,2} \cdot C + y_{1,2}$
parameters region 1	$x = -0.3689$ $y = +2.1666$ $z = -0.2958$	$x_1 = -0.1085$ $y_1 = -0.2427$ -	$x_1 = -0.2340$ $y_1 = -0.3291$ -
parameters region 2	- -	$x_2 = -4.4219$ $y_2 = +5.7116$	$x_2 = -17.1307$ $y_2 = +22.9040$
ΔR at 10^{-5}	1.24	0.42	0.3

In order to use these models, the capacity C^{pd} needs to be computed using estimated channel coefficients $\hat{H}_k$ including errors ΔH_k with a variance determined in (4.3)[2]. For codes whose BER curves exhibit a very high slope, capacity estimation errors lead to dramatic prediction errors of the achievable error rate. This effect shall be investigated now. The channel estimation errors ΔH_k are assumed to be independent and complex Gaussian distributed. Including $\hat{H}_k$ into (4.6) leads to the final estimate $\hat{C}^{\mathrm{pd}} = \frac{1}{N_c}\sum_{k=1}^{N_c} \hat{C}_k^{\mathrm{pd}}$.

An example for the distribution of $\hat{C}^{\mathrm{pd}}$ at a target bit error rate 10^{-5} for FEC code 2 is illustrated in Fig. 4.4a) which corresponds to an average E_s/N_0 of 12.2 dB. Two OFDM pilot symbols are used for channel estimation with subsequent noise reduction. It can be observed that $\hat{C}^{\mathrm{pd}}$ is nearly Gaussian distributed which could be expected since it stems from averaging over 256 subcarriers. The estimates' mean depends on the specific channel realization and is generally close to the true capacity C^{pd} while the variance only depends on the signal to noise ratio of the operating point of the specific BICM scheme. Hence, $\hat{C}^{\mathrm{pd}} > C^{\mathrm{pd}}$ holds in most cases, i.e., the estimate is too optimistic and its application to resource allocation strategies would lead to error rates larger than 10^{-5}. Further investigations revealed that the difference $\Delta C^{\mathrm{pd}} = \hat{C}^{\mathrm{pd}} - C^{\mathrm{pd}}$ becomes not larger than 0.027 bit/s/Hz for 95% of all estimated channels at the specific operating point of code 1, not larger than 0.04 bit/s/Hz for the operating point of code 2 and 0.053 bit/s/Hz for code 3. The distance ΔC^{pd} is getting smaller at high SNRs as can be seen from Fig. 4.4b) illustrating the maximum relative deviation of 95% of all channels. Considering a worst case scenario, an offset reducing $\hat{C}^{\mathrm{pd}}$ by $\max_{95}(\Delta\hat{C}^{\mathrm{pd}})$ guarantees that 95% of all estimated capacities are not larger than C^{pd}. Otherwise, the model causes an outage.

A second influence on the model quality is the variation of the required capacity

[2]It is assumed that the signal to noise ratio is perfectly known although this is a rather optimistic assumption.

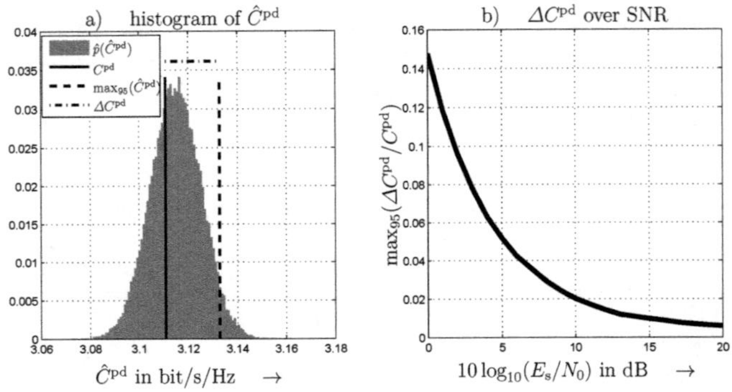

Figure 4.4: a): probability density of $\hat{C}^{\mathrm{pd}}$ and true C^{pd} for operating point of code 2, dashed red line indicates maximum of 95% of all $\hat{C}^{\mathrm{pd}}$, b): maximum of 95% of all $\Delta C^{\mathrm{pd}}/C^{\mathrm{pd}}$

C^{pd} at the target error rate. With respect to Fig. 4.2 the capacity varies at a BER of 10^{-5} by ± 0.25 bit/s/Hz for code 1 and by ± 0.2 bit/s/Hz for code 2. The turbo code shows nearly no variations. Again, a worst case consideration extracts the maximum required capacity to ensure the target BER. Moreover, channel estimation errors at the receiver cause an additional loss which can be modeled by a complementary noise term. This error causes small SNR degradations between 0.2 dB and 0.5 dB and, consequently, a small shift of the curves in Fig. 4.2 towards higher capacities.

4.2.5 Summary and Further Work

Simple generic models have been developed for a BICM-PD OFDM link describing the dependency between the bit-level capacity and the error probability. These models include different coding and modulation schemes related to LTE-specifications but can be easily generalized. Uncertainties due to estimation errors have been analyzed and allow worst case scenarios with which the target error rate can be achieved in a predefined percentage of channels.

In a next step, the OFDM system shall be extended to a MIMO-OFDM system enabling spatial multiplexing and different diversity methods like cyclic delay diversity. This extension requires the analysis of additional spatial parameters characterizing the MIMO channel that have to be integrated into the generic models. Furthermore, automatic-repeat-and-request procedures shall be included into the investigations.

Bibliography

[1] 3GPP, *3rd Generation Partnership Project, Technical Specification Group Radio Access Network, Multiplexing and channel coding (FDD) (Release 9)*, December 2009, available: http://www.3gpp.org/ftp/Specs/html-info/25212.htm.

[2] M. Bossert, *Kanalcodierung*, Teubner, 2 edition, 1998.

[3] V. Kühn, *Wireless Communications over MIMO Channels*, John Wiley & Sons, Ltd, 2006.

[4] C. Stierstorfer and R.F.H. Fischer, "(gray) mappings for bit-interleaved coded modulation," April 2007.

[5] C. Stierstorfer, *A Bit-Level-Based Approach to Coded Multicarrier Transmission*, PhD thesis, University of Erlangen-Nürnberg, Germany, Aug 2009.

[6] 3GPP, *3rd Generation Partnership Project, Technical Specification Group Radio Access Network, Evolved Universal Terrestrial Radio Access (E-UTRA), Physical Channels and Modulation (Release 9)*, December 2009, available: http://www.3gpp.org/ftp/Specs/html-info/36211.htm.

[7] K.-D. Kammeyer and H. Schmidt, "OFDM: An old idea solves new problems," in *International Symposium on Theoretical Electrical Engineering (ISTET 01)*, Linz, Austria, Aug 2001.

[8] H. Zamiri-Jafarian, M.J. Omidi, and S. Pasupathy, "Improved channel estimation using noise reduction for ofdm systems," volume 2, pages 1308 – 1312 vol.2, April 2003.

[9] G. Caire, G. Taricco, and E. Biglieri, "Bit-interleaved coded modulation," *IEEE Transactions on Information Theory*, 44(3):927–946, 1998.

[10] K.-D. Kammeyer, *Nachrichtenübertragung*, Teubner, 3 edition, 2004.

[11] C. Stierstorfer and R.F.H. Fischer, "Adaptive interleaving for bit-interleaved coded modulation," January 2008.

4.3 Resource Allocation Using Broadcast Techniques

M. Bossert, C. Huppert, J.G. Klotz, University of Ulm, Germany

4.3.1 Motivation

In the following we consider the downlink of a multi user communication system. For transmission in such a system the available resources, e.g., bandwidth and power, has to be allocated to the single users. This is often done by means of orthogonal access techniques, e.g., OFDMA, as in the uplink channel. However, the downlink channel is equivalent to the information theoretic broadcast channel since there is only one sender transmitting messages to several receivers. In contrast to the uplink channel which can only be modeled as multiple access channel there is perfect synchronization and coordination for transmitting messages in the downlink channel. Thus, using broadcast techniques as multiple access technique is superior in terms of the achievable rates. This was first shown by Cover in 1972 in [18]. He investigated the achievable rate region in the degraded broadcast channel and showed that all points of this region can be reached by means of superposition techniques.

4.3.2 Resource Allocation Algorithms

In multi-user OFDM systems the available subcarriers and the transmit-power must be allocated to the individual users in a way that certain service requirements are fulfilled. Motivated by the information theoretic superiority of broadcast techniques over orthogonal access techniques, we consider resource allocation schemes for different requirements taking into account the broadcast gain in the following. Furthermore, we do not only consider nodes equipped with a single antenna (SISO), but also multi antenna systems (MIMO). The optimal resource allocation schemes are known for all considered scenarios, however compared to the orthogonal methods it is more complex to determine their solutions and a larger signaling overhead is required to inform the users about the determined allocation. Thus, we propose strategies which achieve a near optimum performance while requiring only a reasonable complexity. These strategies are based on one or more of the following techniques:

- Restriction to at most two users per carrier in order to keep the overhead low.
- Introduction of a new metric to predict interference caused by broadcast techniques (e.g. Eigenvalue Update).
- Usage of hybrid allocation strategies combining orthogonal access with broadcast techniques.

In the following some of the proposed algorithms a briefly described and some simulation results are presented.

Sum-Rate Maximization

First, we consider a scenario where the overall system throughput, defined by the sum over all achievable user rates, is maximized under a transmit power constraint. The proposed algorithm uses eigen-beamforming and dirty paper coding, cf. [13]. The inter-user interference is estimated and the eigenvalues of the affected beams are updated (so-called). Then the optimal power allocation is retrieved by perform water-filling over the adapted eigenvalues, cf. [19, 20]. In Fig. 4.5 the results of the algorithm are compared with the optimal solution. It can be seen that the heuristic algorithm achieves a sum rate of up to 99% of the optimal algorithm for low SNRs. For higher SNRs it still reaches 91%.

Sum-Rate Maximization with Minimum Rate Requirements

Like in the above considered scenario we maximize the sum rate of the system. However an individual minimum rate requirement for each user has to be fulfilled. Such a scheme may be needed in systems where delay critical as well as non-delay critical data should be sent to each user.

Minimum Rate Requirements in SISO-OFDM systems

The proposed algorithm, cf. [12], mainly works in two steps. First, a simple scheduler allocates one user to each carrier aiming in assigning the minimum rates. This scheduler performs the "worst selects" algorithm, i.e., always the instantaneous worst user chooses its best carrier. In the second step, an additional user is added to each suitable carrier by means of broadcast techniques. A modified version of this algorithm avoids irresolvable decoding dependencies in order to make it applicable to code words stretching over several blocks. Some simulation results of these two algorithms, named BC and DEP, are compared to the optimum solution, cf. [21, 22], as well as to a pure scheduling strategy in Fig. 4.6. It can be seen that the proposed algorithm achieves a performance near to the optimum and is clearly superior to the pure scheduler. Furthermore, the results reveal that the modified version still exploit a big part of the possible broadcast gain.

Minimum Rate Requirements in MIMO-OFDM systems

For this problem two different heuristic resource allocation algorithms are proposed, cf. [14], which have a much lower complexity than the existing optimal solution, cf. [23]. The first strategy, extended eigenvalue update (EEU) algorithm, is based on the previously discussed heuristic sum rate maximization algorithm using eigenvalue updates. The second algorithm, the rate based coding (RBC) algorithm, makes use of the duality of uplink and downlink, which allows us to determine the allocation in the dual uplink. The performance of these algorithms for different minimum rates compared to the optimal algorithm is depicted in Fig. 4.7. It can be seen, that the EEU algorithm clearly outperforms the "simple scheduler". The RBC algorithm achieves a better performance than the first algorithm at the cost of more complexity. Actually, it gets very close to the optimal solution for low required minimum rates.

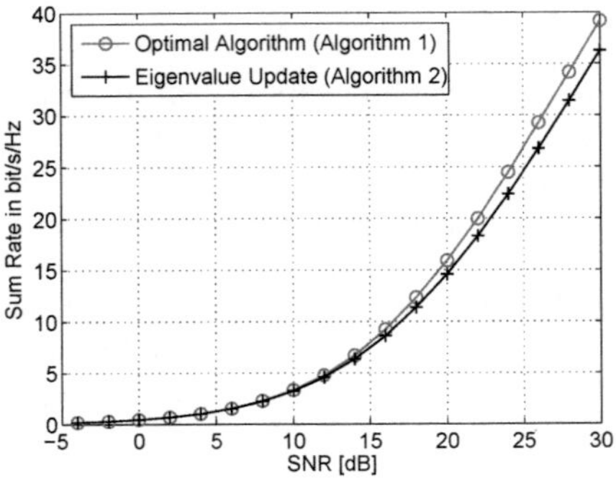

Figure 4.5: Sum-Rate Maximization in MIMO-OFDM (8 tx-antennas, up to 4 rx-antennas, block fading channels, 64 carriers, 40 users, average SNR values are uniformly distributed in a range from 0 dB to 20 dB)

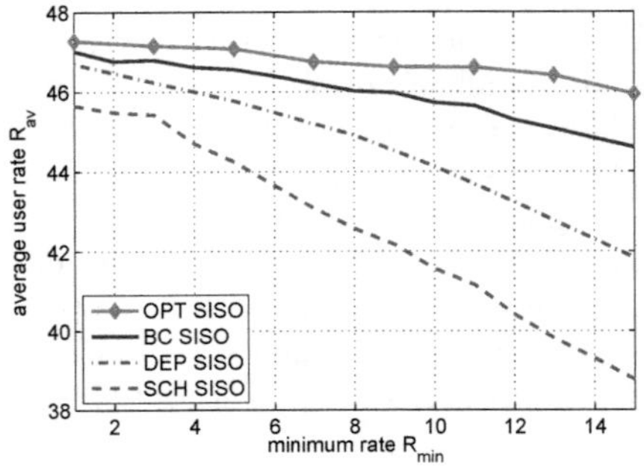

Figure 4.6: Minimum rate requirements in SISO-OFDM (block fading channels, 256 carriers, 40 users, average SNR values are uniformly distributed in a range from 0 dB to 20 dB)

Maximization of the Number of Users

While in the upper two scenarios the number of users, which are served in the system, is constant and the rate of the users is maximized, we maximize in this scenario the number of users which can be served in the system. Each user is provided a fixed rate.

In the following we propose an hybrid algorithm, cf. [11], aiming in maximization of the number of served users. This algorithm works iteratively. In each iteration step it increases the number of users successively until the rate requirements can not be fulfilled anymore. The idea behind this iterative method is that based on a stable system new users should be added. In each iteration all carriers are exclusively assigned to users by the "worst selects" algorithm in a first step. Then, always the instantaneous worst user is added as second user to its best suitable carrier as long as the rate requirements are not fulfilled. Finally, the fraction for the power distribution is determined individually for each carrier. The performance of this algorithms is displayed in Fig. 4.8 and compared to the optimum solution based on [24] as well as to a simple scheduler. It can be seen that for high required rates all algorithms achieve nearly the same performance whereas for lower rate requirements the proposed algorithm clearly outperforms the scheduler by exploiting parts of the broadcast gain.

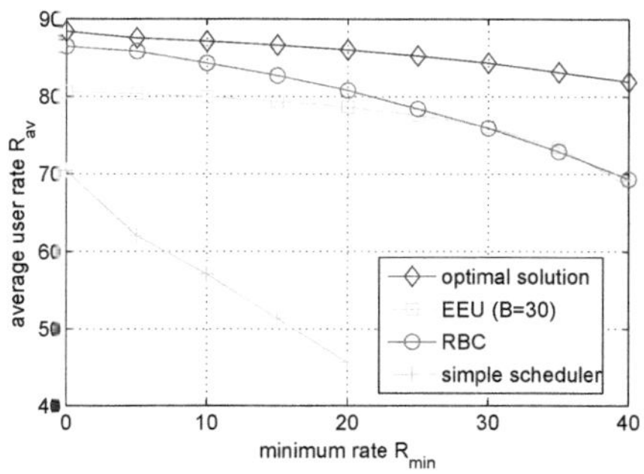

Figure 4.7: Minimum rate requirements in MIMO-OFDM (block fading channels, 64 carriers, 20 users, average SNR values are uniformly distributed in a range from 0 dB to 20 dB)

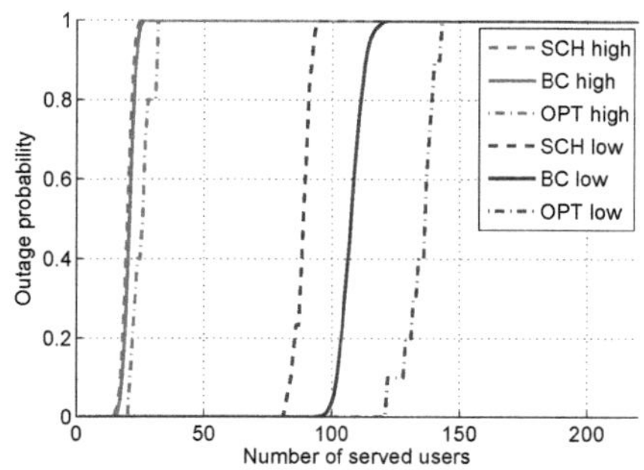

Figure 4.8: User maximization in SISO-OFDM (block fading channels, 256 carriers, average SNR values are uniformly distributed in a range from 0 dB to 20 dB)

Bibliography

Publications Emerged from the Projects:

[1] C. Huppert and M. Bossert, "Delay-limited capacity for broadcast channels," in *11th European Wireless Conference*, pages 829–834, Nicosia, Cyprus, April 2005.

[2] M. Bossert, "Coded modulation for OFDM and the broadcast channel - invited," in *Fifth International Workshop on Multi-Carrier Spread Spectrum*, Oberpfaffenhofen, Germany, September 2005.

[3] C. Huppert and M. Bossert, "Downlink transmission and the broadcast channel - invited," in *RadioTecC*, Berlin, Germany, October 2005.

[4] M. Bossert, "OFDM-Übertragung und der Broadcast-Kanal - invited," in *Öffentliche Diskussionssitzung "'Beyond 3G - Zukünftige Entwicklung mobiler Funksysteme"', ITG*, Ulm, Germany, November 2005.

[5] Carolin Huppert and Boris Stender, "Hybrid scheduling and broadcast with minimum rates in OFDM," in *12th European Wireless Conference*, Athens, Greece, April 2006.

[6] Boris Stender and Carolin Huppert, "Power allocation with constraints over parallel gaussian broadcast channels," in *12th European Wireless Conference*, Athens, Greece, April 2006.

[7] Carolin Huppert and Martin Bossert, "Performance evaluation of a low complex broadcast algorithm for OFDM channels," in *11th International OFDM-Workshop*, Hamburg, Germany, August 2006.

[8] Carolin Huppert, Boris Stender, and Axel Hof, "Resource allocation for OFDM broadcast channels allowing user-wise coding," in *3rd International Symposium on Wireless Communication Systems*, Valencia, Spain, September 2006.

[9] Boris Stender, Carolin Huppert, and Gerd Richter, "Fast broadcasting," In *3rd International Symposium on Wireless Communication Systems*, Valencia, Spain, September 2006.

[10] Carolin Huppert and Martin Bossert, "On achievable rates in the two user AWGN broadcast channel with finite input alphabets," in *IEEE International Symposium on Information Theory*, Nice, France, June 2007.

[11] C. Huppert and M. Bossert, "Heuristic approach of maximizing the number of served users in an OFDM broadcast channel," in *12th International OFDM-Workshop*, Hamburg, Germany, August 2007.

[12] C. Huppert and B. Stender, "Resource allocation with minimum rates for OFDM broadcast channels," *European Transactions on Telecommunications*, 18, No. 6:563–572, October 2007.

[13] J. Klotz, C. Huppert, and M. Bossert, "Heuristic resource allocation for sum rate optimization in MIMO-OFDM systems using eigenvalue updates," in *IEEE International Symposium on Wireless Communication Systems (ISWCS)*, Reykjavik, Iceland, October 2008.

[14] J. Klotz, F. Knabe, and C. Huppert, "Resource allocation algorithms for minimum rates scheduling in MIMO-OFDM systems," in *7th International Workshop on Multi-Carrier Systems & Solutions (MC-SS 2009)*, Herrsching, Germany, May 2009.

[15] C. Huppert, F. Knabe, and J. Klotz, "User assignment for minimum rate requirements in OFDM-MIMO broadcast systems," *IET Electronic Letters*, 45, No. 12:621–623, June 2009.

[16] J. G. Klotz, F. Knabe, and C. Huppert, "Resource allocation algorithms for minimum rates scheduling in MIMO-OFDM systems," *European Transactions on Telecommunications*, 21, No. 5:449–457, August 2010.

[17] C. Huppert and J. Klotz, "Required transmit power applying Tomlinson-Harashima-precoding in scalar and MIMO broadcast systems," *IEEE Transactions on Communications*, 2010. Accepted for publication.

Other References:

[18] T. M. Cover, "Broadcast channels," *IEEE Transactions on Information Theory*, IT-18(1):2–14, January 1972.

[19] T. Michel and G. Wunder, "Optimal and low complex suboptimal transmission schemes for MIMO-OFDM broadcast channels," in *IEEE International Conference on Communications*, Seoul, Korea, May 2005.

[20] R. Böhnke, V. Kühn, and K.-D. Kammeyer, "Fast sum rate maximization for the downlink of MIMO-OFDM systems," in *Canadian Workshop on Information Theory*, Montreal, Canada, June 2005.

[21] T. Michel and G. Wunder, "Minimum rates scheduling for OFDM broadcast channels," in *IEEE International Conference on Acoustics, Speech and Signal Processing*, volume 4, pages 41–44, Toulouse, France, May 2006.

[22] G. Wunder and T. Michel, "Optimal resource allocation for parallel Gaussian broadcast channels: Minimum rate constraints and sum power minimization," *IEEE Transactions on Information Theory*, 53(12):4817–4822, December 2007.

[23] G. Wunder and T. Michel, "Minimum rates scheduling for MIMO-OFDM broadcast channels," in *9th IEEE International Symposium on Spread Spectrum Techniques and Applications*, Manaus, Brazil, August 2006.

[24] D. N. C. Tse, "Optimal power allocation over parallel Gaussian broadcast channels," in *IEEE International Symposium on Information Theory*, Ulm, Germany, June 1997.

4.4 Rate Allocation for the 2-user Multiple Access Channel with MMSE Turbo Equalization

M. Grossmann, R. Thomä, Ilmenau University of Technology, Germany

4.4.1 Introduction

Recently, iterative (turbo) techniques have been recognized as practical solutions to multi-user detection/equalization problems in coded communication systems. In [1], utilization of the optimal a posteriori probability (APP) equalizer in combination with the APP-based decoder is considered for turbo equalization in frequency-selective fading channels. In [2], the APP detector is replaced by a less computational complex detector that performs soft canceling and minimum mean squared error (SC MMSE) filtering. The turbo equalization technique for block-transmissions over multiple-access channels (MACs) in [3] performs the equivalent signal processing in frequency domain (FD), further reducing the computational complexity.

The convergence of turbo systems can be analyzed by extrinsic information transfer (EXIT) charts [4]. Ashikhmin *et al.* [5] showed that for any code with rate R, the area under its corresponding EXIT function is $1-R$. Based on this area property of the EXIT chart, it has been shown in [6]- [8], that the problem of rate allocation in the *single-user* or *equal-rate multi-user* turbo case, reduces to a simple curve-fitting problem of the two-dimensional (2D) EXIT curves of the detector and decoder.

In this contribution, we consider the problem of rate allocation for the 2-user MAC in the presence of frequency-selective fading employing the low-complexity SC FD-MMSE turbo equalizer [3]. Specifically, we show that for such a turbo system, the equalizer EXIT characteristic is given by multidimensional surfaces, and thus, rate allocation to the users is no longer a simple 2D matching of the EXIT curves. Moreover, we derive an upper bound for the rate region of the turbo system, and study the problem of maximizing the sum rate of both users.

4.4.2 Turbo Equalization

Consider the cyclic prefix (CP) assisted U-user uplink system in Fig. 4.9, where a base station having M receive antennas receives signals from U active users, each equipped with K transmit antennas. For the ease of analysis, we assume in the following $U = M = 2$ and $K = 1$. However, the extension to more generic cases ($U > 2$, $M > 2$, $K > 1$) is rather straightforward. The transmission scheme of the u-th user ($u = 1, 2$) is based on bit interleaved coded modulation, where the information bit sequence is independently encoded by a binary encoder, randomly bit-interleaved, binary phase-shift keying (BPSK) modulated, and grouped into several blocks that are transmitted over the frequency-selective multiple-access MIMO fading channel.

At the receiver side, iterative processing for joint equalization and decoding is performed. The receiver consists of an SC FD-MMSE equalizer and two single-user APP decoders. Within the iterative processing, the *extrinsic* LLR sequences $\{\lambda_u(n)\}$ and $\{\zeta_u(n)\}$ of the coded bit sequences $\{b_u(n)\}$ are exchanged between the

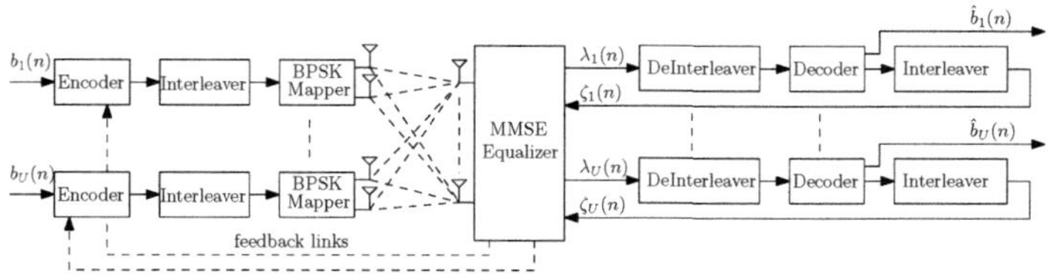

Figure 4.9: Structure for a coded multiuser MIMO system with turbo equalization.

equalizer and both decoders, each separated by the interleaver and deinterleaver in their iteration loop, following the turbo principle [2].

The receiver also selects the code to be used for each user, for the channel realization given, from an available code set, where the users are notified of the codes selected through separated feedback links. We assume zero-delay and error-free feedback links.

4.4.3 Rate Allocation using EXIT Charts

In the 2-user case the convergence characteristic of the equalizer is defined by two EXIT functions [3],

$$\mathbf{f}_e : \mathbf{I}_d \to \mathbf{f}_e \equiv \big(f_{e,1}(\mathbf{I}_d), f_{e,2}(\mathbf{I}_d)\big) \in [0,1]^2,$$

which depend on the mutual information (MI) $\mathbf{I}_d \equiv (I_{d,1}, I_{d,2}) \in [0,1]^2$ with $I_{d,u}$ being the MI between the transmitted bits $b_u(n)$ and the LLRs $\zeta_u(n)$. Similarly, the convergence characteristics of the two decoders are defined by two EXIT functions[3] $f_{d,k} : I_{d,k} \to f_{d,k}(I_{d,k}) \in [0,1]$. An example of the equalizer EXIT vector-functions $\mathbf{f}_e$ and the decoder EXIT function $f_{d,1}$ is shown in Fig. 4.10 (a). Also shown is a possible decoding trajectory (plotted as a projection onto the plane region $\mathcal{U} \equiv \{\mathbf{I}_d : \mathbf{I}_d \in [0,1]^2\}$) of the MI exchange over the iterations, and a region $\mathcal{D}$, which is referred to as the *feasible region* of the EXIT functions $\mathbf{f}_e$, $f_{d,1}$, and $f_{d,2}$. Note that $f_{d,2}$ (not shown) is drawn in the $I_{d,2}$-coordinate. For the computation of the decoding trajectory, the codes of both users were in this case assumed to be identical, and hence the shapes of their EXIT functions are exactly the same.

The monotonicity of the EXIT functions imply that the decoding trajectory converges monotonically to a unique limit point [9]. Convergence of turbo equalization is achieved, when the decoding trajectory attains the maximum point $a_2 \equiv (1,1)$. This is possible for the number of turbo iterations being sufficiently large, if the following two constraints hold: $\mathcal{D}$ is pathwise connected and $a_2 \in \mathcal{D}$.

Let $A_{\mathcal{D}} \equiv \iint_{\mathcal{D}} \mathrm{d}\mathbf{I}_d$ be the area of $\mathcal{D}$. Assume now that each decoder EXIT function $f_{d,k}$ is matched to the corresponding equalizer EXIT function $f_{e,k}$ so that only an infinitesimally small open tube between the four EXIT surfaces remains. Note that such decoder EXIT functions imply 1) an ideally designed code for each user of infinite block length to achieve a nearly zero BER and 2) an infinite number of

[3]Note that $f_{d,k}$, denoted here as the decoder EXIT function, corresponds to the inverse decoder EXIT characteristic defined in [4].

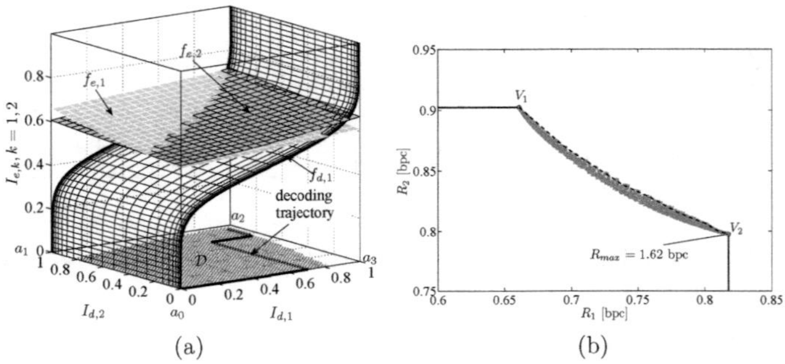

Figure 4.10: (a) Equalizer EXIT functions $f_{e,1}$ and $f_{e,2}$ for a single random Rayleigh channel realization having $L = 10$ taps and decoder EXIT function $f_{d,1}$ for a rate-1/2 convolutional code. (b) Rate region of the 2-user MAC with SC FD-MMSE turbo equalization for a single random channel realization at $E_s/N_0 = 5$ dB, numerically computed by generating a large number of different admissible convergence curves (a gray dot corresponds to one curve).

turbo iterations. Under this assumption, the size of the area $A_{\mathcal{D}}$ is close to zero and the feasible region $\mathcal{D}$ can be characterized by a single curve, which is referred to as convergence curve.

Let $\mathcal{P}$ be the set of admissible convergence curves in the plane region $\mathcal{U}$, as defined in [10]. With the the area property of EXIT functions [5], we derive in a straightforward manner an upper bound for the rate region of both users, as

$$\mathcal{R} \equiv \bigcup_{\mathbf{p}\in\mathcal{P}} \left\{ (R_1, R_2) : R_k < \int_{\mathbf{p}} f_{e,l}(\mathbf{I}_d) \mathrm{d}I_{d,k}, k = 1, 2 \right\}. \tag{4.10}$$

Figure 4.10 (b) illustrates an example of the rate region in (4.10), where $f_{e,1}$ and $f_{e,2}$ have been computed for a random channel realization using the algorithm in [3]. Note that the rate region in (4.10) is non-convex, in general, where the dominant face of this region strongly depends on the particular realization of the equalizer EXIT vector-function $\mathbf{f}_e$. Using the rate bound in (4.10), the problem of maximizing the sum rate of both users can be formulated as $R_{max} \equiv \max_{\mathbf{p}\in\mathcal{P}} \sum_{k=1}^{2} \int_{\mathbf{p}} f_{e,k}(\mathbf{I}_d) \mathrm{d}I_{d,k}$. Specifically, we show in [10] that this optimization problem can be efficiently solved by using the Euler-Lagrange formalism [11]. As a result, the optimal decoder EXIT curves for both users with respect to the maximum sum rate of the turbo system, are obtained.

For practical reasons, however, each transmitter is restricted to have only a finite number of codes with fixed rates. For this scenario assumption, we propose in [10] a simple *code selection* algorithm for the rate allocation that maximizes the coding rate of each user, given the optimal decoder EXIT curve obtained from (4.10), while satisfying the constraints for successful decoding.

Figure 4.11 shows the average total throughput of the 2-user turbo system with the proposed *code selection* approach for frequency-selective Rayleigh fading channels having $L = 32$ taps. We assume sufficient long CP for inter-block interference to be negligible. The binary-input sum capacity of the 2-user system is shown as a reference. Also shown is the average total throughput performance for an automatic repeat-request scheme with fixed coding rates of both users. For clarity, the rates shown do not include the fractional rate loss incurred by the CP. As observed in Fig. 4.11, substantial throughput gain is obtained with the proposed *code selection* algorithm over the fixed rate case. Further, we find that the throughput performance is only 2 dB away (in the high E_s/N_0 region) from the theoretical limit.

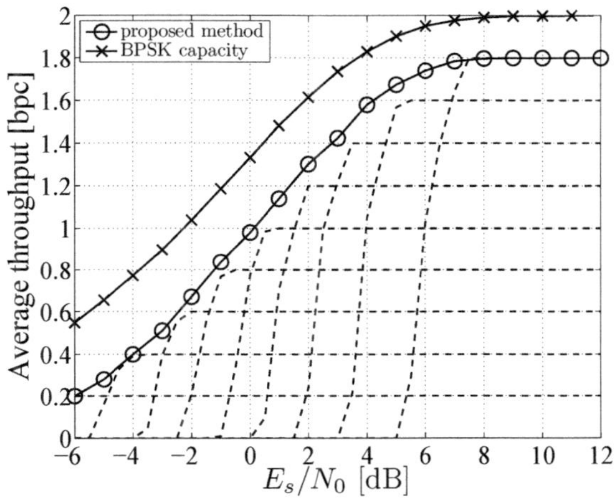

Figure 4.11: Average total throughput of both users versus E_s/N_0 for the proposed rate allocation scheme using the rate-compatible punctured serial concatenated convolutional codes in [12], and for automatic repeat-request with fixed coding rates $r_{1/2} = 0.1 \cdot n$ with $n = 1, .., 9$ (dashed curves, from bottom to top).

4.4.4 Summary

In this contribution, we consider the problem of rate allocation in the frequency-selective 2-user Gaussian multiple-access fading channel employing a low complexity MMSE turbo equalizer. We derive an upper bound on the rate region of the turbo system and study the problem of maximizing the sum rate of both users. In addition, numerical results of a simple code selection algorithm for rate allocation to both users are presented.

Bibliography

[1] C. Douillard, et al., "Iterative correction of intersymbol interference: Turbo equalization," *European Trans. Telecomm.*, vol. 6, no. 5, pp. 507-511, Sept. 1995.

[2] X. Wang, and H. V. Poor, "Iterative (turbo) soft interference cancellation and decoding for coded CDMA," *IEEE Trans. Commun.*, vol. 47, no. 7, pp. 1046-1061, July 1999.

[3] K. Kansanen, and T. Matsumoto, "An analytical method for MMSE MIMO turbo equalizer EXIT chart computation," *IEEE Trans. Wireless Commun.*, vol. 6, no. 1, pp. 59-63, Jan. 2007.

[4] S. ten Brink, "Convergence behavior of iteratively decoded parallel concatenated codes," *IEEE Trans. Commun.*, vol. 49, no. 10, pp. 1727-1737, Oct. 2001.

[5] A. Ashikhmin, G. Kramer, and S. ten Brink, "Extrinsic information transfer functions: model and erasure channel properties," *IEEE Trans. Inform. Theory*, vol. 50, no. 11, pp. 2657-2673, Nov. 2004.

[6] S. ten Brink, G. Kramer, and A. Ashikhmin, "Design of low-density parity-check codes for modulation and detection," *IEEE Trans. Commun.*, vol. 52, no. 4, pp. 670-678, April 2004.

[7] A. Sanderovich, M. Peleg, and S. Shamai (Shiz), "LDPC Coded MIMO Multiple Access With Iterative Joint Decoding," *IEEE Trans. Inform. Theory*, vol. 51, no. 4, pp. 1437-1450, April. 2005.

[8] A. Roumy, and D. Declercq, "Characterization and Optimization of LDPC Codes for the 2-User Gaussian Multiple Access Channel," *EURASIP Journal on Wireless Commun. and Netw.*, vol. 2007, Article ID 74890, 10 pages, 2007.

[9] F. Brännström, L.K. Rasmussen, and A.J. Grant, "Convergence analysis and optimal scheduling for multiple concatenated codes," *IEEE Trans. Inform. Theory*, vol. 51, no. 9, pp. 3354-3364, Sept. 2005.

[10] M. Grossmann, T. Ortlepp, T. Matsumoto, "Rate Allocation for 2-User MAC with MMSE Turbo Equalization," *IEEE Trans. on Wireless Commun.*, vol. 5, pp. 1033-1043, March 2010.

[11] J. Jost, and X. Li-Jost, "Calculus of variations," *Cambridge Studies in Advanced Mathematics*, vol. 64, Cambridge University Press, Cambridge, 1998.

[12] M. Tüchler, "Design of serially concatenated systems depending on the block length," *IEEE Trans. Commun.*, vol. 52, no. 2, pp. 209-218, Feb. 2004.

4.5 Coexistence of Systems

F. K. Jondral, U. Berthold[4], Karlsruhe Institute of Technology (KIT), Germany
M. Schnell, S. Brandes, DLR, Germany

The *efficiency of spectrum usage* for a radio communications system employing a fixed bandwidth is measured as the average number of bits that are transmitted per second and per Hertz in a certain area. For simplicity reasons we will use the notion *spectrum efficiency* instead of efficiency of spectrum usage here. The results of measurement campaigns recently accomplished, pointed out that, even in heavily occupied bands, usually not more than fifteen percent of the spectral resource is effectively employed. This leads to the justifiable assumption that dynamic access upon the transmission resource will at least contribute to a reduction of spectrum scarcity.

The TAKOKO[5] project within TakeOFDM dealt with the utilization of resources left idle in the frequency band allocated to a primary system of authorized users (so-called spectrum holes in the time-frequency-plane) by an overlay system of secondary users. A coexistence strategy is investigated that explicitly harnesses the flexibility of the OFDM (orthogonal frequency-division multiplexing) method with respect to spectrum utilization. I.e. an OFDM-based overlay system is installed in order to enhance the spectrum efficiency within the frequency region allocated to an authorized primary user system.

As a starting point of our investigations, frequency regions as well as primary user systems appearing capable for additional use by an overlay system were identified. These are essentially systems employing the resource by FDMA (frequency division multiple access) or TDMA (time division multiple access). Table 4.2 provides an overview over the scenarios adopted.

Table 4.2: Overviews over scenarios adopted

	Scenario 1	Scenario 2	Scenario 3
Frequency region	VHF band	960 - 1215 MHz	1800 MHz
Primary system	Aircraft radio	Distance Measuring Equipment (DME)	GSM
Overlay system	WLAN for communication at airports	Flexible mobile communication system	Metropolitan Area Network (MAN)
Special aspects	Interference suppression	Detection of spectrum holes, suppression of jammers	Utilization of spectrum holes

Reliable detection of idle resources in the time-frequency-plane, i.e., the detection of primary user signals by the overlay system, as well as the signalization of the results within the overlay systems are of paramount importance for the acceptance of overlay systems by primary users.

[4]Dr.-Ing. Ulrich Bertold is now an employee of COMSOFT GmbH, Karlsruhe.

[5]TAKOKO is an acronym for technologies, algorithms, and concepts for OFDM systems coexisting with authorized systems in the same frequency band

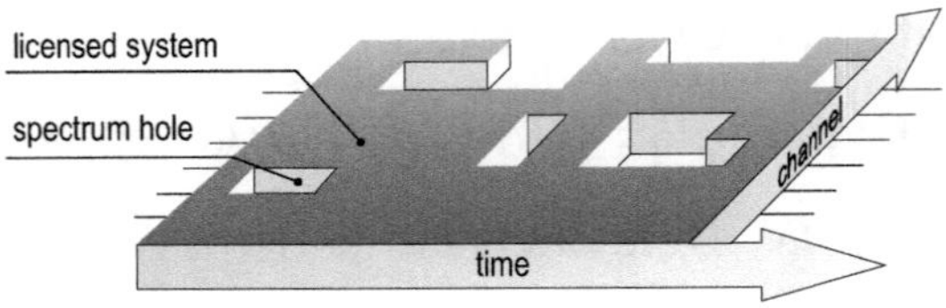

Figure 4.12: Idle resources represented by spectrum holes

The impact that the overlay has on the primary system has to be kept minimum. Therefore, detection as well as false alarm probabilities for the detection of primary users' signals by the overlay system were derived. In the course of the work it became evident that only the application of a detection algorithm distributed over the overlay system's stations is able to guarantee sufficiently high detection probabilities and at the same time lead to tolerable false alarm rates.

Moreover, an efficient signaling within the overlay system of the spectrum utilization by the primary system proves to be important in order to optimally make use of the spectrum holes.

Employing OFDM in overlay systems, of course, also implicates some disadvantages. Among these disadvantages are the strong spurious transmissions that may lead to disturbances in the primary system. To reduce them, in addition to well known methods, like windowing of the transmitted signal in the time domain or the introduction of guard bands, subcarrier weighting, the application of cancellation carriers as well as a combination of windowing and cancellation carriers were investigated in TAKOKO. In scenario 1, for example, it was shown that the application of this combination may reduce the energy level contained in a secondary lobe to a value of 50 dB below the OFDM signal energy level. The new techniques for disturbance suppression were integrated into the simulation model of an adaptive OFDM transmitter.

At the receiver side of an overlay system's user the desired signal is superimposed by interferences generated by the primary system. Although the channels occupied by primary system's users are generally not employed by the overlay system, the overlay system's performance is influenced by these interferences. In this connection, interference reduction methods concerning narrow band interferers were investigated with respect to scenario 1. Here time domain windowing of the received signal as well as the frequency domain estimation of the interfering signal were applied. High quality results were achieved with a combination of time domain windowing and leakage compensation in the frequency domain. The interference minimization techniques were integrated into the overlay system's receivers. To facilitate a dynamic adaptation to varying interference situations, the OFDM frame structure had to be modified by introducing observation carriers such that their position may be adjusted to the actual spectrum occupancy.

Further essential aspects of the TAKOKO project were the investigation of the requirements on the MAC (medium access control) layer as well as the specification of an adequate MAC protocol. Here the essential assumption was that the overlay system operates as an ad-hoc network, i.e., without central coordination. Therefore, all individual stations are equal and the MAC protocol has to incorporate the system's distributed character. In order to be able to initiate a data transmission the stations essentially need a) knowledge about the frequency occupation by the primary system (overlay feature) and b) information about those stations in their environment to which it is possible to establish a connection (ad hoc feature). For determination of the primary system's frequency occupation measurements are performed periodically. The measurement periods for the overlay system's stations have to be coordinated because no overlay system's stations should radiate signals then to guarantee a reliable detection of primary system signals. Exactly this coordination is assumed by the AHO-MAC (ad hoc overlay MAC) protocol developed in the TAKOKO project. Similar to PRMA (packet reservation multiple-access), as many time slots for data transmission as mini slots for acknowledgments are provided. Moreover, special time slots for occupancy measurements and for the mutual synchronization of the overlay stations are defined.

The representation of the frequency utilization by the primary system in form of the occupancy vector plays an important role within the overlay system and affects the PHY (physical) as well as the MAC layer. Therefore, an approach of cross layer optimization was chosen. Here the layered model was kept but the interfaces between the layers were extended by the definition of more detailed as well as more specific parameters.

In summary promising methods were developed in TAKOKO which show that in principle OFDM based overlay systems may be implemented. Although the TAKOKO approaches were considered with respect to preferably realistic overlay scenarios (Table 4.2), further investigations have to be performed before OFDM based overlay systems may be practically implemented. Those investigations could not be accomplished within the project because of the given time and effort constraints and because they were beyond the project's objectives. Moreover, the real implementation of OFDM based overlay systems requires fundamental decisions in the political as well as in the regulatory area in order to define the principle framework. The scientific results acquired in the TAKOKO project are essentially summarized in the doctoral dissertations [1] and [2][6].

Bibliography

[1] U. Berthold, *Dynamic Spectrum Access Using OFDM-based Overlay Systems*, Dissertation, Forschungsberichte aus dem Institut für Nachrichtentechnik der Universität Karlsruhe (TH), Band 21, Karlsruhe, 2009.

[2] S. Brandes. *Suppression of Mutual Interference in OFDM Based Overlay Systems*, Dissertation, Forschungsberichte aus dem Institut für Nachrichtentechnik der Universität Karlsruhe (TH), Band 22, Karlsruhe, 2009.

[6] Both dissertations may be downloaded from www.cel.kit.edu

4.6 System Design for Time-Variant Channels

P. Klenner, S. Vogeler, K.-D. Kammeyer, University of Bremen, Germany
L. Reichardt, S. Knörzer, J. Maurer, W. Wiesbeck, University of Karlsruhe, Germany

The following two subsections summarize the cooperation between the Arbeitsbereich Nachrichtentechnik (ANT) at the University of Bremen and the Institut für Hochfrequenztechnik und Elektronik (IHE) at the University Karlsruhe. Both partners were engaged in two projects within the DFG program "TakeOFDM". The project titles coincide with the names of the following sections. The focus is in Section 4.6.1 on SIMO (Single-Input-Multiple-Output)-OFDM and in Section 4.6.2 on MIMO (Multiple-Input-Multiple-Output)-OFDM in rapidly fading channels, which are modeled by ray-optic methods to include realistic propagation effects.

4.6.1 Multicarrier Systems for Rapidly Moving Receivers

In this section, the use of multicarrier systems and, in particular, OFDM for high-rate wireless data transmission and for fast moving receivers mounted on high-speed trains, cars, or airplanes is considered. Realizing high data rates at simultaneously high mobility is a difficult feat for any communication system. If broadband signaling for a high data rate causes the sample duration to become smaller than the channel's maximal echo duration, then OFDM can simplify the resulting equalization effort. But a frequency-selective channel is only separated into orthogonal parallel subcarriers in the case of a time-invariant channel, i.e., if the channel remains unchanged during one OFDM symbol. High mobility environments, however, are characterized by time-invariant channels, which leads to a loss of the subcarriers' orthogonality. Intercarrier interference (ICI) occurs, the equivalent to intersymbol interference in time-domain.

In the simplest case, ICI can be modeled as a source of noise in addition to the thermal noise. Neglecting it results in a degradation of the achievable bit-error rate. More dramatic is the impact, if the channel is changing so fast, that the violation of the sampling theorem prevents a reliable channel estimation. This case does not necessarily occur due to extreme speed, but also due to a large pilot symbol spacing.

The observations conducted here refer to a system for a high-speed train in a single frequency network. The channel model described in Section 2.2.2 is applicable to different frequencies and is transferable to other types of mobiles. Similarly, the considerations apply to the Doppler effect for other types of fast moving receivers. For an extensive investigation under realistic channel conditions, the project is based on modeling the propagation conditions via ray-tracing methods by which a series of channel impulse response is computed.

Subsequently, three main strategies will be discussed with the aim to solve the conflict between the transmission channel's high time and frequency selectivity: The application of a pre-equalizer to shorten the channel impulse response, the use of soft impulse shaping for a non-orthogonal multicarrier system, and multi-antenna concepts to reduce the Doppler spread.

During the project, parameters of an OFDM system were defined taking into account the main characteristics of the channel that was determined by ray-tracing simulations. The system's performance limits were established through Monte Carlo simulations. The system parameters have been dimensioned such that the time-bandwidth conflict caused by the interference is completely focused on the frequency domain, while time domain intersymbol interference is avoided. Thus, the procedures to counter the Doppler effect can be performed after the DFT in the receiver.

By consolidating the effects of multipath and Doppler spread in the form of a channel matrix, a linear equalization of both effects according to the MMSE criterion can be carried out. The disadvantage of this procedure, however, is a relatively high computational effort, while the knowledge of the signal alphabet is not explcited. Therefore, two well-known algorithms for low-complex equalization were implemented: a successive linear MMSE equalizer and a successive equalizer with decision feedback structure, which were modified in accordance to the requirements of the scenario. Both methods, however, did not provide satisfactory simulation results, which in the decision-feedback method is due to the suboptimal detection sequence. This called for the sorted QR decomposition (SQRD) employed in V-BLAST to improve the detection sequence. This resulted in the best detection performance, but at the cost of a comparably high computational effort. Through the gradual decomposition of the channel matrix in submatrices the effort can be reduced significantly, yielding a trade-off between high performance and low complexity equalization. This successive SQRD is detailed in [11, 12].

As part of the investigations of alternative multicarrier concepts the subcarrier spectra of OFDM were replaced by soft non-orthogonal pulse shapes, i.e., time-limited Gaussian pulses. In the frequency domain, this leads to a concentration of interference power in the vicinity of the main diagonal of the channel matrix. This in turn has a positive effect on the performance of the aforementioned equalization methods, particularly for the low-complex variants such as the successive SQRD. The soft impulse shaping also allowed for Maximum Likelihood detection with reasonable effort, because the concentration of the interference power allows the use of a time-variant Tailbiting-MaxLogMAP algorithm with a reduced number of states. This procedure resulted in the best compromise between complexity and performance of the transmission system [13].

Shortening the channel impulse response potentially allows for a shorter OFDM symbol, which is beneficial in rapidly time-varying environments. Numerous well-known non-blind and blind algorithms for computing equalizer coefficients were investigated regarding their applicability in the scenarios. It turned out that these methods produce either ill-conditioned frequency responses or that the convergence speed is not sufficient in case of the blind methods.

Multi-antenna concepts open up the spatial domain to procure Doppler spread compensation. Sectorizing antennas exploit the relationship between the impinging waves' angle of incidence and their resulting Doppler shift. By restricting the range of angles of incidences a sectorizing antenna simultaneously reduces the effective Doppler spread compared to a omnidirectional antenna without any restriction of the incidence angle. One practical implementation is beamforming via a Uniform Linear Array (ULA), and the beams are formed such that the Doppler spread is equally distributed among all beams [2]. Furthermore, a ULA can be used for a method called spatial interpolation, in which the signals of the individual antenna elements are filtered by a time-variant interpolator such that the filter output resembles the output of a time-invariant channel [3].

The two scenarios of a noise barrier and vegetation are considered as typical high speed lines. The radio channel has very different properties, therefore, these scenarios are used as a reference. Details can be found in [6,8,9]. For the simulation of wave propagation, a three dimensional ray-propagation model is employed. It provides characteristic parameters for each multipath, from which the sequence of channel impulse responses and channel characteristics are extracted. The wave propagation model takes into account the propagation phenomena of reflection, diffraction and scattering. The approach to modeling the distribution of vegetation is described in 2.2.2.

Figure 4.13 shows the BER performance for a SIMO-OFDM system with receive diversity in a rapidly fading channel (one transmit and two or four receive antennas, $N_t = 1$, $N_r = 2, 4$). For omnidirectional reception widely spaced antennas are used, which yield completely uncorrelated channels per antenna. Without any means of Doppler compensation this results in the worst performance. Sectorizing antennas and spatial interpolation based on a ULA with interantenna spacing d/λ yield a better performance. Sectorization is here achieved by forming two beams in and against the direction of the movement. Spatial interpolation is achieved by computing two virtually non-moving antennas, which fix the channel over the middle of the respectively received OFDM signal. The received signals on the resulting virtual antennas are correlated. Taking this correlation into account by means of a whitening matched filter (WMF) improves the performance over the matched filter (MF).

4.6.2 Highly Mobile MIMO-OFDM-Transmission in Realistic Propagation Scenarios

Doppler spread compensating antenna structures at the receiver offer an efficient approach to solve the problem of the channel's time-variance already in time-domain. The focus now is on the two antenna configuration discussed in Section 4.6.1, whose function ultimately is to affect a less rapidly fading channel prior to the DFT. Unlike the previous discussion, where a single transmit antenna was employed, in the following the focus is shifted to the transmitter side.

The MIMO philosophy is based on two pillars: Spatial transmit diversity is provided by Space-Time codes, and spatial multiplexing allows to increase the data rate by transmitting independent data streams. The development of these techniques is shaped by the assumption of a time-invariant channel, and most methods even demand that. In the following, the mutual benefit of MIMO-signaling and Doppler spread compensation measures is described. If the Doppler spread is small, then the loss of time diversity can be made up by spatial diversity, and if the Doppler spread is large, then the Doppler compensating antennas can effect a less time-invariant channel such that the MIMO-signal processing upholds its validity [1, 4].
Rapid channel fluctuations in conjunction with multiple transmit antennas pose the

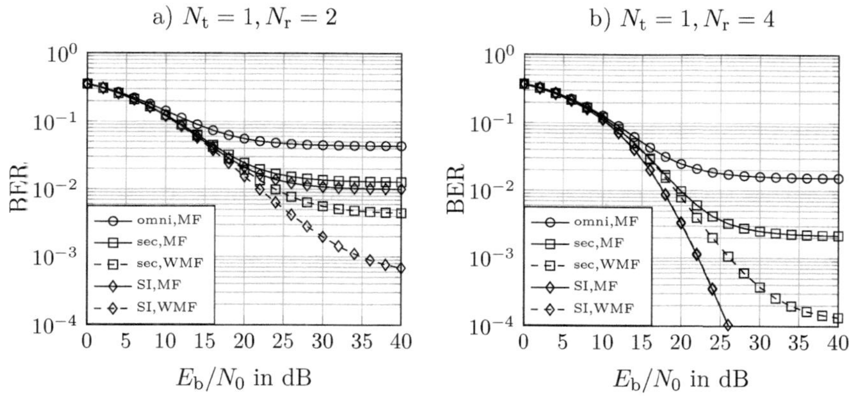

Figure 4.13: BER-performance for SIMO-OFDM with a single transmit antenna and different receiver configurations: omnidirectional antennas with widely spaced antennas, sectorizing antennas and spatial interpolation based on a ULA, parameters: 256 subcarriers, 16QAM, exp. power delay profile, antenna spacing $d/\lambda = 0.25$, Doppler spread $f_D T_s = 0.2$, raised cosine filters ($r = 0.18$), zero forcing channel est.

possible disadvantage of decreased bandwidth efficiency, in that frequent training becomes necessary. However, since the channel is forced to become less rapidly fading by Doppler compensating measures the pilot spacing in time direction can become larger. Similarly, noncoherent ST codes can profit from a channel that does not change too much during a signaling interval.

During the course of the project it is investigated how the functions of Doppler compensation and the separation of data streams using multiple receiving antennas can be combined. One possibility is the use of multiple spatially separated antenna arrays, each used for Doppler compensation, followed by the recovery of the transmitted data streams.

Parameter estimation is a further important issue. In particular, the channel's autocorrelation function needs to be estimated, since the receiver requires it in order to perform the spatial interpolation. The idealized assumption of the Bessel function as autocorrelation holds only in isotropic scattering environments. Other scattering conditions and mutual antenna coupling lead to different correlations. Realistic channels already studied lead to an SNR loss. It is possible to find an estimate of the ACF based on the cyclic OFDM signal structure without spending extra training data [5].

Furthermore, the optimization of antenna structures for the purpose of minimizing the mutual antenna coupling is considered. The algorithms used for data processing in the baseband yield a better performance if the individual antennas for sectorization or spatial interpolation are decoupled. This condition is violated in practical systems, since the individual antenna elements are positioned very close to each other, so that inevitably crosstalk of the antennas occurs. The effect of mutual coupling and its minimization and compensation are investigated, so that the idealized assumptions of the baseband processing can be maintained.

The consideration of realistic channel models as a benchmark of these approaches represents a further focus of the project. Two scenarios are considered, a high-speed train and vehicle-to-vehicle environments. In the former, the propagation conditions are dominated by a major incident direction (by the base stations along the railroad tracks), while in the latter, in principle, all incidence directions are possible. Regarding the car-to-car scenario, a further challenge is to find Doppler compensating antenna structures, which are not impaired by the car structure.

Figure 4.14 shows the BER performance for a (2×2)-MIMO OFDM systems with the receiver configurations known from Fig. 4.13. In Fig. 4.14a transmit diversity is provided by the Alamouti Space-Time code, whereas in Fig. 4.14b the V-BLAST is employed, i.e., independent data streams are transmitted from both transmit antennas. For a fair comparison identical data-rates are used, i.e., 16QAM for the Alamouti scheme, and QPSK for V-BLAST. Comparing the two-fold transmit diversity scheme in Fig. 4.14a with the single transmit diversity case in Fig. 4.13a shows that except for spatial interpolation the gains promised by two-fold transmit diversity are small. This can be attributed to the channel estimation which needs to determine twice as many channel parameters. On the other hand, the gains for V-BLAST in employing Doppler compensating antenna structure are more impressive.

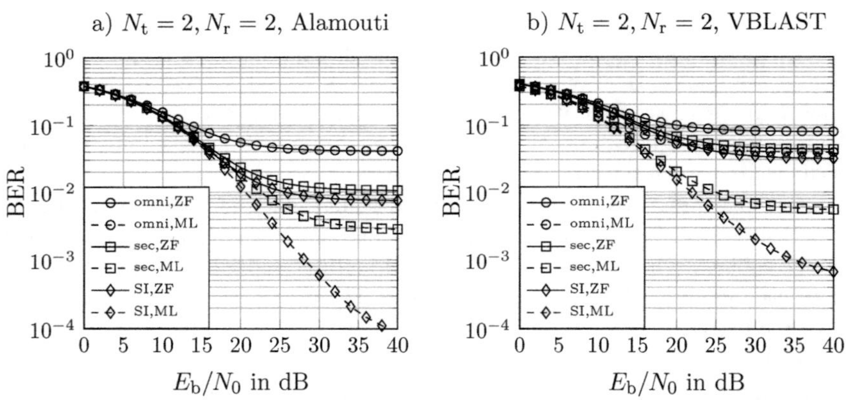

Figure 4.14: BER-performance for (2×2)-OFDM and different receiver configurations: omnidirectional antennas with widely spaced antennas, sectorizing antennas and spatial interpolation based on a ULA, parameters: 256 subcarriers, 16QAM (a) and QPSK (b), exp. PDP, $d/\lambda = 0.25$, $f_D T_s = 0.2$, RC-filters ($r = 0.18$), ZF channel est.

Bibliography

[1] K.-D. Kammeyer and P. Klenner, "Space-Time Coded OFDM Transmission with Spatial Interpolation in the Presence of Severe Doppler Spread," in *3rd*

International Symposium on Communications, Control and Signal Processing (ISCCSP 08), St. Julians, Malta, March 2008.

[2] P. Klenner and K.-D. Kammeyer, "Doppler-Compensation for OFDM-Transmission by Sectorized Antenna Reception," in *6th International Workshop on Multi-Carrier Spread Spectrum (MCSS 07)*, Herrsching, Germany, May 2007.

[3] P. Klenner and K.-D. Kammeyer, "Spatially Interpolated OFDM with Channel Estimation for Fast Fading Channels," in *IEEE Vehicular Technology Conference 2007 (VTC2007-Spring)*, Dublin, Ireland, April 2007.

[4] P. Klenner and K.-D. Kammeyer, "Performance of Space-Time-Coded OFDM with Sectorized Antenna Reception," in *International ITG/IEEE Workshop on Smart Antennas (WSA 08)*, Darmstadt, Germany, Feb. 2008.

[5] P. Klenner and K.-D. Kammeyer, "Temporal Autocorrelation Estimation for OFDM with Application to Spatial Interpolation," in *Asilomar Conf. on Signals, Systems and Computers*, Monterey, CA, Oct. 2008.

[6] S. Knörzer, *Funkkanalmodellierung für OFDM-Kommunikationssysteme bei Hochgeschwindigkeitszügen*, PhD thesis, University Karlsruhe, Germany, 2009.

[7] S. Knörzer, J. Maurer, T. Fügen, and W. Wiesbeck, "Wave Propagation Modeling for Communication between Moving Vehicles," in *National Radio Science Meeting*, Boulder, USA, Jan 2005.

[8] S. Knörzer, J. Maurer, S. Vogeler, K.-D. Kammeyer, and W. Wiesbeck, "Channel Modeling and Characterization for High-Speed Train OFDM Systems," in *COST 273 TD(05)086*, Leuven, Belgium, June 2005.

[9] S. Knörzer, J. Maurer, S. Vogeler, K.-D. Kammeyer, and W. Wiesbeck, "Channel modeling for a high-speed train ofdm communication link supporting high data rates," in *Proc. 5th Int. Conf. on ITS Telecomm. 2005*, pages 333–336, Brest, France, June 2005. Best Student Paper Award.

[10] J. Maurer, S. Knörzer, and W. Wiesbeck, "Ray Tracing in Rich Scattering Environments for Mobile-to-Mobile Links," in *Proc. of the Int. Conf. on Electromagnetics in Advanced Applications*, Italy, Sept. 2005.

[11] S. Vogeler, *Verfahren zur Kompensation von Doppler-Einflüssen in Mehrträger-Übertragungssystemen*, PhD thesis, University of Bremen, Germany, 2006.

[12] S. Vogeler, L. Brötje, P. Klenner, V. Kühn, and K.-D. Kammeyer, "Intercarrier Interference Suppression for OFDM Transmission at Very High Velocities," in *9th International OFDM-Workshop (InOWo 2004)*, Dresden, Germany, Sept. 2004.

[13] S. Vogeler, P. Klenner, and K.-D. Kammeyer, "Multicarrier Transmission for Scenarios with High Doppler Influence," in *10th International OFDM-Workshop (InOWo 2005)*, Hamburg, Germany, Aug. 2005.

4.7 Combination of Adaptive and Non-Adaptive Multi-User OFDMA Schemes in the Presence of User-Dependent Imperfect CSI

A. Kühne, A. Klein, Technische Universität Darmstadt, Germany

4.7.1 Introduction

The Orthogonal Frequency-Division Multiple Access (OFDMA) transmission scheme is a promising candidate for future mobile networks. It offers an efficient adaptation to the channel conditions by performing a time-frequency scheduling of the different subcarriers to the different users. However, Channel State Information (CSI) is required at the transmitter in order to perform such an adaptive scheduling in an optimal way. Having perfect CSI for all users at the Base Station (BS), the use of frequency adaptive OFDMA schemes achieves very good performances by exploiting multiuser diversity . Having no CSI at all at the BS, the use of frequency non-adaptive OFDMA schemes exploiting frequency diversity independently from any CSI is the best strategy, however, not achieving the performance of adaptive schemes with perfect CSI . For the case of imperfect CSI, only pure adaptive OFDM-based systems have been studied in the literature but not a combination of adaptive and non-adaptive transmission modes. A comparison of adaptive and non-adaptive multiuser OFDMA schemes in the presence of imperfect channel knowledge has been investigated assuming the same degree of CSI imperfectness for each user [1]- [5]. It appears that at a certain level of CSI imperfectness, it is beneficial to switch from adaptive to non-adaptive transmission, i.e., depending on the quality of the channel knowledge, either all users apply the adaptive or non-adaptive transmission mode. In a realistic scenario however, the level of CSI imperfectness differs from user to user, i.e., for some users, the CSI is only slightly corrupted, whereas for other users, the CSI is totally inaccurate. Hence, we propose a hybrid OFDMA scheme where both adaptive and non-adaptive transmission schemes co-exist at the same time and show how to optimally combine these transmission schemes in the presence of user-dependent imperfect CSI [6].

4.7.2 Combining Transmission Schemes

In the considered hybrid OFDMA scheme, different users are served either adaptively or non-adaptively sharing the available bandwidth. Applying the non-adaptive OFDMA transmission scheme, a fixed modulation and subcarrier allocation is performed. This scheme does not rely on instantaneous CSI but exploits frequency diversity. Applying the adaptive OFDMA transmission scheme, an adaptive subcarrier allocation together with an adaptive modulation based on the instantaneous CSI is performed. The goal is to achieve a maximum system data rate under the constraint of a minimum user data rate and target Bit Error Rate (BER). Hence, the question arises, which user shall be served adaptively or non-adaptively taking into account user-dependent imperfect CSI and furthermore how to choose the signal-to-noise-ratio (SNR) thresholds for the applied modulation schemes in order to maximize the system data rate while guaranteeing a certain minimum user data rate and target BER. Since the performance of an adaptive users strongly depends on the total number of adaptive users in the system due to the selection process and the multi-user diversity involved, the decision whether a user is served adaptively or non-adaptively cannot be made userwise independently from the other users but has to be done jointly considering all users, resulting in a combinatorial problem.

For the order of allocating the available subcarriers to the adaptive and non-adaptive users, two possibilities are considered. Firstly, the subcarriers of non-adaptive users are allocated in a first step and the remaining subcarriers are then allocated to the adaptive users in a second step referred to as Non-Adaptive First (NAF). Secondly, first the subcarriers of the adaptive users are allocated followed by the allocation of the subcarriers of the non-adaptive users referred to as Adaptive First (AF).

To solve the described optimization problem, it is split up into two subproblems without losing optimality, namely the SNR threshold problem and the user serving problem. The SNR threshold problem can be solved applying a Lagrange multiplier approach leading to the optimized user data rate for each possible combination of serving the different users. In order to do so, analytical expressions for user data rate and BER have been derived taking into account user-dependent imperfect CSI. Solving the combinatorial user serving problem, it appears that it is not necessary to check all 2^U possible user serving combinations with U denoting the number of users in order to find the optimal combination maximizing the system performance subject to the mentioned data rate and BER constraints. Utilizing the fact that the data rate of an adaptive user does not depend on which users are adaptively served but only on the total number of adaptive users, the order of complexity can be reduced to $\mathcal{O}(U^3)$ without loosing optimality. Taking into account the characteristics of the user data rate as a function of the number U_A of adaptive users, the complexity can be further reduced to $\mathcal{O}(U^2)$.

4.7.3 Numerical Results

In Fig. 4.15a, the system data rate of a single cell OFDMA system with $N = 125$ subcarriers and $U = 25$ users in the downlink is depicted as a function of the average Mobile Station velocity $\bar{v}$ in the cell for the different transmission schemes. The target BER is set to 10^{-3} while each user shall achieve at least the data rate achievable applying the Pure Non-adaptive Transmission Scheme (PNTS). As one can see, PNTS achieves a constant system data rate since it does not depend on the reliability of the CSI. In case of $\bar{v} = 0$ km/h, the Pure Adaptive Transmission Scheme (PATS) and the hybrid transmission schemes achieve the same system data rate and outperform PNTS. However, when increasing $\bar{v}$ in the cell and, thus, the unreliability of the CSI, the performances of PATS dramatically decrease, especially for the naive approach, where the SNR thresholds are calculated assuming perfect CSI for all users, since now, due to the imperfect CSI, wrong users and modulation schemes are selected for transmission. This results in a BER which no longer fulfills the target BER requirements. For PATS which is aware of the imperfect CSI, the

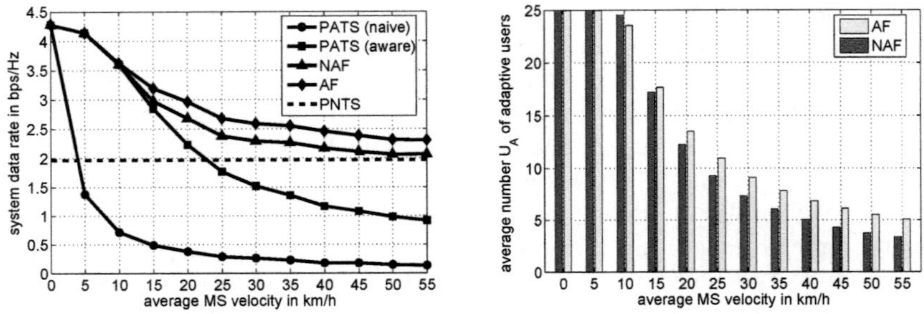

(a) (b)

Figure 4.15: (a) System data rate and (b) Number U_{A} of adaptive users vs. $\bar{v}$

decrease is less dramatical. However, at some point the system performance is worse than using PNTS. Applying the hybrid schemes NAF and AF for an increasing $\bar{v}$ in the cell, the system performance is always equal or better than both PNTS and PATS, where AF outperforms NAF due to the more exclusive resource selection. For large velocities, the performances of the hybrid schemes converge to the one of PNTS, since now all users in the hybrid scheme are served non-adaptively due to the totally outdated CSI. This effect is also shown in Fig. 4.15b, where the number U_{A} of adaptively served users is depicted as a function of $\bar{v}$. It can be seen that for low velocities, almost all users are served adaptively. When increasing $\bar{v}$, more and more users are served non-adaptively.

Bibliography

[1] A. Kühne and A. Klein, "An analytical consideration of imperfect CQI feedback on the performance of a Multi-user OFDM-system," in Proc. *12th International OFDM-Workshop (InOWo'07)*, Hamburg, Germany, August 2007.

[2] A. Kühne and A. Klein, "Adaptive subcarrier allocation with imperfect channel knowledge versus diversity techniques in a multi-user OFDM-system," in Proc. *Proc. International Symposium on Personal, Indoor and Mobile Radio Communications (PIMRC'07)*, Athens, Greece, September 2007.

[3] A. Kühne and A. Klein, "Throughput analysis of Multi-user OFDMA-Systems using imperfect CQI feedback and diversity techniques," *IEEE Journal of Selected Areas in Communications*, vol. 26, no. 8, pp. 1440-1451, October 2008.

[4] A. Kühne and A. Klein, "Adaptive MIMO-OFDM using OSTBC with imperfect CQI feedback," in Proc. *International ITG Workshop on Smart Antennas (WSA'08)*, Darmstadt, Germany, February 2008.

[5] A. Kühne, A. Klein, X. Wei, and T. Weber, "Transmit Antenna Selection with

imperfect CQI feedback in Multi-user OFDMA systems," in Proc. *Proc. 13-th International OFDM-Workshop (InOWo'08)*, Hamburg, August 2008.

[6] A. Kühne and A. Klein, "Combining adaptive and non-adaptive Multi-user OFDMA schemes in the presence of user-dependent imperfect channel knowledge," *IEEE Transactions on Wireless Communications.* (submitted)

4.8 Integration of COFDM Systems with Multiple Antennas and Design of Adaptive Medium Access Protocols

D. Martini, B. Wolz, B. Rembold, B. Walke, RWTH Aachen University, Germany

4.8.1 Abstract

Besides medium access in time and frequency domain, the space and polarization domain can be exploited using multiple antenna (MIMO) systems for space division multiple access. SDMA combined with OFDM are the key technologies for future wireless and mobile communication systems as specified in standards such as IEEE 802.16d, IEEE 802.11n and 3GPP LTE. This work focuses on the development and prototype implementation of Medium Access Control (MAC) protocols of a demonstration system that uses multiple antennas. The goal is to serve all user terminals (UT) best by taking their instantaneous channel condition into account. A number of candidate channel-adaptive MAC protocols were designed to be implemented in a stochastic event driven simulator based on the openWNS tool. The protocol stack under study operates in between a multi-media load generator and an interference engine that calculates the current SINR at the receiver dependent on the direction of arrival (DoA). The contribution of a IEEE 802.16 specific transceiver chain is fed-in from link-level simulation results providing DoA specific bit-error characteristics. The openWNS tool is mainly employed to compare the performance of the MAC protocols, the usefulness of the interface between MAC and PHY and the constraints of an SDMA enabled scheduler. Besides simulation studies to understand the performance critical parameters, an analytical validation of the used simulator and a complexity analysis of a number of SDMA based service disciplines was performed. The demonstration system was completely implemented using a modular FPGA system and two high-performance power PCs that to perform MAC in real time. The contribution of MIMO technologies to improve the system capacity was evaluated by means of the SINR values that results from the studied antenna configurations. While the FPGA hardware was taken from a former project, the bidirectional transceiver chains including multiport antennas were new designed and implemented, resulting in a realtime 4x4 bidirectional MIMO testbed.

4.8.2 MAC Frame for SDMA Operation and Spatial Grouping

A multi-cellular scenario served by 120° antennas each serving 15 UTs was chosen operated with re-use one. Transmit power is 1 W and UTs control their power between 200 mW and 1 W. The scheduler decides based on data volume in MAC- and PHY layer, signaling overhead per UT and SINR of an UT, estimated based on the beamforming algorithm information. Figure 4.16 shows an example two-dimensinal MAC frame to serve two spatially separated data streams. Above the

MAC frame the transmit situation is shown using abstract antenna diagrams. The first part of the frame is transmitted omnidirectional to implement broadcast mode. DL-MAP (blue) and UL-MAP (green) and arrows are shown pointing to the time instants contained in the MAPs, where up to four radio bursts can be transmitted, spatially separated. Similar applies to the UL-MAP.

UTs cannot be arbitrarily scheduled for concurrent SDMA transmissions since their spatial separability by the beam-forming antenna array depends on their relative spatial positions. Spatially separated concurrent transmissions cannot be assumed to be orthogonal. Therefore, a hierarchical scheduling algorithm is introduced that first computes a spatial grouping of UTs that can be well separated by the base station's (BS) beamforming antenna. The result of this grouping is a set of groups of UTs. Users of the same spatial group can be separated and thus be served at the same time. Users in different groups must be separated in the time domain. Structuring of the scheduling process into two hierarchical steps adds flexibility and simplicity to the scheduling process. The grouping process is independent from the TDMA scheduling and vice versa. Thus, spatial grouping and group scheduling procedures can be freely combined and interchanged according to the specific needs of the system. Spatial grouping of UTs has been proposed earlier [2] and hierarchical grouping and scheduling in an SDMA enhanced IEEE 802.16 systems is proposed first in [4]. Since an optimal grouping is far too complex to be applied, a greedy algorithm to sort spatial groups according to achievable throughput was used resulting in grouping gain slightly below the optimal grouper's gain. A tree-based heuristic algorithm only estimating the most promising spatial groups appear well suited to reduce run-time complexity to be applicable in real-time condition, with a grouping gain comparable to that of the greedy algorithm.

Simulation Results

Figure 4.17 shows the aggregate DL cell throughput versus offered traffic. Saturation is reached where the throughput deviates to grow linearly with offered traffic. Clearly, with omnidirectional it is lower than with SFIR or SDMA transmission. Under SDMA with up to 4 parallel transmissions, the throughput curve climbs from a value of about 22 MBit/s, where the first UTs reach saturation, to a final maximum of 30 MBit/s. SDMA transmission with up to 4 beams achieves about 440% gain compared to omnidirectional transmission. Most of the results on MIMO based MAC protocols, grouping etc. gained in this project are available from [3]. The scheduler algorithms developed for multiple antenna systems were implemented on a real-time hardware platform.

4.8.3 Hardware Implementation of COFDM Systems with Multiple Antennas

A realtime testbed was developed and used to validate the scheduler algorithms in "real world" scenario. The demonstrator architecture had to be adapted resulting in redesign of the analog frontend as well as the digital transceiver. The testbed is based on OFDM transmission with 256 subcarriers for each UT. A WLAN transceiver chip

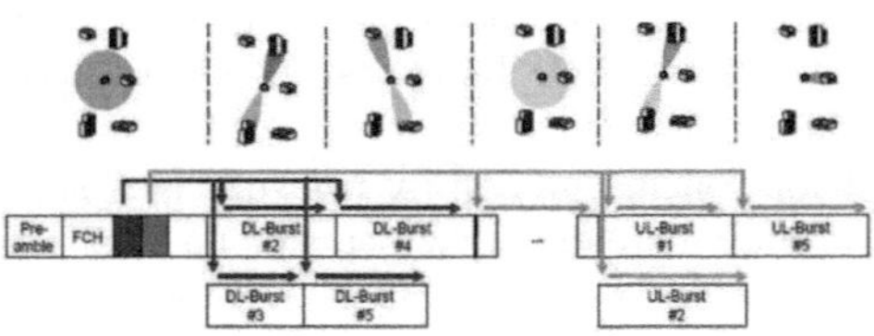

Figure 4.16: A MAC frame supporting SDMA operation

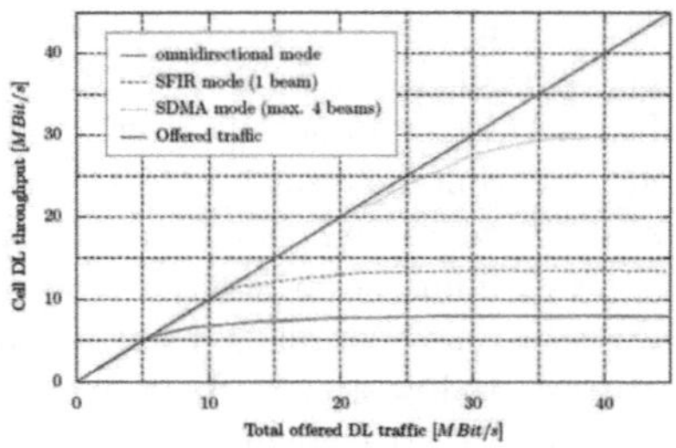

Figure 4.17: DL throughput vs. total offered DL traffic for omnidirectional, SFIR, and SDMA transmission modes

is used for the analog frontend operating at a radio frequency of 5.6 GHz with an RF bandwidth of 25 MHz. The modulation of the radio link is adaptive and is BPSK, QPSK, 16- or 64-QAM, and can be defined independently for each user. In total the system can serve four BS antennas and four UTs resulting in a 4x4 MIMO system. Each antenna combination between 1x1 (SISO) and 4x4 (MIMO) with or without diversity can be operated. Up- and downlink are separated in time division duplex (TDD). Digital signal processing performs the linear equalization algorithms ZF and MMSE in real time [1]. Furthermore, pre-equalization at the transmitter is possible using channel information from uplink transmission.

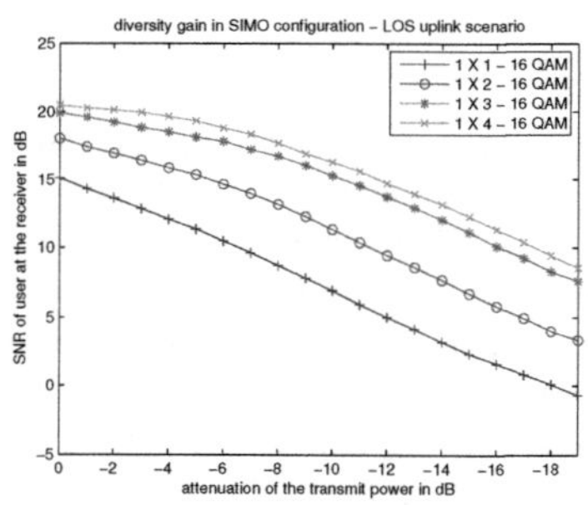

Figure 4.18: Diversity gain under SIMO with different number of antennas

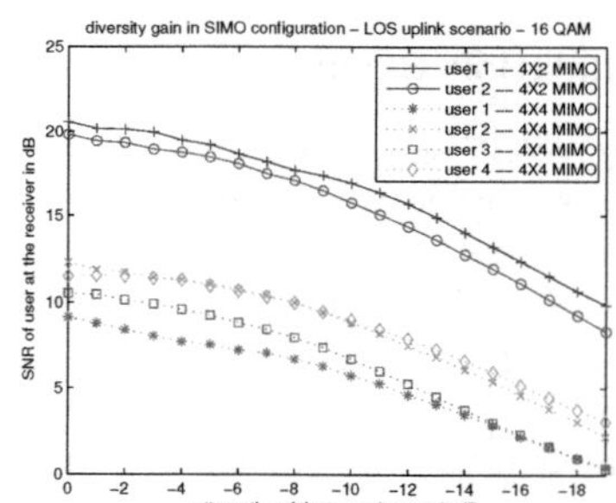

Figure 4.19: SNR per user with 4x2 and 4x4 MIMO

As an example Fig. 4.18 presents the measured diversity gain that is obtained using more than one antenna. Regarding MIMO scenarios diversity also plays an important role as shown in Fig. 4.19. All user can be served with smaller SNR. Regarding two users the SNR and therefore the data rate is much higher. The measurements were taken at the entrance hall of the IHF building. The BS was equipped with a compact four-port-inverted-F antenna and the UT was fit with another compact four-port antenna using four orthogonal radiation modes [5].

Bibliography

[1] L. Bruehl, *Parallele Prozessorarchitektur fuer die Raum-Zeit-Entzerrung in breitbandigen Funksystemen mit adaptiven Gruppenantennen*, PhD thesis, RWTH Aachen, 2005, Shaker ISBN 3-8322-4523-5.

[2] M. Fuchs, G. Del Galdo, and M. Haardt, "A Novel Tree-based Scheduling Algorithm for the Downlink of Multi-user MIMO Systems with ZF Beamforming," in *Proc. IEEE Internat. Conf. on Acoustics, Speech, and Signal Proc.(ICASSP*, 2005.

[3] C. Hoymann, *IEEE 802.16 Metropolitan Area Network with SDMA Enhancement*, PhD thesis, RWTH Aachen University, Department of Communication Networks, Faculty 6, July 2008, http://www.comnets.rwth-aachen.de.

[4] C. Hoymann, H. Meng, and J. Ellenbeck, "Influence of SDMA-Specific MAC Scheduling on the Performance of IEEE 802.16 Networks," in *Proceedings of 12th European Wireless Conference 2006*, Athens, Greece, April 2006, http://www.comnets.rwth-aachen.de.

[5] C. Oikonomopoulos-Zachos and B. Rembold, "A 4-Port Antenna for MIMO Channels," in *IEEE Conf. of Antennas and Propagartion (EUCAP)*, pages 1–4, 2007.

4.9 Large System Analysis of Nearly Optimum Low Complex Beamforming in Multicarrier Multiuser Multiantenna Systems

C. Guthy, W. Utschick, Technische Universität München, Germany

4.9.1 Introduction

In general, analytical expressions for the average sum rate in the Multiple-Input Multiple-Output (MIMO) Broadcast Channel (BC), resulting from signal processing algorithms requiring perfect channel state information at the transmitter, are difficult to obtain. That is true for the optimum sum capacity [11] as well as near optimum algorithms such as the Successive Encoding Successive Allocation Method (SESAM) [9] and its variant with Minimum Mean Square Error (MMSE) transmit filters [8, Ch. 4.2]. Thus, performance evaluation of these algorithms is only possible via simulation results. In the large system limit however, i.e., when at least two system parameters go to infinity at a fixed and finite ratio [6, 10], the performance of many algorithms becomes deterministic, although random variables are used. That is because the eigenvalues of random matrices often converge to an asymptotic limit, which can be obtained from the asymptotic eigenvalue distribution (a.e.d.) and which is independent of the current realization of the random matrix. The most prominent example is the Marčenko-Pastur distribution [5] for Wishart matrices. Furthermore the analytical expressions in the large system limit often serve as a good approximation of the system performance also with finite system parameters, which makes large system analysis an interesting tool for the analysis of signal processing algorithms.
In this chapter we will present analytical approximations of the ergodic sum rates achievable with SESAM with MMSE transmit filters [8] and zero-forcing transmit filters [9] based on large system results with Gaussian i.i.d. channel matrices and an infinite number of transmit and receive antennas. It will be shown that the results also serve as good approximations for the average sum rate in systems with finite parameters.

4.9.2 System Model

We consider a multi-user multi-carrier MIMO system with C carriers, one base station with N antennas and K non-cooperative users with R antennas. The k-th user's channel matrix on carrier c, denoted as $\boldsymbol{H}_{k,c} \in \mathbb{C}^{R\times N}$, consists of uncorrelated Gaussian entries with zero mean and variance $1/R$. Perfect knowledge of the matrices $\boldsymbol{H}_{k,c}$ at the transmitter is assumed. The additive noise at each receiver is assumed to be white Gaussian with zero mean and unit covariance matrix.

4.9.3 Description of Algorithms

As the optimum solution, both variants of SESAM rely on the principle of Dirty Paper Coding (DPC). Furthermore they successively allocate data streams to users, where in each step filters and user allocation of previously allocated data streams are kept fixed and the next data stream is allocated to that user such that the increase in the objective function becomes maximum. The allocation stops, if no increase in the objective function is possible with a new allocation. Denoting the number of totally allocated subchannels on carrier c with M_c, the achievable sum rate computes according to

$$R_{\mathrm{sum}} = \sum_{c=1}^{C}\sum_{i=1}^{M_c} \log_2(1 + p_{i,c}\lambda_{i,c}).$$

In the following we will shortly describe how the subchannel powers $p_{i,c}$ and the channel gains $\lambda_{i,c}$ compute for the two algorithms. For notational convenience, we assume that the user allocation is given in the following, where $\pi_c(i)$ denotes the user to which the data stream encoded at i-th place has been allocated to on carrier c. Certainly, when the algorithm is run, potential subchannel gains have to be computed for every user and carrier to select the most suitable user and carrier for each data stream.

SESAM with MMSE Filters

With MMSE filters, the problem of maximizing sum rate under a total power constraint can be solved almost optimally at drastically reduced computational complexity. In this case a simplified power allocation is assumed such that $p_{i,c} = \frac{P_{\mathrm{Tx}}}{CM_c}$[7], where P_{Tx} denotes the available transmit power. The subchannel gains $\lambda_{i,c}$ then compute according to

$$\lambda_{i,c} = \rho_1\left(\boldsymbol{H}_{\pi_c(i),c}\left(\boldsymbol{I} + \sum_{j=1}^{i-1}\frac{P_{\mathrm{Tx}}}{CM_c}\boldsymbol{H}^{\mathsf{H}}_{\pi_c(j),c}\boldsymbol{t}_{j,c}\boldsymbol{t}^{\mathsf{H}}_{j,c}\boldsymbol{H}_{\pi_c(j),c}\right)^{-1}\boldsymbol{H}^{\mathsf{H}}_{\pi_c(i),c}\right), \tag{4.11}$$

where $\rho_1(\boldsymbol{A})$ denotes the principal eigenvalue of the matrix $\boldsymbol{A}$. $\boldsymbol{t}_{j,c}$ denotes the transmit filter in the dual uplink for the j-th data stream on carrier c and is equal to the unit-norm eigenvector belonging to the principal eigenvalue of the matrix

$$\boldsymbol{H}_{\pi_c(i),c}\left(\boldsymbol{I} + \sum_{j=1}^{i-1}\frac{P_{\mathrm{Tx}}}{CM_c}\boldsymbol{H}^{\mathsf{H}}_{\pi_c(j),c}\boldsymbol{t}_{j,c}\boldsymbol{t}^{\mathsf{H}}_{j,c}\boldsymbol{H}_{\pi_c(j),c}\right)^{-1}\boldsymbol{H}^{\mathsf{H}}_{\pi_c(i),c} \tag{4.12}$$

[7]For SESAM with MMSE filter the $p_{i,c}$ correspond to the powers in the dual uplink. As the rates in the broadcast and the dual uplink are the same [8], we consider the uplink rates for the SESAM MMSE algorithm in this chapter.

SESAM with Zero-Forcing Filters

With zero-forcing filters the subchannel gains $\lambda_{i,c}$ are given by

$$\lambda_{i,c} = \rho_1 \left(\boldsymbol{H}_{\pi_c(i),c} \left(\boldsymbol{I} - \sum_{j=1}^{i-1} \boldsymbol{t}_{j,c} \boldsymbol{t}_{j,c}^{\mathsf{H}} \right) \boldsymbol{H}_{\pi_c(i)}^{\mathsf{H}} \right), \tag{4.13}$$

where here the $\boldsymbol{t}_{j,c}$ are the downlink transmit filters for the j-th data stream on carrier c and equal to the unit-norm eigenvectors corresponding to the principal eigenvalues of the matrices

$$\left(\boldsymbol{I} - \sum_{\ell=1}^{j-1} \boldsymbol{t}_{\ell,c} \boldsymbol{t}_{\ell,c}^{\mathsf{H}} \right) \boldsymbol{H}_{\pi_c(j),c}^{\mathsf{H}} \boldsymbol{H}_{\pi_c(j),c} \left(\boldsymbol{I} - \sum_{\ell=1}^{j-1} \boldsymbol{t}_{\ell,c} \boldsymbol{t}_{\ell,c}^{\mathsf{H}} \right).$$

Each matrix $\boldsymbol{I} - \sum_{\ell=1}^{j-1} \boldsymbol{t}_{\ell,c} \boldsymbol{t}_{\ell,c}^{\mathsf{H}}$ projects into the nullspace of the beamformers of the previously allocated data streams on carrier c and therefore assures that the $\boldsymbol{t}_{j,c}$ to do not interfere with the previously allocated subchannels on carrier c. Interference on later allocated subchannels is suppressed by DPC. With this kind of zero-forcing, several Quality-of-Service (QoS) constrained optimization problems can be solved at reduced computational complexity, where the optimum can be achieved quite closely. Such optimization problems can be the weighted sum rate maximization under minimum and maximum rate requirements with a transmit power constraint or the transmit power minimization for the fulfillment of minimum rate requirements. As zero-forcing filters are applied together with DPC, the subchannels are interference-free and the optimum powers $p_{i,c}$ can be determined by water-filling alike solutions. For the details of the power and user allocation the reader is referred to [9] for the pure sum rate maximization, to [4] for the weighted sum rate maximization under a power constraint and [1] for QoS constrained optimization problems.

4.9.4 Approximation of the Ergodic Sum Rate with Large System Analysis

In this section we will present approximations for the subchannel gains $\lambda_{i,c}$ with results from large system analysis. For the large system analysis the number transmit antennas and receive antennas go to infinity at a finite fixed ratio β, i.e., $N \to \infty$, $R \to \infty$, $\beta = \frac{N}{R}$, finite. Then the empirical eigenvalue distributions of the matrices in (4.13) and (4.11) converge almost surely to an asymptotic limit. For finite systems, we propose to use the a.e.d. $f_{\boldsymbol{A}}(x)$ of a matrix $\boldsymbol{A}$ to find an approximation for the i-th strongest eigenvalue as follows. Let $\boldsymbol{A}$ be a $L \times L$ matrix with finite L, then the i-th eigenvalue of this matrix is approximated by these two implicit equations.

$$\rho_i(\boldsymbol{A}) = L \int_{m_{i-1}}^{m_i} x f_{\boldsymbol{A}}(x) \, \mathsf{d}\, x, \quad \text{where} \int_0^{m_i} f_{\boldsymbol{A}}(x) \, \mathsf{d}\, x = \frac{L-i+1}{L}.$$

Hence the x domain is divided into L intervals, where in each interval $[m_{i-1}; m_i]$ the integral $\int_{m_{i-1}}^{m_i} f_{\boldsymbol{A}}(x')\,\mathsf{d}\,x'$ is equal to $1/L$. The centroid of each interval then represents one eigenvalue. Using these approximations for the computation of eigenvalues in (4.11) and (4.13), the user and power allocation is afterwards done as originally proposed for finite systems. The a.e.d.s $f_{\boldsymbol{A}}(x)$ can be computed from the Stieltjes transformation $S_{\boldsymbol{A}}(z)$ as described for example in [10]. In the following we will therefore present equations for the Stieltjes transforms.

SESAM with MMSE Filters

As a direct computation of the Stieltjes transforms of the matrices in (4.11) seams to be difficult, we first introduce the following approximation of the subchannel gains.

$$\lambda_{i,c} \approx \rho_{n_{i,c}+1}\left(\boldsymbol{H}_{\pi_c(i),c}\left(\boldsymbol{I}+\frac{P_{\text{Tx}}}{CM_c}\boldsymbol{V}_{n_{i,c}}\boldsymbol{V}_{n_{i,c}}^{\mathsf{H}}\right)^{-1}\boldsymbol{H}_{\pi_c(i),c}^{\mathsf{H}}\right), \tag{4.14}$$

where $n_{i,c}$ denotes the number of subchannels assigned to the user $\pi_c(i)$ in previous steps on carrier c. $\boldsymbol{V}_{n_{i,c}} \in \mathbb{C}^{N\times i-1-n_{i,c}}$ is a matrix with orthonormal columns, i.e., $\boldsymbol{V}_{n_{i,c}}^{\mathsf{H}}\boldsymbol{V}_{n_{i,c}} = \boldsymbol{I}$, independent of $\boldsymbol{H}_{\pi_c(i),c}$. Hence the effect of the interfering subchannels allocated to other users is taken into account by the matrix $\boldsymbol{V}_{n_{i,c}}$, which consists of as many orthonormal columns as subchannels assigned to other users than user $\pi_c(i)$ interfere with the i-th subchannel on carrier c. The effect of interference of subchannels allocated to the same user is considered by taking the $n_{i,c}+1$-th largest eigenvalue in (4.14), denoted as $\rho_{n_{i,c}+1}$. Due to the independence between the matrices $\boldsymbol{H}_{\pi_c(i),c}$ and $\boldsymbol{V}_{n_{i,c}}$, the Stieltjes transform $S_{\tilde{\boldsymbol{A}}_{i,c}}(z)$ of the matrix $\tilde{\boldsymbol{A}}_{i,c} = \boldsymbol{H}_{\pi_c(i),c}\left(\boldsymbol{I}+\frac{P_{\text{Tx}}}{CM_c}\boldsymbol{V}_{n_{i,c}}\boldsymbol{V}_{n_{i,c}}^{\mathsf{H}}\right)^{-1}\boldsymbol{H}_{\pi_c(i),c}^{\mathsf{H}} \in \mathbb{C}^{R\times R}$ can be derived from [10, Theorem 2.39] and is given by the implicit equation

$$\beta = \frac{1+zS_{\tilde{\boldsymbol{A}}_{i,c}}(z)}{1-\frac{1}{1+S_{\tilde{\boldsymbol{A}}_{i,c}}(z)}\frac{N-(i-1-n_{i,c})}{N}-\frac{1+\frac{P_{\text{Tx}}}{CM_c}}{1+\frac{P_{\text{Tx}}}{CM_c}+S_{\tilde{\boldsymbol{A}}_{i,c}}(z)}\frac{i-1-n_{i,c}}{N}}.$$

For details the reader is referred to [3].

SESAM with ZF Filters

As the matrix $\boldsymbol{H}_{\pi_c(i),c}\left(\boldsymbol{I}-\sum_{j=1}^{i-1}\boldsymbol{t}_{j,c}\boldsymbol{t}_{j,c}^{\mathsf{H}}\right)\boldsymbol{H}_{\pi_c(i),c}^{\mathsf{H}}$ has the same nonzero eigenvalues as $\boldsymbol{A}_{i,c} = \boldsymbol{V}_{i-1,c}^{\mathsf{H}}\boldsymbol{H}_{\pi_c(i),c}^{\mathsf{H}}\boldsymbol{H}_{\pi_c(i),c}\boldsymbol{V}_{i-1,c}$, where $\boldsymbol{V}_{i-1,c}$ contains the $N-i-1$ orthonormal basis vectors of the projector $\boldsymbol{I}-\sum_{j=1}^{i-1}\boldsymbol{t}_{j,c}\boldsymbol{t}_{j,c}^{\mathsf{H}}$, we derive an implicit equation for the Stieltjes transform $S_{\boldsymbol{A}_{i,c}}(z)$ of the latter matrix in the following. The computation of $S_{\boldsymbol{A}_{i,c}}(z)$ works recursively, as it requires the a.e.d. $f_{\boldsymbol{A}_{\ell_{i,c},c}}(x)$, where $\ell_{i,c}$ denotes the last step before step i, in which a subchannel has been assigned to the user $\pi_c(i)$ on carrier c. $S_{\boldsymbol{A}_{i,c}}(z)$ is given implicitly by

$$\int_0^{n_1}\frac{f_{\boldsymbol{A}_{\ell_{i,c},c}}(x)}{1-\tilde{\beta}+(x-z)\tilde{\beta}S_{\boldsymbol{A}_{i,c}}(z)}\,\mathsf{d}\,x = \frac{N-\ell_{i,c}-1}{N-\ell_{i,c}},$$

where

$$\int_0^{n_1} f_{\boldsymbol{A}_{\ell_{i,c},c}}(x)\,\mathrm{d}\,x = \frac{N-\ell_{i,c}-1}{N-\ell_{i,c}}.$$

For details the reader is referred to [2].

4.9.5 Numerical Results

Figure 4.20 exhibits the sum rate averaged over 1000 channel realizations versus the number of transmit antennas, where the sum rate maximization under a transmit power constraint is considered. The ratio β has been fixed to $\beta = 2$. Each cross for the SESAM approaches corresponds to the sum rate achieved with one channel realization and the line goes through the average sum rates. Figure 4.20 (a) com-

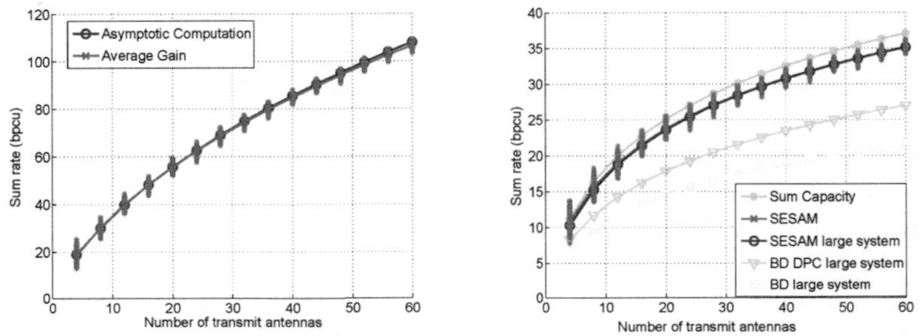

(a) Sum rate averaged over 1000 channel realizations compared to asymptotic sum rate in a system with $K = 2$ users, $\beta = N/R = 2$

(b) Large system approximation of sum rates for different algorithms with $K = 5$, $P_{\mathrm{Tx}} = 10$, $\beta = 2$

Figure 4.20: Numerical Results

pares the approximated sum rate with the average sum rate for SESAM with MMSE filters in a system with 2 users, one carrier, and at a transmit power of $P_{\mathrm{Tx}} = 100$, which corresponds to a transmit SNR of 20dB. In Fig. 4.20 (b) the same comparison is made with SESAM and ZF filters at 10 dB and in a system with 5 users. Additionally, Fig. 4.20 (b) exhibits the average sum capacities and the large system approximations of Block Diagonalization [7] with and without DPC, which have also been derived in [2]. From both figures we can conclude that the presented approximations match the average sum rate quite well, already with reasonable numbers of transmit and receive antennas.

Bibliography

[1] C. Guthy, W. Utschick, and G. Dietl, "Spatial Resource Allocation for the Multiuser Multicarrier MIMO Broadcast Channel - A QoS Optimization Perspective," in *Proc. of International Conference on Acoustics, Speech and Signal Processing (ICASSP)*, 2010.

[2] C. Guthy, W. Utschick, and M.L. Honig, "Large System Analysis of Projection Based Algorithms for the MIMO Broadcast Channels," submitted to *International Symposium on Information Theory (ISIT)*, 2010.

[3] C. Guthy, W. Utschick, and M.L. Honig, "Large System Analysis of the Successive Encoding Successive Allocation Method for the MIMO BC," In *Proc. of ITG/IEEE Workshop on Smart Antennas*, 2010.

[4] C. Guthy, W. Utschick, R. Hunger, and M. Joham, "Efficient Weighted Sum Rate Maximization With Linear Precoding," accepted for publication in *IEEE Transactions on Signal Processing*, December 2009.

[5] V. Marčenko and L. Pastur, "Distribution of Eigenvalues for Some Sets of Random Matrices," *Math USSR Sbornik*, 1:457–483, 1967.

[6] M.J.M. Peacock, I.B. Collings, and M.L. Honig, "Advances in Multiuser Detection," volume 1 of *Wiley Series in Telecommunications and Signal Processing*, chapter 4 Performance with Random Signatures. John Wiley & Sons, 2009.

[7] Q.H. Spencer, A.L. Swindlehurst, and M. Haardt, "Zero-forcing Methods for Downlink Spatial Multiplexing in Multiuser MIMO Channels," *IEEE Transactions on Signal Processing*, 52(2):461–471, February 2004.

[8] P. Tejera, *Principles and Algorithms for Transmission in Multiple-Input Multiple-Output Systems*, PhD thesis, Technische Universität München, 2008.

[9] P. Tejera, W. Utschick, G. Bauch, and J. A. Nossek, "Subchannel Allocation in Multiuser Multiple Input Multiple Output Systems," *IEEE Transactions on Information Theory*, 52(10):4721–4733, October 2006.

[10] A.M. Tulino and S. Verdu, "Random Matrix Theory and Wireless Communications," Number 1 in Foundations and Trends in Communications and Information Theory. Now Publishers, 2004.

[11] H. Viswanathan, S. Venkatesan, and H. Huang, "Downlink Capacity Evaluation of Cellular Networks With Known-Interference Cancellation," *IEEE Journal on Selected Areas in Communications*, 21(6):802–811, June 2003.

4.10 Combined Radar and Communication Systems Using OFDM

M. Braun, C. Sturm, F. Jondral, T. Zwick, Karlsruhe Institute of Technology (KIT), Germany

4.10.1 Introduction

In the current technological development, the radio frequency front-end architectures used in radar and digital communication technology are becoming more and more similar. In both applications more and more functions that have traditionally been accomplished by hardware components are now being replaced by digital signal processing algorithms. Moreover, today's digital communication systems use frequencies in the microwave regime for transmission, which are close to the frequency bands traditionally used for radar applications. This technological advancement opens the possibility for the implementation of joint radar and communication systems that are able to support both applications with one single platform and even with a common transmit signal. A typical application area for such systems would be intelligent transportation networks, which require the ability for inter-vehicle communications as well as the need for reliable environment sensing.

First concepts of joint radar and communication systems have been primarily based on spread spectrum techniques. Recently, OFDM signals have gained a lot of attraction for this purpose. This is motivated by two facts: First, most currently released communications standards, e.g., IEEE 802.11p, employ OFDM signals. Second, in the radar community OFDM signals recently have attracted general interest and their suitability for radar applications has been proven. Hence, OFDM signals currently seem to be the ideal basis for joint radar and communication implementations.

A possible application scenario of an OFDM based joint radar and communication system is illustrated in Fig. 4.21. The OFDM signal (colored in green) is transmitted from the car on the left side and transports information to distant receivers. At the same time this signal is reflected at objects, also cars, in the neighborhood (reflected signal depicted in gray). The OFDM system observes the echoes of its own transmit signal and calculates a radar image by applying suitable processing algorithms. Also, communication with base stations could be included in this concept.

In the following, detailed considerations regarding an optimum parametrization of the OFDM signals for simultaneous radar and communication operation as well as optimum radar processing strategies will be discussed. Moreover, a fully operational system demonstrator and verification measurement results will be presented.

4.10.2 Signal Design

A major challenge is the design of the OFDM signals. It must be guaranteed that the communication link is reliable over mobile communication channels with severe fading, and that the radar imaging algorithm is not negatively affected by the signal

Table 4.3: Channel limitations for the OFDM parameters

Property	Urban	Autobahn
RMS excess delay	0.102 µs	0.122 µs
Maximum Doppler spread	7.24 kHz	5.23 kHz
Coherence bandwidth	2246.1 kHz	1269.53 kHz
RMS Coherence time	0.401 ms	0.46 ms

design. A large variety of parameters can be changed, ranging from sub-carrier distance to channel coding. This section will explain the most important parameters. A more detailed description can be found in [1, 2].

Physical Parameters

Physical parameters of an OFDM frame are sub-carrier distance, guard interval duration, total bandwidth and the frame duration. Their choice depends on the required radar accuracy and the quality of the mobile propagation channels.

In [1] and [2], we analyzed the effects of the mobile propagation channels. Basis for the analysis were RayTracing channels, i.e., simulations of real traffic scenarios where the channels between transmitter and receiver were obtained by optical methods [3]. A database of 10567 channels from a total of eight urban traffic and two highway ("Autobahn") scenarios was used to test and evaluate parameterizations. These channels were analyzed for time and frequency variance to obtain limits for the physical OFDM parameters. In particular, the sub-carrier spacing is limited by the coherence time and the Doppler spread; the guard length duration depends on the excess delay [1]. Table 4.3 shows the results of the analysis.

The requirements of the radar accuracy also impose constraints on the parametrization of the signal. In particular, the resolution in range and Doppler domain set minimum limits on bandwidth B and frame duration T_F, which can be estimated

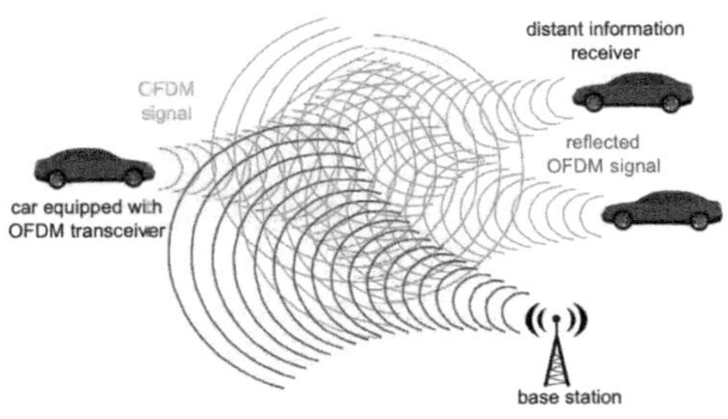

Figure 4.21: Application scenario for a combined radar and communication system

by the following equations:

$$B \geq \frac{c_0}{2\Delta d_{\max}}, \quad T_F \geq \frac{c_0}{2\Delta v_{\max} f_c} \tag{4.15}$$

For a range resolution of 1.5 m, the bandwidth must therefore be on the order of 100 MHz. This in turn has other side effects, such as a low power density.

A less obvious design criterion is the effect of the parametrization on the OFDM radar processing algorithm. The maximum likelihood estimator presented in the following section is prone to threshold effects. In [4], we present a method to test the range in which a given set of parameters works without threshold effects.

Channel Coding

A suitable channel coding is an important choice in the frame design process. Vehicular applications in particular have high demands regarding reliability of the data transmission. At the same time, the signal's low power density and the highly frequency- and time-variant channels make error-free data transmission very difficult. On the other hand, the large bandwidth allows for high data rates, which might not be necessary. It therefore makes sense to sacrifice raw data rate for lower bit error rates by using robust codes with low coding rates.

Although several types of codes can satisfy these requirements, Reed-Muller codes appear particularly suitable. Their big advantage is the possibility to use sub-sets of codes which exhibit a low peak-to-average power ratio (PAPR) [5, 6]. This is gives the whole system a new degree of freedom, since the low PAPR values exhibit fewer requirements towards the amplifiers. Simulations have shown the codes to perform well under adverse channel effects with a fixed PAPR of 3 dB [5].

4.10.3 The Radar Subsystem

The radar subsystem deals with the task of estimating range and relative speed of other objects in the vicinity. When transmitting, it receives and analyses the backscattered signal and processes it to gain information about the surrounding objects. In order to identify a suitable estimation algorithm we must first analyze the effects the backscattering has on OFDM frames.

In the following, one OFDM frame shall be described by an $N \times M$-matrix $\mathbf{F}_{\mathrm{Tx}} \in \mathbb{C}^{N \times M}$, where every element $(\mathbf{F}_{\mathrm{Tx}})_{k,l}) = \pm 1$ is a modulation symbol from the BPSK modulation alphabet. Every row in $\mathbf{F}_{\mathrm{Tx}}$ corresponds to an OFDM sub-carrier; every column corresponds to an OFDM symbol. $s(t)$ denotes the transmitted time domain signal and is created from $\mathbf{F}_{\mathrm{Tx}}$ by the usual OFDM modulation process of calculating an IFFT of every column and prepending the result with a cyclic prefix.

During transmission, a receiver is active. The received signal $r(t)$ consists of the Doppler shifted and time-delayed reflected signals. In the case of H reflecting targets, the relation between transmitted and received signals is

$$r(t) = \sum_{h=0}^{H-1} b_h s(t - \tau_h) e^{j2\pi f_{D,h} t} + w_{\sigma^2}(t). \tag{4.16}$$

The Doppler shift and roundtrip propagation time of the h-th target are denoted $f_{D,h}$ and τ_h, respectively. $b_h = |b_h|e^{j\tilde{\varphi}_h}$ is the corresponding complex attenuation factor. The received signal is a linear superposition of reflected signals from every target, plus complex white Gaussian noise $w_{\sigma^2}(t)$ with variance σ^2.

For the development of the estimation algorithm it is useful to analyze the effects of Doppler shift and time delay on $\mathbf{F}_{\mathrm{Tx}}$. For simplicity, we will analyze the effect of a single reflecting target with Doppler shift of f_D and a roundtrip delay of τ. In this case, the Doppler shift causes an oscillation of the matrix rows by $e^{j2\pi l T_O f_D}$. This assumes a constant Doppler shift over all frequencies, which is a valid approximation if the bandwidth is much smaller than the signal's center frequency. The delay causes a phase shift on every sub-carrier of $e^{-j2\pi(f_0+k\Delta f)\tau_0}$. Without loss of generality, $|b_0|$ can be assumed to be of unit value. Representing the received signal in the same matrix notation as $\mathbf{F}_{\mathrm{Tx}}$ can thus be done as

$$(\mathbf{F}_{\mathrm{Rx}})_{k,l} = (\mathbf{F}_{\mathrm{Tx}})_{k,l} \cdot e^{j2\pi l T_O f_{D,0}} e^{-j2\pi k \tau_0 \Delta f} e^{j\varphi_0} + (\mathbf{W})_{k,l}. \tag{4.17}$$

$\mathbf{W}$ is a matrix representation of the additive white Gaussian noise (AWGN); its entries are i.i.d. random values from a circular, complex, zero-mean normal distribution with variance σ^2. All phase shifts which are constant for the entire frame are summarized into the phase term φ. Before estimation, the known modulation symbols inside $\mathbf{F}_{\mathrm{Tx}}$ can be eliminated from $\mathbf{F}_{\mathrm{Rx}}$ by simple element-wise division. The resulting matrix is thus

$$(\mathbf{F})_{k,l} = \frac{(\mathbf{F}_{\mathrm{Rx}})_{k,l}}{(\mathbf{F}_{\mathrm{Tx}})_{k,l}} = e^{j(2\pi(lT_O f_{D,0} - k\tau_0\Delta f)+\varphi_0)} + \frac{(\mathbf{W})_{k,l}}{(\mathbf{F}_{\mathrm{Tx}})_{k,l}}. \tag{4.18}$$

All estimation is now performed on $\mathbf{F}$, which consists of two orthogonal oscillations and AWGN. It must be noted that the statistics of the noise are not affected by the division if the BPSK symbols are not correlated, since in this case, the division is nothing but a random phase rotation by either π or zero of the rotationally invariant noise. Therefore, the estimation of f_D and τ is equivalent to estimating the frequencies of the two orthogonal oscillations within the matrix $\mathbf{F}$ and is therefore very similar to an identification of spectral components.

Finally, the target parameters must be estimated from $\mathbf{F}$. We have chosen a maximum likelihood estimate (MLE) approach, which was originally introduced in [7] and [8]. The MLE is obtained by calculating [4]

$$C(m,n) := \left| \underbrace{\mathrm{IFFT}(n) \left\{ \underbrace{\mathrm{FFT}(m)\,\{\mathbf{F}\}}_{\text{FFT over every row of }\mathbf{F}} \right\}}_{\text{IFFT over all columns of the FFT result}} \right| \tag{4.19}$$

and finding the values $\hat{m}$, $\hat{n}$ which maximize $C(m,n)$. The MLE for Doppler shift and propagation time is then

$$\hat{\tau} = \frac{\hat{n}}{N_{\text{IFFT}} \Delta f}, \quad \hat{f}_D = \frac{\hat{m}}{M_{\text{FFT}} T_O}, \tag{4.20}$$

where N_{IFFT} and M_{FFT} are the lengths of the IFFT and the FFT, respectively. It must be noted that the complexity of such an approach is smaller than the classical approach of correlating in frequency and time domain. Moreover, additional investigations have been conducted regarding the estimation of the direction of arrival (DoA) with a multiple antenna receiver. It has been shown that standard DoA estimation techniques can be applied directly to the output of the range and velocity estimators in Eq. (6). Detailed results have been published in [9].

4.10.4 Measurements

Demonstrator Setup

The demonstrator system consists of three main hardware components: a Rohde & Schwarz (R&S) SMJ100A vector signal generator, a R&S FSQ26 signal analyzer, and optionally a R&S SMR40 microwave signal generator. The SMJ100A is limited to a maximum carrier frequency of 6 GHz but offers higher output power than the SMR40. Therefore it has been decided to implement two different configurations, one with the SMJ100A only in order to achieve high output power at 6 GHz and another one with both SMJ100A and SMR40 in order to generate a signal at the intended carrier frequency of 24 GHz but with reduced transmit power. All instruments are connected through an Ethernet link and controlled from a computer via the MatLab Instrument Control Toolbox. All signals are generated and processed in MatLab. The OFDM system parameters that have been applied for the measurements are summarized in Table 4.4. These parameters have been obtained through a theoretical study described in [10] and verified with ray tracing simulations in [1].

Table 4.4: OFDM system parameters

Symbol	Quantity	Value
f_c	Carrier frequency	6 GHz / 24 GHz
N_c	Number of subcarriers	1024
Δf	Subcarrier spacing	90.909 kHz
T	Elementary OFDM symbol duration	11 µs
T_p	Cyclic prefix length	1.375 µs
B	Total signal bandwidth	93.1 MHz

The first configuration of the system setup is shown in Fig. 4.22. The transmit signal is generated in MatLab, transferred to the signal generator, converted to the carrier frequency and radiated. The signal analyzer is synchronized in phase through a 10 MHz reference signal and in time through a trigger signal. The signal analyzer samples the I and Q components of the received signal after conversion

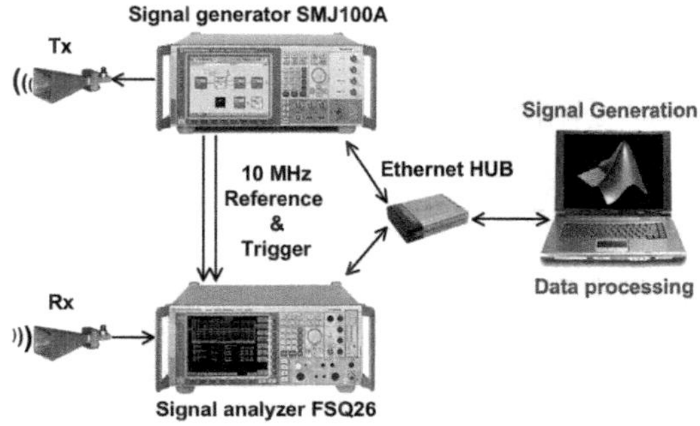

Figure 4.22: OFDM system setup for a maximum carrier frequency of 6 GHz

to the baseband and transfers them back to the computer. The signal generator provides a maximum carrier frequency of 6 GHz and a maximum peak power of 20 dBm. Since with the chosen parameters the OFDM signal shows a relatively stable PAPR of approx. 10 dB, a maximum mean transmit power of 10 dBm is available for uncoded transmission. The employed horn antennas at the transmitter and at the receiver have a gain of 18.5 dBi each.

In order to carry out measurements with a carrier frequency of 24 GHz an additional mixer is required. In that case a slightly different second setup is used, in which the output signal of the SMJ100A signal generator is fed to the external modulation signal input of the SMR40 signal generator. With an intermediate frequency of 200 MHz at the input of the SMR40 and a local oscillator frequency of 23.85 GHz, the center frequency of the upper sideband occurs at 24.05 GHz. The radiation of a lower sideband cannot be suppressed in this configuration, however the receiver is tuned only to the upper sideband, which spans from 24.0 GHz to 24.1 GHz. The external modulation input of the SMR40 does not allow for output power control. When driving the SMR40 with an average input signal power of 0 dBm a total average output power of only -12 dBm is available in the upper sideband. With an additional medium power amplifier the output power can be increased to 10 dBm. Also in this setup horn antennas are employed, which have a gain of 22 dBi each.

Measurement Results

In order to verify the developed algorithms a dynamic scenario with at least one moving object is required. Therefore in the scenario shown in Fig. 4.23a measurements have been taken with the system demonstrator. The scenario consists of a corner reflector with a radar cross section of $\sigma_{\mathrm{RCS}} = 16.3\ \mathrm{dBm}^2$ at 24 GHz and a car moving towards the radar with a velocity of approximately 25 km/h. The measurement was taken at the instant when the car was at the same distance of $R =$ 20 m as the reflector. The result obtained from the Doppler estimation algorithm is shown in Fig. 4.23b. In the FFT processing a Hamming window has been applied for both Doppler and range processing. It can be observed that in the distance of

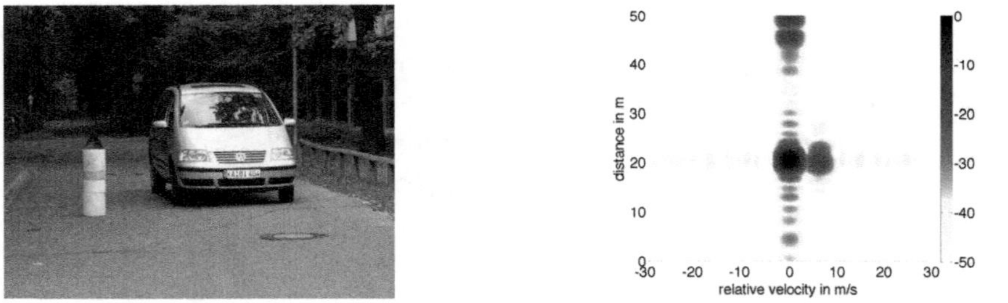

(a) Investigated scenario

(b) Measured radar image (normalized, in dB)

Figure 4.23: Verification measurement in a dynamic scenario

20 m both a high peak from the reflector at zero velocity and an additional peak at approximately 7 m/s corresponding to the speed of the car appear in the image. The reflection from the car is around 15 dB weaker than the signal scattered from the reflector. In the radar image additional reflections from ground clutter and objects in the background appear at zero velocity. The measurement result proves that both objects can be clearly identified and separated with the proposed processing algorithm.

In order to completely characterize the system performance also the SNR of the radar image after the processing has to be analyzed. The estimator described in Eq. (4.20) provides an SNR gain equivalent to the product of the number of subcarriers N and the number of evaluated OFDM symbols M. The expected *radar image SNR* amounts to

$$\mathrm{SNR_{image}} = \frac{P_{Tx} N M G_{Tx} G_{Rx} \lambda^2 \sigma_{\mathrm{RCS}}}{P_N (4\pi^3) r^4} \tag{4.21}$$

with P_{Tx} being the transmitted signal power, G_{Tx} and G_{Rx} being the transmit and receive antenna gain, λ being the wavelength, and σ_{RCS} denoting the radar cross section of the reflector.

In order to verify that this relation applies for practical OFDM radar measurements with the proposed estimator, additional measurements have been carried out with the system setup for the 24 GHz ISM band. In these measurements radar images of the trihedral reflector have been taken for three different distances of r = 4, 10, 20 m without using the amplifier. For each measurement result the ratio between the peak caused by the reflector and the average background noise level has been determined, assuming that this value represents SNR_{image} from (4.21). The measured values have been compared to the theoretically expected values, taking into account the receiver noise power level specified by the manufacturer to -143 dBm/Hz corresponding to a total noise power of -64 dBm. The results are reported in Table 4.5.

With increasing distance the measured SNR values approach the expected ones.

Table 4.5: Radar image SNR for P_{Tx} = -12 dBm

Distance to the reflector in m	4	10	20
Measured SNR in dB	49.8	41.4	30.8
Expected SNR in dB	60.7	44.5	32.8

For the distance of 4 m the discrepancy is caused by the fact that the reflector is not yet in the far field. For higher distances of the reflector there is only a minor discrepancy between the measured and the expected SNR, which results form the SNR degradation caused by the Hamming window. From the measurement results it is evident that the proposed estimator provides the gain claimed in (4.21). A detailed report on the measurements can be found in [11].

4.10.5 Summary

In this project a detailed concept for a combined radar and communication system based on OFDM signals has been elaborated and evaluated. A suitable estimator has been developed that allows for performing range and Doppler measurements with OFDM signals without any negative impact of the simultaneously transmitted user information. Both measurements and simulations show that OFDM radar is an interesting and feasible new technology with some interesting qualities.

Acknowledgement

We would like to thank Rohde & Schwarz for providing the measurement equipment.

Bibliography

[1] M. Braun, C. Sturm, A. Niethammer, and F. K. Jondral, "Parametrization of Joint OFDM-based Radar and Communication Systems for Vehicular Applications," *20th IEEE Symposium on Personal Indoor and Mobile Radio Communications*, 2009.

[2] M. Braun, C. Sturm, and F. K. Jondral, "On the Frame Design for Joint OFDM Radar and Communications," *15th International OFDM Workshop, Hamburg*, 2010.

[3] C. Sturm, M. Braun, and W. Wiesbeck, "Deterministic propagation modeling for joint radar and communication systems," in *Proceedings 2010 International Symposium on Electromagnetic Theory*, Aug 2010.

[4] M. Braun, C. Sturm, and F. K. Jondral, "Maximum Likelihood Speed and Distance Estimation for OFDM Radar," *Radar Conference, IEEE International*, 2010.

[5] M. Braun, Y. Koch, C. Sturm, and F. K. Jondral, "Signal Design and Coding for High-Bandwidth OFDM in Car-to-Car Communications," *Vehicular Technology Conference, 2010. 72nd IEEE*, September 2010.

[6] J. A. Davis and J. Jedwab, "Peak-to-mean power control in OFDM, Golay complementary sequences and Reed-Muller codes," *IEEE Transactions on Information Theory*, 45(7), November 1999.

[7] C. Sturm, E. Pancera, T. Zwick, and W. Wiesbeck, "A Novel Approach to OFDM Radar Processing," *Radar Conference, IEEE*, May 2009.

[8] C. Sturm, M. Braun, T. Zwick, and W. Wiesbeck, "A Multiple Target Doppler Estimation Algorithm for OFDM based Intelligent Radar Systems," *7th European Radar Conference*, 2010.

[9] C. Sturm, L. Reichardt, T. Zwick, and W. Wiesbeck, "Evaluation of beamforming algorithms for automotive OFDM signal based radar," in *Proceedings of the European Radar Conference EuRAD 2009*, pages 141–144, Rome, Sept 2009. 2009.09.30.

[10] C. Sturm, T. Zwick, and W. Wiesbeck, "An OFDM System Concept for Joint Radar and Communications Operations," *Vehicular Technology Conference, 2009. 69th IEEE*, April 2009.

[11] C. Sturm, M. Braun, T. Zwick, and W. Wiesbeck, "Performance Verification of Symbol-Based OFDM Radar Processing," *Radar Conference, IEEE International*, 2010.

5 System Level Aspects for Multiple Cell Scenarios

5.1 Link Adaptation

R. Kays, O. Hoffmann, TU Dortmund University, Germany

5.1.1 Motivation

OFDM based systems provide many options and parameters to set up an efficient radio transmission for given requirements and channel conditions. System design is very demanding especially if many transmissions share a limited bandwidth and have to be optimized without a central coordination instance. Self-organizing networks are an example for such demanding system design tasks. Especially if high requirements on quality of service are given in an environment with many active transmission nodes, only careful selection of link adaptation methods can yield the required performance. In any case, the basic OFDM design has to be suited to the transmission task. System design is based on the selection of general parameters like transmission frequency, bandwidth per channel, maximum transmit power (typically limited by regulation) as well as the definition of the fundamental OFDM parameters like number of carriers and the duration of the guard interval. In order to specify a practical system, further specifications are needed to allow for synchronization. Transmission standards therefore include the definition of pilot carriers and synchronization symbols. Link adaptation means the dynamic choice of individual link parameters within a given OFDM-framework. Typically, this process has to be based on the observation of the transmission conditions which influence a single radio link as well as a complete network. As the question of optimum link adaptation is closely related to the environment of the links, many different options exist. In order to restrict our discussion to a limited complexity, we will focus on a certain application scenario which includes many aspects of general interest. This scenario is sketched in the following chapter.

5.1.2 Example of a Multiple Link Scenario

For the purpose of discussing different aspects of link adaptation, a scenario should be selected which poses demanding requirements on transmission efficiency and quality of service. The transmission of live media streams in home environments may serve as such an example. Home environments are as individual as people are, and they will differ between regions and cultures. Assuming a typical household in many

H Rohling (ed.), *OFDM: Concepts for Future Communication Systems*,
Signals and Communication Technology, DOI: 10.1007/978-3-642-17496-4_5,

industrial countries one might assume that a critical network situation is the simultaneous use of different video services by several members of the household. There may be different local storage servers containing audio-video (AV) material bought by the consumer (like DVD or CD collections), recorded (broadcast time shift), downloaded or produced (like personal video and still images). Furthermore, different access points to external networks are available - satellite reception equipment, broadband coax cable or twisted pair cable carrying speech communication as well as DSL Internet access. Different devices may also link to terrestrial mobile communication (GSM, UMTS) or digital broadcast networks like DVB-T, DVB-H, or DAB. Figure 5.1 shows a typical scenario. Network access devices and local storage servers are located in different rooms, and manifold AV devices are used in several places. Services are provided based on individual links between data sources and media renderers. The system has to assure that every user has convenient access to every service - the vision of anything, anywhere, anytime should become reality for the personal environment of people. From the system design point of view, an efficient network is required, allowing for data rates in the order of 5 - 15 Mbit/s per media stream. Although the network should appear as one logical instance from the users' perspective, it consists of a number of independent links with individual transmission parameters. Each link is competing for the scarce transmission resources.

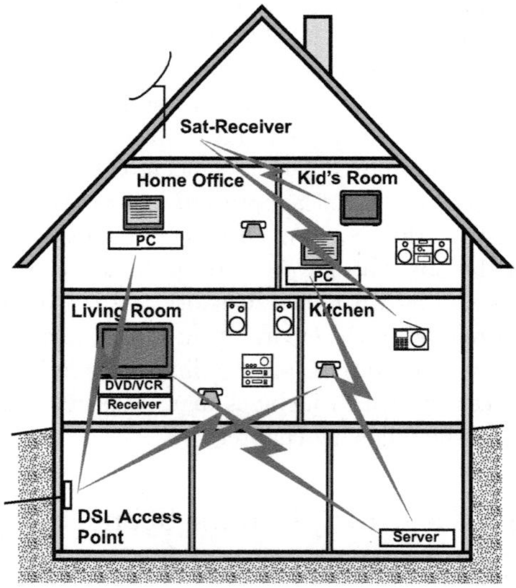

Figure 5.1: Typical wireless network in a detached house

Optimization can start by selection of suitable parameters for each individual link. The task becomes more demanding if several links are operated in parallel. It can be expected that such network concepts will be introduced to many households. This fact makes the competition for scarce resources more demanding, because a limited number of wireless channels has to be shared between neighboring logical networks.

Figure 5.2 shows the levels of system organization. The final goal of link adaptation will be the best selection of parameters for all links such that overall efficiency will be maximized.

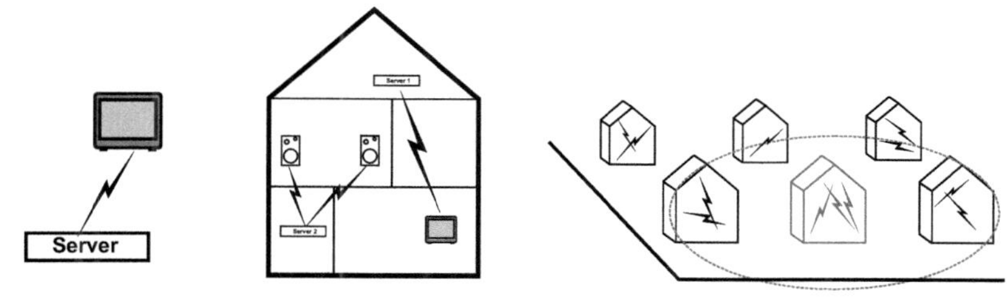

Figure 5.2: Levels of system organization

As already mentioned, OFDM based transmission is very well suited to optimize the transmission efficiency for each link. In order to discuss these aspects, some basic parameters should be defined. A parameter set given by the OFDM transmission modes of standardized wireless LAN (IEEE 802.11a/g) is a good platform. We therefore use it as a starting point for the discussion of the general aspects of link adaptation in a multiple link environment as given by our system example. Basic transmission parameters of WLAN in Europe are [1]:

- channel bandwidth: 20 MHz
- transmission frequencies: 2.4 - 2.483 GHz, 5.15 - 5.35 GHz, 5.47 - 5.725 GHz, 4 + 19 non-overlapping channels
- total number of carriers: 53 (48 data carriers + 4 continuous pilots, center carrier set to zero amplitude)
- net symbol duration: 3.2 microseconds, yielding a carrier spacing of 312.5 kHz
- guard interval: 0.8 microseconds
- modulation scheme: BSPK, QPSK, 16-QAM, 64-QAM
- forward error correction: convolutional code, constraint length 7, rate 1/2, puncturing for rate 2/3 and 3/4 in certain standardized transmission modes

Regulation with respect to allowed transmit power and power density slightly differs between frequency bands and countries. Limits for the 2.4 GHz band in Europe may serve as a reference: Maximum equivalent isotropically radiated power (EIRP) is 20 dBm, maximum power density is 10 mW/MHz.

Efficient link adaptation requires knowledge about the channel, e.g., transfer function or packet transmission statistics. In order to simulate and discuss the options,

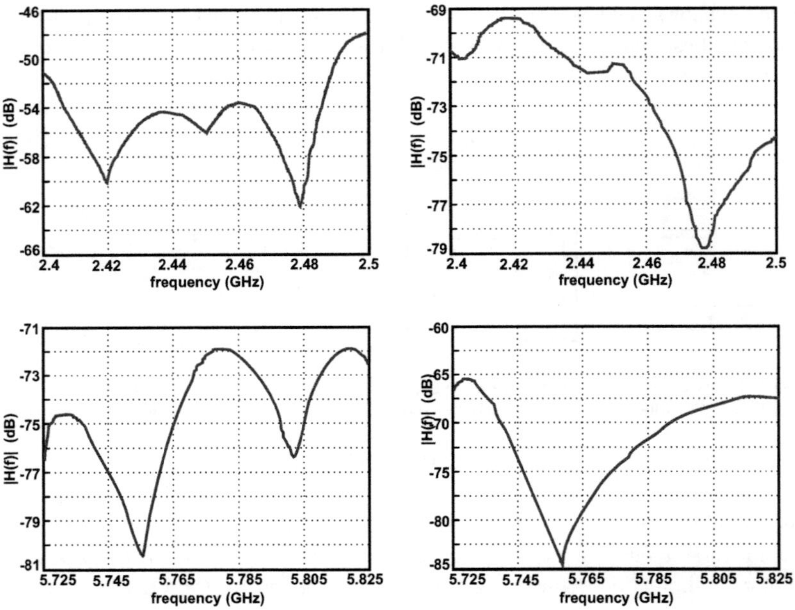

Figure 5.3: Typical transfer functions in a home environment (non-line-of-sight) for the 2.4 and 5 GHz bands, respectively. One WLAN transmission channel is a 20 MHz portion of the spectrum.

a channel model is required to deliver the required information. In our scenario, the coherence bandwidth is large compared to the carrier spacing. Furthermore, we assume precise channel estimation due to the pilot carriers and synchronization sequences in WLAN. Therefore, it will be sufficient to know the amplitude transfer function of typical transmission channels in home environments. Indoor transmission has been analyzed in many publications. However, most of these channel measurements relate to office environments. Therefore, a large number of channel transfer function measurements in typical homes have been carried out within the scope of this analysis [2]. Figure 5.3 gives examples of typical transfer functions. These have been measured using a channel sounder which provided a radio signal of 100 MHz bandwidth. Measurements have been taken for the ISM bands at 2.4 GHz and 5.75 GHz. One WLAN channel covers a portion of 20 MHz of these bands. The results of the measurements will be used within the following chapters.

5.1.3 Adaptation of Physical Link Parameters

Assuming the basic parameter set of IEEE 802.11a/g, the options for link adaptation on the physical layer are discussed in the following. We start the discussion for one single link, thus neglecting aspects of cross-layer optimization and interference with competing transmissions for the moment. If channel bandwidth, duration of guard interval, number of carriers, and carrier spacing are fixed according to the standard, link adaptation in the physical layer means individual, dynamic selection of the following parameters:

1. transmission channel (within the number of available channels)
2. total transmit power
3. modulation and coding scheme (and thus the physical data rate)
4. transmit power/modulation per carrier

Some of these options can be used within the boundaries of existing standards, others require extensions.

1. *Dynamic Frequency Selection*

   In off-the-shelf WLAN products, the transmission frequency is selected once (e.g., by default parameters coming with the product or during setup procedure by manual selection). A dynamic selection of the transmission frequency, adapted to the transmission environment, can improve the system performance significantly. Selecting a channel with an advantageous transmission function (low attenuation) allows the maximization of received power and thus SNR. In a network with high node density, the amount of competing transmissions in the radio channels has to be taken into consideration, too. By measuring the occupation of the channel, a second criterion for the dynamic selection of the channel is given. Let us first consider the transmission function. Generally, the potential for improvements depends on the scenario. In a home environment, the transmission function of the wireless channel remains nearly constant, if there are no movements of persons or furniture. Long term measurements of habitations have shown that the transmission function of the channel is influenced significantly by persons moving around. However, in many measurements the best channel remains the optimum choice even when habitants changed their position. Even if adaptation to the best transmission frequency is required, this can be organized with limited communication overhead, because changes happen comparably slow. The evaluation of a large number of measured transmission channels has shown the potential of a dynamic frequency selection in our use case. Measurements have been carried out for a bandwidth of 100 MHz (2.4 - 2.5 GHz), thus covering the complete ISM band. A WLAN channel can be characterized by a portion of 20 MHz within this band. The 13 WLAN channels specified for Europe have carrier frequencies of 2.412 - 2.472 GHz with a 5 MHz distance. Therefore, the available channels overlap. Calculating the average received power of these specified channels and comparing it for the different possible choices, and evaluating this figure for a large number of measurements, a statistical measure for improvements in a single link can be given. The frequency selection gain illustrated in Fig. 5.4 is specified as the difference of the received average power between the worst and the best channel. In the 2.4 GHz band this gain is up to 20 dB. Furthermore, it turns out that in 50% of all cases this gain is at least 5.4 dB for a line-of-sight (LOS) transmission and 7.2 dB for a non-line-of-sight (NLOS) transmission.

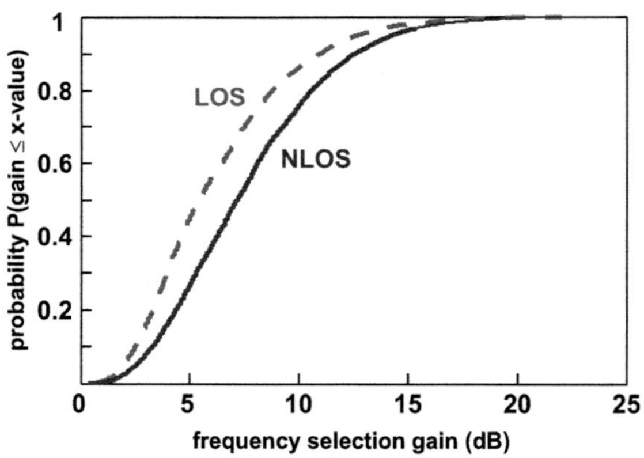

Figure 5.4: CDF for improvements realized by dynamic frequency selection (2.4 GHz)

The frequency selection gain in real systems can be even higher considering the reduction of interference and load balancing, for example. However, the quantification of this gain largely depends on the topology, environment, load, number of active nodes, and further system parameters.

2. *Transmit Power Control (TPC)*

The transmit power determines the range in which packets can be received at a certain error rate. Therefore, the transmit power strongly impacts the performance and the efficiency of a network, e.g., in terms of energy consumption, throughput, delay, and interference. By means of TPC the transmit power can be adapted collectively for all subcarriers of the OFDM signal. Besides reducing energy consumption of the devices, TPC can also be used to minimize the interference effects in a given topology. However, implementation of TPC requires a change of paradigms. Many state-of-the-art systems, also current WLANs, exploit the limits defined by regulation and therefore waste network resources at the expense of competing links. In the home environment, there are a lot of device arrangements that require significantly lower transmit power to fulfill their transmission task. Considering a wireless link between a set-top-box and a presentation device located in the same room, a transmit power of 20 dBm will not only be wasting radio resources, but may be even critical due to overdriving the RF-input of the receiver. If adaptation of the transmit power, i.e., reduction to the required level, were introduced, the network nodes would increase the performance of the whole network, because interference would be limited. In this respect, TPC is more than an improvement method for a single link. TPC aims to increase the performance of all nodes in a network. In this way, neighboring nodes cooperate to enable the fulfillment of as many transmission tasks as possible under the constraint of limited resources. However, TPC is used today mainly in multi-hop networks or for meeting regulative requirements, e.g., in IEEE 802.11a systems. In order to reduce the energy consumption and increase the throughput, two general principles are deployed: transmission in time and in space multiplex. In space multiplex, the transmit power is adapted in a way that each link covers only the minimum range, so that all the network resources can be reused in minimum distance. Thus, it is possible that other stations can simultaneously transmit on the same channel without causing significant interference to other stations. In time multiplex, stations compete for medium access as usual (e.g., via CSMA/CA). When obtaining medium access, the stations can use the transmit opportunity to transmit data packets with reduced transmit power. Time multiplex is feasible, for example, when space multiplex is not possible - e.g., because for space multiplex the PHY data rate would be so low that the required throughput could not be fulfilled or because legacy devices do not allow for space multiplex. Of course, improvement of overall system performance cannot be characterized by simple E_b over N_0 curves. Again, the gain largely depends on the topology, environment, number of active nodes etc. In order to judge this approach, a link efficiency measure as derived in chapter 5.1.3 is required.

3. *Modulation and FEC*

Once the channel transfer characteristics are given, the selection of the modulation and FEC parameters has to be done. WLAN according to the OFDM modes of IEEE 802.11 allows for different combinations (modulation BPSK, QPSK, 16-QAM, 64-QAM, convolutional coding with rates 1/2, 2/3, or 3/4). According to the standard, power is allocated equally to all subcarriers, and all of them use the same modulation and FEC scheme. The selection of a suitable transmission scheme leading to a certain gross data rate is done by a process called "dynamic rate adaptation" (DRA). By this process, the link can be adapted to different SNR ratios on the channel. When adapting modulation and coding rate, there is always a trade-off between achievable throughput and resulting packet error rate. Therefore, the transmission mode comprises an important variable in efficient link management. In many practical implementations, this selection is based on evaluation of the transmission statistics. In literature, different improved schemes for DRA can be found (see, e.g., [3–6]). Existing adaptation schemes can roughly be distinguished into two categories, depending on the way channel quality is estimated. On the one hand, MAC layer-based adaptation schemes adapt the data rate on the basis of packet error rate (PER) calculated at MAC layer, as in [3]. In contrast, PHY layer based adaptation schemes evaluate current channel conditions, both at transmitter and receiver side, and adapt data rate according to the SNR, as in [4]. Typically, these single-stage approaches exhibit at least one main drawback. For a single link, evaluating channel conditions on the basis of packet error rate might perform reasonably. Since packet loss is considered only as caused by degrading channel conditions, these methods fail in wireless networks with high node density. In case of packet loss induced by collisions, the adaptation scheme reduces the transmission rate. Thus, channel occupancy and collision probability are increased even more. Deriving channel conditions from the current SNR is not trivial and highly fault-prone. A solution to these problems might be the introduction of a two-stage concept, which comprises advantages of both approaches and compensates their disadvantages. Examples of a two-stage approach can be found in [5, 6].

4. *Transmit Power/Modulation per Carrier*

   A more sophisticated approach is the adaptation of parameters individually for different subcarriers. In this case, improvements can be expected for frequency-selective channels. Figure 5.5 shows the transmission function of two representative channels that have been selected for the simulations in the following.

   Five adaptation methods have been compared in the simulation series described below. Perfect channel knowledge is assumed in all cases. Figures 5.5 and 5.6 each give five curves A-E for the two fading channel examples. The simulations were carried out for the 54 Mbit/s mode (64-QAM, R=3/4). Curve A represents a standard transmission using a simple receiver concept as described in [7]. In this case, the transmission function of the channel is compensated by inverse filtering at the receiver side (zero forcing). In the next step, demodulation and Viterbi decoding is done, based on a three bit soft decision (as often proposed in literature). This configuration reflects the state-of-the-art and is used as reference. As zero forcing and calculation of soft information is done separately, performance is lost. Curve B is based on a standard transmission combined with perfect soft decoding. Here, the receiver combines zero forcing and calculation of soft information. In order to overcome the problem of receiver side filtering, one approach that can also be realized in standardized

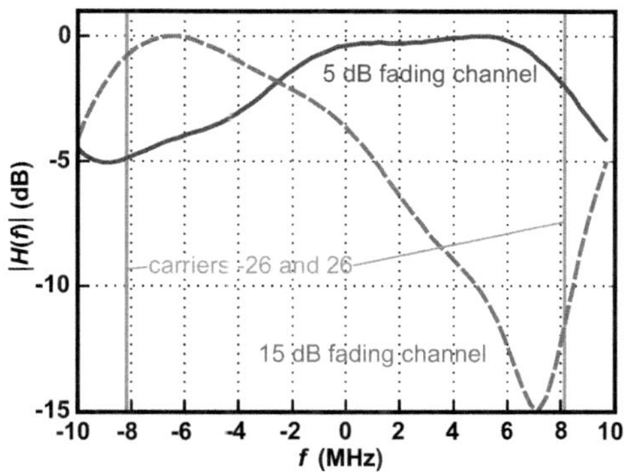

Figure 5.5: Sample channels used in simulations

systems is a channel inversion at the sender. Carriers which are subject to high attenuation are amplified at the transmitter side, thus yielding a flat spectrum at the receiver input. If attenuation is high, a high amount of power is used for weak carriers. An extension of this approach is the option to completely leave out a certain number of carriers (in our simulation six of 48) and use the saved power to boost the remaining carriers. As the transmission is using forward error correction, this appears to be a specific kind of puncturing. Curves C and D give the simulation results for the options channel inversion at transmitter and channel inversion with suppressing the six weakest carriers at transmitter, respectively. It turns out that this approach is worse than other options, because even in the case of carrier suppression, with the given transmission functions much energy is wasted for transmitting information in inefficient parts of the frequency band. The information theoretic point of view shows us that this naive adaptation approach has to be suboptimal. Therefore, processing as simulated in curve B is the best choice among these options, as can be clearly seen in the figures. However, careful evaluation of the characteristics for different channel characteristics has shown that the interleaver specified in IEEE 802.11 is not perfectly suited to the application scenario discussed here. As the interleaver was selected to match primarily the characteristics of office environments, it can be further optimized for the home scenario. Thus, a modified interleaver has been developed [8]. Curve E shows the simulation results when assuming perfect soft decision as in curve B in combination with the improved interleaver.

The performance comparison yields similar results for other channel profiles and transmission parameters (modulation and convolutional encoder). In general, the difference between the modes A - E increases with fading depth and data rate. Regarding channel profiles with deep fades, this modification yields a gain of more than 2 dB in contrast to the wide spread simplified solution of a serial connection of equalizer and Viterbi decoder with 3 bit quantization. However, the gain which can generally be achieved by these link adaptation methods has to be seen in combination with other methods. If dynamic chan-

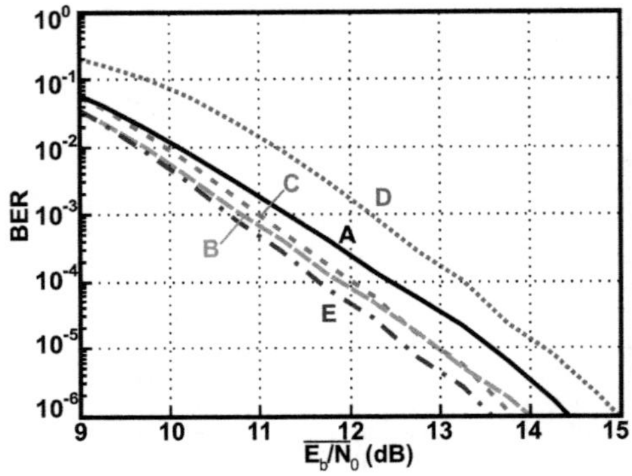

Figure 5.6: BER versus averaged E_b/N_0 in a typical home channel with a single 5 dB fade

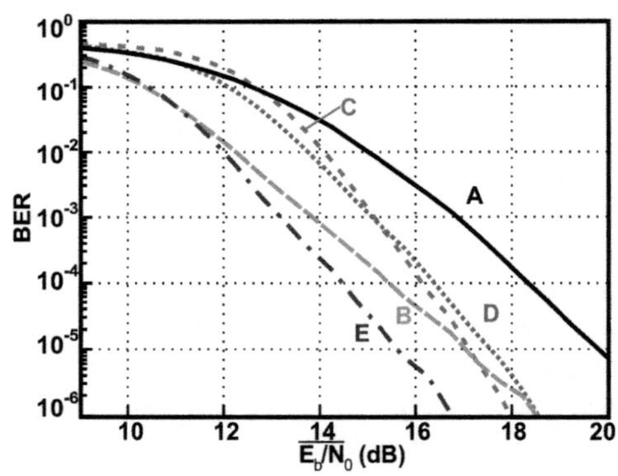

Figure 5.7: BER versus averaged E_b/N_0 in a typical home channel with a single 15 dB fade

nel selection is implemented, only channels with low fading depths will be selected in most situations. The possible improvements are thus very limited. Furthermore, it should be considered that organization of adaptation on a subcarrier basis needs communication overhead for channel measurements. Another practical disadvantage is the fact that the options are not within the line of existing standards.

Link Efficiency as Optimization Criterion

Generally speaking, link efficiency of transmission systems should be optimized so that all transmission tasks are fulfilled completely while consuming a minimum of network resources. A comparison related to E_b/N_0 figures does not cover all important aspects of network optimization. Therefore, an enhanced efficiency measure is needed. Discussing in terms of the ISO/OSI model, this link efficiency measure should focus on the lower two layers. A fair comparison of different transmission concepts should be possible, including options of improved modulation and error correction coding schemes, use of multiple antennas as well as modified MAC methods. In order to derive a generic definition, we define the link efficiency by the ratio between the achieved transmission performance and the consumed resources. The resources spent to achieve the transmission performance comprise the occupied bandwidth, duration of channel occupation, and the transmit power. On this basis, the efficiency of a packet transmission can be calculated as follows. The basic element of the transmission is one symbol. However, efficiency analysis requires the inclusion of protocol overhead (of the lower two ISO/OSI layers in our case) as well as an investigation of the trade-off between forward error correction and automatic repeat request (ARQ) methods. Therefore, we start with the definition of the packet transmission efficiency η_P:

$$\eta_P = \frac{PL}{t_p \cdot B \cdot \log_2\left(1 + \frac{P_T \cdot a}{(B \cdot N_0 + P_{Int}) \cdot I}\right)} \tag{5.1}$$

PL is the number of transmitted payload bits in the packet. PL equals zero, if the packet is not transmitted successfully (i.e., either the data packet or the acknowledgment packet associated with it is in error). Retransmissions will be considered when analyzing the transmission of a packet sequence.

t_p is the duration of channel occupancy. It is including the transmission time of the data packet and the acknowledgment as well as the contention time and the interframe spaces.

P_T is the transmit power.

a is the channel attenuation factor ($0 \leq a \leq 1$) and includes free space attenuation and slow fading.

N_0 is the spectral density of noise. As white noise is assumed, multiplication by the bandwidth B gives the received noise power.

P_{Int} covers interference from concurring transmissions in the same band.

I allows introducing an implementation margin. Thus, the performance of different implementations of network nodes can be compared. A typical value for IEEE 802.11 systems is 15 dB (10 dB noise figure plus 5 dB implementation loss).

During the transmission of each single packet, the bandwidth B, noise spectral density N_0, implementation margin I, transmit power N_0, and the interference power P_{Int} are assumed to be constant. During a transmission attempt there are several phases (contention, transmission, interframe spaces), where different transmit powers are used and different interference levels may occur. A temporal average of the different levels seems to be inadequate when analyzing realistic network situations. For the sake of simplicity, it is proposed to assume that P_T equates to the transmit power during the data packet transmission phase and P_{Int} equals the maximum interference level during this phase. The difference to the exact values will be low. Efficiency analysis should reflect the gain when the transmission conditions are improved, e.g., by methods to combat fast fading like diversity reception. Therefore, the severity of the hurdle of the wireless channel (as indicated by the factors a, P_{Int} and N_0) is averaged over the duration of the link and the whole frequency band. If, for example, dynamic frequency selection is applied and a channel switch enables higher throughput because of better transmission conditions, the effect of this link adaptation would be directly reflected by η_P. The reason is that the ratio of PL/t_p increases, whereas a, P_{Int}, and N_0 are constant.

In our analysis, the transmission of a sequence of data packets between two specific network nodes is regarded as one link. Typically, there will be at least one source of traffic at the transmitter side, generating packets to be sent to a receiving station. An efficiency measure on link level can be derived by using the packet transmission efficiency as defined above and averaging over the duration of link existence. For each transmitted packet, the transmission efficiency η_P can be calculated. For the reasons explained above, only the parameters PL, t_p, and P_T are calculated on packet basis and therefore indexed with i. Due to packet retransmission, the number of generated packets N_G generally will differ from the number N_T of transmitted packets and the number of successfully decoded packets N_D. To obtain the efficiency of a single link, $\eta_{P,i}$ is averaged over the N_T transmitted packets. This leads to the following definition of link efficiency η_L:

$$\eta_L = \frac{1}{N_T} \sum_{i=1}^{N_T} \frac{PL_i}{t_{p,i} \cdot B \cdot \log_2 \left(1 + \frac{P_{T,i} \cdot a}{(B \cdot N_0 + P_{Int}) \cdot I}\right)} \tag{5.2}$$

The link efficiency takes into account the resources spent to achieve the performance as well as the complexity of the transmission task. It is topology independent and can be applied to different use cases. Thus, it allows a fair comparison of systems and improvement methods.

5.1.4 Cross-Layer Adaptation

Besides the adaptation of the physical transmission parameters as discussed above, optimization also has to cover aspects of the MAC layer. Typical networks transmit packets containing payload data and overhead for medium access, synchronization, and addressing. The complete bit sequence is then mapped to a number of OFDM symbols, giving rise to another loss due to padding bits that are required to generate complete symbols. The size of the packet payload is crucial for achieving a high efficiency. Figure 5.8 shows the maximum throughput of one link for different gross data rates as a function of packet size. IEEE 802.11 allows for the insertion of a request to send / clear to send sequence to reduce the loss of capacity due to collisions with hidden stations. Both options, with and without RTS/CTS sequence, are plotted. It is obvious that under these assumptions the maximum packet length yields the highest MAC throughput of 19.8 Mbit/s for the 24 Mbit/s mode without RTS/CTS, for example.

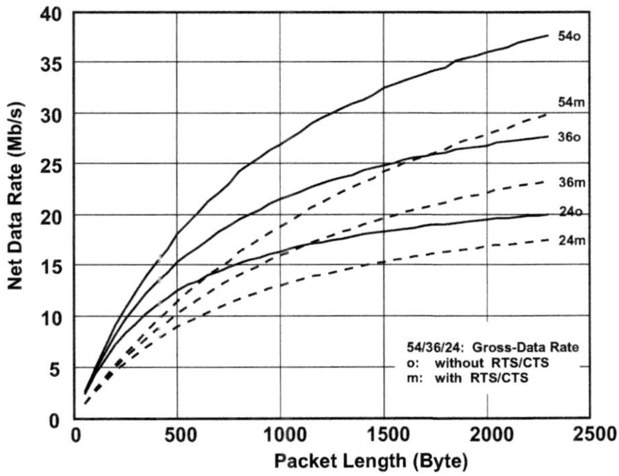

Figure 5.8: MAC throughput as a function of packet length

Up to now, the throughput has been calculated under the assumption that no collisions occur. A practical network comprising several transmitters will be degraded due to collisions. The amount of collisions largely depends on the number of nodes and the kind of data traffic. If several nodes wait for a transmission opportunity, the random process decides which node will be the first to transfer. If more than one node starts with the same backoff number a collision occurs. It turns out that the MAC algorithm as defined in the standard is not well suited for our scenario. A good idea for a modified approach is to specify deterministic parameters for media streams which require high grades of service. Data transfers for PCs which have low requirements with respect to delay and guaranteed data rate are assigned low priorities. The differentiation at the MAC layer is achieved by suitable selection of a fixed and a variable amount of the contention window. It turns out that overall efficiency can be increased by avoiding collisions by such an approach. The achievable improvement depends on the scenario (number of active nodes, data rates). Simulations have shown that up to 20% increased net data rates can be achieved [9].

5.1.5 Multiple Link Network - Overall Adaptation

Some general aspects of link adaptation have been discussed so far, focusing on the system scenario of media transmission in a dense environment. It turns out that optimization is a complex task. Many parameters, which influence each other, have to be selected. A general link efficiency measure has been derived which allows a fair comparison of all options. Link adaptation may start on a per link basis. Even in this simple approach, several parameters exhibit significant interdependence. Defining optimum parameters in a complex environment is even more demanding. Furthermore limitations due to standards and regulation have to be considered. Different strategies are possible. In the home media network described above, optimization may start by the selection of the best channel. In a given setup, this can be achieved by selection of the best combinations of available antenna constellation and transmission frequency. In the next step, the usage of this channel can be optimized by careful selection of modulation and coding scheme. In our practical example, characterized by typical amplitude transmission functions as described above, it turns out that selecting a common scheme for all subcarriers is close to an optimum solution. Improvements by adaptation of parameters per subcarrier would require additional communication overhead. A practical problem might be the fact that such techniques are not allowed in existing or upcoming standards. Further room for significant improvements is given by selecting suitable packet sizes and MAC procedures.

Bibliography

[1] IEEE 802.11-2007, *Wireless LAN Medium Access Control (MAC) and Physical Layer (PHY) Specifications*, Jun. 2007.

[2] K. Jostschulte, R. Kays, W. Endemann, "Enhancement of Wireless LAN for Multimedia Home Networking," *IEEE Trans. on Consumer Electronics*, Vol. 51, No. 1, pp. 80-86, Feb. 2005.

[3] M. Lacage et al.,"IEEE 802.11 Rate Adaptation: A Practical Approach," *Proc. ACM International Conference on Modeling, Analysis and Simulation of Wireless and Mobile Systems (MSWiM)*, Oct. 2004.

[4] J. Pavon et al., "Link Adaptation Strategy for IEEE 802.11 WLAN via Received Signal Strength Measurement," *Proc. IEEE International Conference on Communications (ICC)*, May 2003.

[5] I. Haratchev et al., "Automatic IEEE 802.11 Rate Control for Streaming Applications," *Wireless Communications and Mobile Computing*, Vol. 5, No. 4, pp. 421-437, Jun. 2005.

[6] H. Koetz, R. Kays, "Efficient Dynamic Data Rate Adaptation for Wireless Home Media Networks," *Proc. International OFDM Workshop*, Aug. 2008.

[7] J. Heiskala, J. Terry, *OFDM Wireless LANs: A Theoretical and Practical Guide*, SAMS Publishing, Sep. 2002.

[8] K. Jostschulte, R. Kays, W. Endemann, "Optimisation of IEEE 802.11 Based WLAN for Home Media Networks," *Proc. International OFDM Workshop*, Aug./Sept. 2005.

[9] R. Kays, K. Jostschulte, W. Endemann, "Wireless Ad-Hoc Networks with High Node Density for Home AV Transmission," *IEEE Trans. on Consumer Electronics*, Vol. 50, No. 2, pp. 463-471, May 2004.

5.2 System Concept for a MIMO-OFDM-Based Self-Organizing Data Transmission Network

C. Fellenberg, R. Grünheid, H. Rohling, Hamburg University of Technology, Germany

5.2.1 Introduction

Future wireless communication networks will provide numerous different services with user-specific Quality-of-Service (QoS) demands. For this reason, flexibility as well as adaptivity have to be considered in the system concept design, the data transmission technique and the medium access scheme, respectively. An efficient resource allocation procedure in the multiple access scheme should be realized to fulfill the individual user requirements and different data rates. The Orthogonal Frequency-Division Multiplexing (OFDM) transmission technique offers this required flexibility and adaptivity even in multipath propagation situations and in frequency-selective radio channels. In the OFDM case, the total bandwidth is divided into many subchannels which are simultaneously modulated and transmitted in a superimposed way. This allows for exploiting multi-user diversity (MUD) by the resource allocation procedure.

Furthermore, in cellular networks, the OFDM transmission technique has the additional advantage that all base stations and mobile terminals can be synchronized in time and carrier frequency. In doing so, a so-called Single Frequency Network (SFN) is considered for the proposed system concept. An SFN offers high flexibility in resource allocation as the resources can be provided to those base stations which have the highest traffic load. This concept is very promising, particularly for unequal user distributions.

The OFDM transmission technique is very suitable for broadband radio channels due to its ability of equalization with low computation complexity. A single subcarrier is affected by a narrowband channel and allows for multiple antenna techniques (MIMO). Therefore, in this chapter, a MIMO-OFDM based transmission technique is considered to achieve very high flexibility in digital transmission technique and multiple access scheme as well as a high system performance. Especially, beamforming techniques are treated, allowing for multiple users to be served on the same time frequency resource. Thus, the resources to be allocated are separated in space, time, and frequency. Moreover, depending on the channel knowledge at the transmitter, different antenna diagrams and beam patterns are applicable.

The resource allocation scheme is integrated into the scheduler. Based on the optimization objectives (fairness, capacity, ...), the scheduling procedure is organized in a cross-layer approach by taking the QoS parameters of the data link control (DLC) layer as well as the radio channel knowledge of the physical layer into account. The goal of this chapter is to design and develop a system concept for future wireless cellular communication networks, based on the MIMO-OFDM transmission technique and related multiple access schemes. The resource allocation is performed in a self-organized procedure.

The chapter is organized as follows. First, the different beamforming concepts are introduced. Afterwards, the system concept is described, subdivided into allocation of new resources and reallocation of already assigned resources. The first point depicts the main idea of the proposed system concept, while the reallocation of resources explains how the system performance can be further improved.

5.2.2 Beamforming Concepts

Beamforming is an efficient technique which uses appropriate weighting of multiple antenna elements to concentrate the energy in a desired direction. The determination of the weighting is dependent on the channel knowledge. In the following two different beamforming techniques are described, which are summarized in Table 5.1.

Table 5.1: Beamforming Techniques

channel knowledge	beamforming technique
no	fixed beams
full	zeroforcing beams

The base stations are equipped with Uniform Linear antenna Arrays (ULAs), which consist of several antenna elements. In order to ensure roughly the same performance in each azimuth angle direction, the service area of a base station is subdivided into sectors of 120° width.

As mentioned before, multiple antenna techniques allow for exploiting space division multiple access (SDMA) gains by serving multiple users simultaneously on the same frequency and time resource.

Fixed Beams

Fixed beams are uniquely precalculated without considering the current position of the users inside the cell. The weighting coefficients are generated using the Dolph-Chebychev method to achieve a constant sidelobe magnitude. To cover the sector, the number of beams is determined so that a user between two beams has a degradation of 3dB compared to users in the center of a beam. To perform resource allocation, the measured Channel Quality Indicator (CQI) by the user is signaled back to the base station. In general, the antenna diagrams for multiple users are not orthogonal and crosstalk can occur.

Zeroforcing Beams

In contrast to the fixed beam concept, zeroforcing beams are calculated user-specifically. Therefore, full channel knowledge is required at the base station. For zeroforcing beams, ideal channel knowledge is assumed to be known at the transmitter. Consequently, better performance is expected, compared to fixed beams. The calculation of the weighting coefficients w_k is given in Eq. (5.3),

$$W = H^*(HH^*)^{-1} \tag{5.3}$$

where the matrix H comprises the channel vectors h_k of the considered users and matrix W consists of the corresponding weighting vectors w_k. By means of zeroforcing beams, intra-sector interference is completely avoided for the considered users.

For zeroforcing beams the question arises which users should be served simultaneously on the same frequency resource. Here, heuristic grouping approaches are used, exploiting, e.g., the orthogonality of the channel vectors [4]. In [5] new algorithms were proposed to exploit SDMA as well as MUD gains.

5.2.3 System Concept

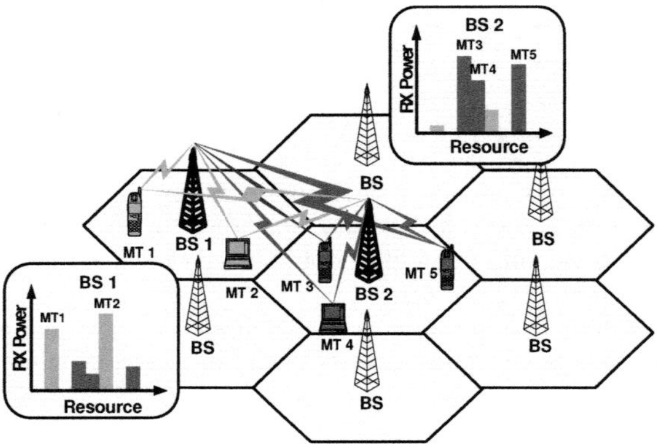

Figure 5.9: Cellular Network

This section covers the design of the proposed system concept for a MIMO-OFDM transmission system in the described context. The scheduling of resources is based on the CQI and the QoS parameters of the users. In addition, the cellular network is assumed to be totally synchronized in time and frequency to fulfill the requirement of single frequency networks supporting point-to-point connections (Fig. 5.9). In [3] it is shown that the synchronization in an OFDM based cellular network is feasible. Due to the carrier synchronization, the interference of neighboring cells is only observed as co-channel interference. The base stations are equipped with multiple antennas, while the mobile terminals make use of a single antenna. A time division duplex (TDD) scheme is applied to separate the down- and uplink for data transmission.

The philosophy of the system concept is to incorporate high flexibility into the system while keeping the interference on a low level. The self-organizing approach is based on measurable and predictable interference situations in the network. In this way, the mobile terminals as well as the base stations are able to measure the interference situation and allocate appropriate resources. While the interference is different for the base station and the mobile terminal, the useful power is the

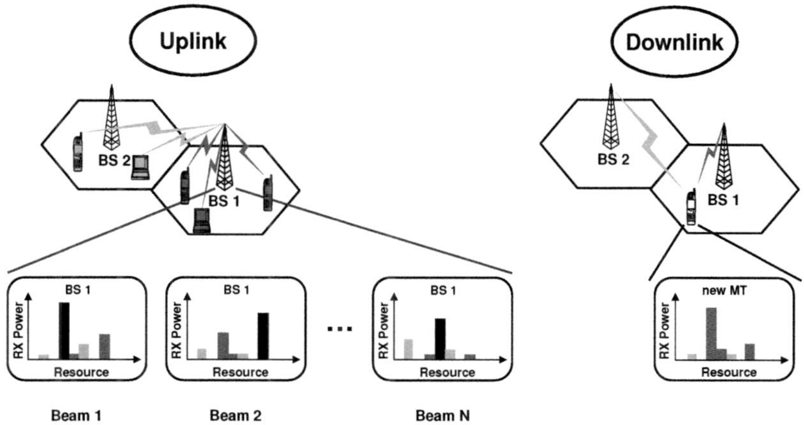

Figure 5.10: Interference Measurement

same due to the TDD scheme. By allocating the resources for long durations, the interference situation becomes predictable.

Compared with the classical network design, there are two main and important additional system functions. The carrier and time synchronization is important for the SFN functionality and the self-organized resource allocation gives high flexibility in the network.

Allocation of New Resources

New resources are requested by new mobile terminals coming into the cellular network. Alternatively, all users who need a higher data rate ask for additional resources. The resource allocation procedure for users is based on the interference power measured by the base station and the mobile terminal. The interference power at the base station is measured continuously in frequency and space on resources during the uplink phase. In Fig. 5.10 the interference power is marked in grey scales, while the occupied resources are marked in black. The interference level measured at the mobile terminal during the downlink phase are signaled in a free time slot to the base station for further consideration. If no collisions occur within this time slot, the useful power is obtained at the base station as well. With this information, the mobile terminal is assigned by the base station to suitable frequency-space resources with low interference. The underlying assumption is that the interference is predictable. However, due to the fluctuation of the interference situation within the network, caused by allocation of new resources and mobility of the users, the measurement of useful and interference power is updated after each frame in order to perform link adaptation.

Results

For quantitative analysis, a cellular network consisting of 19 cells is considered. The wrap-around technique is used to ensure the same number of neighboring cells for each cell. As mentioned before, a cell is subdivided into three sectors, each of which served by 8 antennas at the base station. In the following, the downlink from the base station to the mobile terminal is evaluated, in which the number of receive antennas at the mobile terminal is set to one. Moreover, the performance for one Time-Frequency block (TF block) is assumed to be constant. Further parameters for the simulation are listed in Table 5.2. Additionally, the QoS parameters in Table 5.3 are assumed for the downlink.

Table 5.2: Simulation

Parameters	Value / Assumption
Cell radius	750 m
User distribution	uniform
Carrier frequency	5 GHz
Bandwidth	20 MHz
Duplex scheme	TDD
Channel model	WINNER C2 [6]
MAC-frame duration	2 ms
OFDM-Symbol duration	60 μs
Number of subcarriers	336
Time span of one TF block	24 OFDM symbols
Frequency span of one TF block	8 subcarriers
Number of bits per packet	832
Bandwidth efficiency of PHY modes	1/3, 2/3, 4/3, 8/3, 12/3, 16/3 [bits/s/Hz]

Table 5.3: QoS Parameters

max. delay	100 ms
max. PER	0.01
data rate	1 packet per MAC-frame (416 kbit/s)

Figure 5.11 shows the performance for the different beamforming techniques and the reference scenario. The percentage of unsatisfied users is plotted versus the average number of users in a cell. A user is called unsatisfied if his request for a link connection is denied or his QoS parameters are not fulfilled. The single antenna case (SISO) serves as a reference. Here, 6 users can be served in average for 5% unsatisfied users in the system. If fixed beams are applied, the number of users which can be served is considerably increased to 34. This is due to the following reasons:

- increased receive power at the mobile terminal due to steering of the transmit power

- SDMA gains
- better interference situation within the system

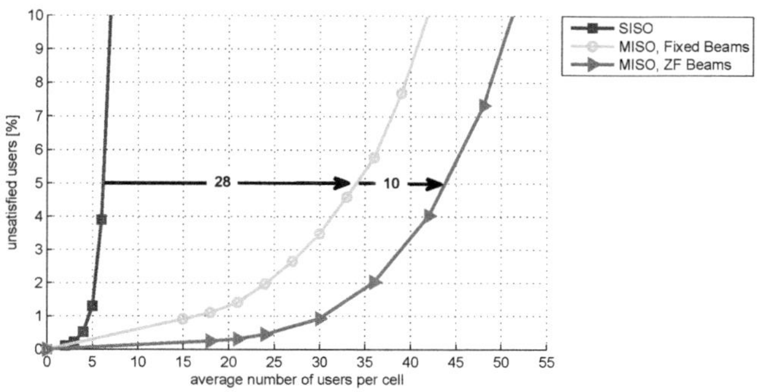

Figure 5.11: Allocation of New Resources

The last point is visualized in Fig. 5.12, where a cellular system and a single subcarrier are considered. The receive power is plotted for all base stations from one mobile terminal localized at the green cross. The dark blue color represents the noise power while the dark red color represents high signal power. Obviously, for the MISO system lots of more resources are still usable for every base station compared to the SISO system.

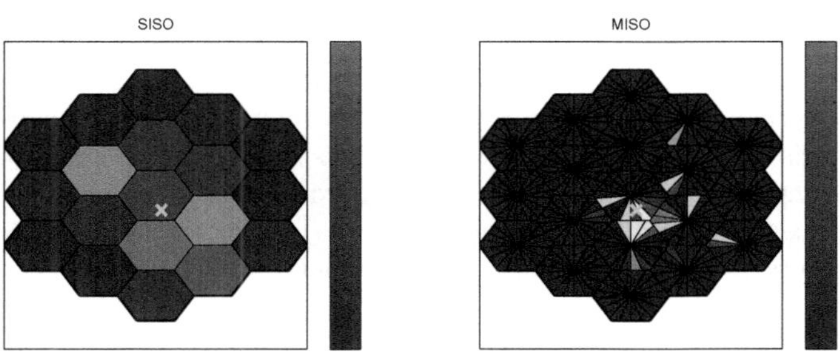

Figure 5.12: Interference at BS

Expectedly, the zeroforcing beams achieve the highest performance, yielding an increase of 10 users compared to fixed beams. Beside the advantages of the MISO system pointed out before, the ideal channel knowledge is exploited so that intra-sector interference is fully avoided. Additionally, the steering of transmit power towards the users is enhanced.

Reallocation

Due to the fluctuation of the interference power level within the network, the CQI values are only valid for a short duration. Therefore, the measurement of the useful and interference power is updated after each frame. In doing so, each user measures the power on the resources occupied by the base station and signals it back. In other words, a subset of the entire resources is available for the scheduler at the base station to perform reallocation, while the interference situation for users in other cells remains the same. By reallocation, the scheduler is capable of balancing the QoS demands between users with high CQI and those with low CQI.

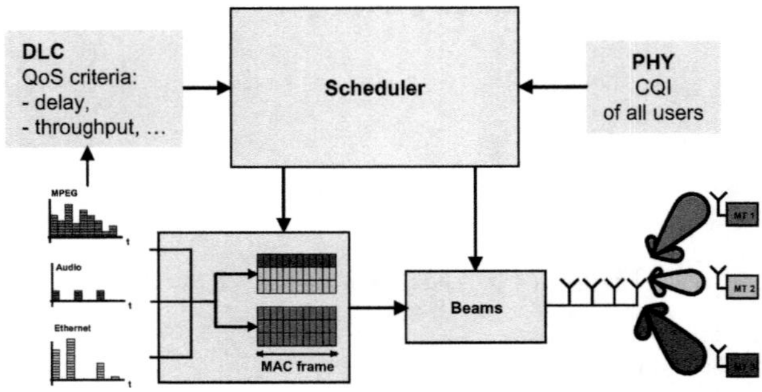

Figure 5.13: Crosslayer approach, incorporating scheduling and beamforming

The scheduler is responsible for the resource allocation. Optimization criteria for the scheduler could be fairness in the system or maximizing the number of satisfied users. A user is called satisfied if all of his QoS demands are fulfilled. To do so, the scheduler can make use of the CQI of the physical layer as well as the QoS parameters of the DLC (see Fig. 5.13), a concept called Cross-Layer design [1,2,7].

As mentioned before, the measurement of useful and interference signal power is updated periodically on resources which are occupied by the base station. On the basis of this information, well-known schedulers can be utilized. A promising technique for the scheduler is the utility-based approach, which transforms the challenge of the scheduler into an optimization task [8]. This technique offers a compromise between fairness and maximizing the throughput.

Utility-Based Scheduling

Utility-based scheduling aims at maximizing the sum utility U_S,

$$U_S = \max_{D_i, i \in M} \sum_{i \in M} U_i(a_i) \tag{5.4}$$

where U_i is the utility function of user i and M is the number of users. The argument a_i is a function of all assigned resources D_i of user i, which is a subset out of the available resources D. The optimization task is subject to the following conditions:

$$\bigcup_{i=1}^{M} D_i \subseteq D$$

$$D_i \bigcap D_j = \emptyset, \, i \neq j$$

An appropriate choice of the utility function for each user is dependent on the optimization task (e.g., capacity, fairness, QoS demands of the users). In the scope of this project, different utility functions have been considered.

Data Rate-Based Utility Function

For the data rate based utility function, the argument a_i in Eq. (5.4) represents the overall data rate r_i for one MAC-frame, assigned to user i. The overall data rate is given by the sum of data rates $c_{i,d}$ on single resources d.

$$a_i = r_i = \sum_{d \in D_i} c_{i,d}$$

So far, solving the optimization task yields the resource allocation for one MAC-frame. However, fairness and QoS demands are related to time. Therefore, a low-pass filter is used for the sum of data rates in order to exploit the time dimension.

$$\bar{r}_i(n) = \alpha \bar{r}_i(n-1) + (1-\alpha) r_i(n)$$

As a matter of fact, the argument a_i in Eq. 5.4 is replaced by $\bar{r}_i(n)$. Fairness among the users is achieved by a logarithmic slope of the utility function. In this way, users with a relatively low data rate are preferred for the resource allocation. In literature this kind of scheduling is well-known as proportional fair scheduling (PFS) [9]. Certainly, PFS evaluates pure channel state information. In order to maximize the number of satisfied users, it is self-evident that information from the DLC has to be incorporated into the progression of the utility function. This is, e.g., achieved with modified largest weighted delay first (M-LWDF) scheduling [10] by weighting the logarithmic slope of the utility function with the waiting time of the first packet in the queue T_i. Thus, also users with a large waiting time of the first packet in the queue are preferred for resource allocation. The resulting utility function is given in Eq. (5.5).

$$U_i(\bar{r}_i(n)) = \begin{cases} T_i log(\bar{r}_i(n)) & , \bar{r}_i(n) \leq r_{max,i}(n) \\ T_i log(r_{max,i}(n)) & , \bar{r}_i(n) > r_{max,i}(n) \end{cases} \tag{5.5}$$

The parameter $r_{max,i}(n)$ ensures that the assigned data rate for user i is not further increased if there are no more packets in his queue. Fig. 5.14 shows the characteristics of the utility function for different users and waiting times of the first packets. The optimization task is expressed as follows:

$$U_S = \max_{D_i, i \in M} \sum_{i \in M} U_i(\bar{r}_i(n))$$

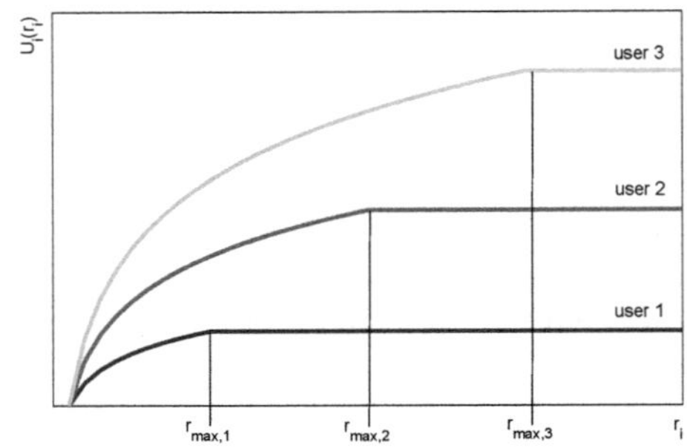

Figure 5.14: Data Rate-based Utility Function

Delay-Based Utility Function

The idea of the delay based utility function is to minimize the waiting time of packets in the queue $W_i(n)$ [11], which is

$$W_i(n) = \frac{Q_i(n)}{\lambda_i},$$

where $Q_i(n)$ is the number of packets in the queue at time n and λ_i is the arrival rate of new packets in the queue. The number of packets in the queue can also be expressed as follows

$$Q_i(n) = Q_i(n-1) + \lambda_i - p_i(n),$$

where $p_i(n)$ is the number of packets transmitted within the current MAC-frame. For the same reason mentioned before, low-pass filtering is used for the actual number of packets in the queue.

$$\bar{Q}_i(n) = \alpha \bar{Q}_i(n-1) + (1-\alpha) Q_i(n)$$

Therefore, the average waiting time $\bar{W}_i(n)$ becomes

$$\bar{W}_i(n) = \frac{\bar{Q}_i(n)}{\lambda_i}$$

and the optimization task is expressed as follows.

$$U_S = \max_{D_i, i \in M} \sum_{i \in M} U_i(\bar{W}_i(n))$$

In order to prefer users with large waiting time, the contribution to the sum utility for large waiting times have to be higher than for low ones. This behavior is achieved by the following utility function

$$U_i(\bar{W}_i) = \begin{cases} -\frac{1}{\gamma_i} \bar{W}_i^{\gamma_i} & , \bar{W}_i \leq W_{max,i} \\ a + b\bar{W}_i & , \bar{W}_i > W_{max,i} \end{cases},$$

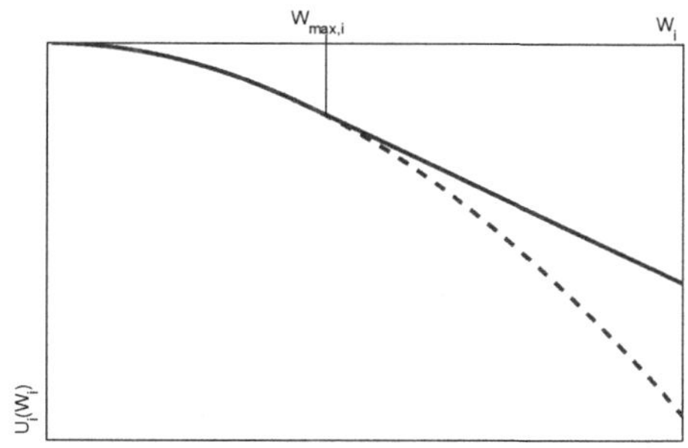

Figure 5.15: Delay-based Utility Function

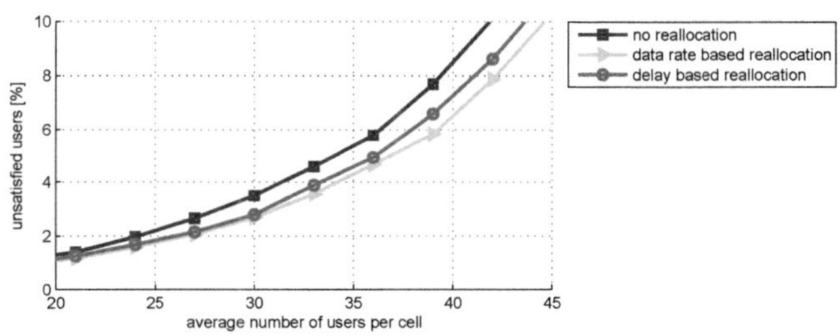

Figure 5.16: Reallocation, Fixed Beams

where $\gamma_i \geq 1$. The values of a and b are $a = W_{max,i}^{\gamma_i}(1 - \frac{1}{\gamma_i})$, $b = -W_{max,i}^{\gamma_i - 1}$ so that the utility function is differentiable. $W_{max,i}$ is determined according to the QoS parameters of the user and is applied to reduce the influence of users with very poor channel gains. Figure 5.15 depicts the delay based utility function. The dashed line represent the progression of the utility function $-\frac{1}{\gamma_i}\bar{W}_i^{\gamma_i}$ for values of $\bar{W}_i > W_{max,i}$.

Results

The same setup and parameters of Table 5.2 and Table 5.3 are used for the results. For the evaluation of the reallocation, fixed beams are considered. For the uppermost curve in Fig. 5.16, no reallocation is performed. This curve serves as the reference for the utility-based scheduling.

Both utility functions achieve a performance gain compared to the reference curve. On average, 2 respectively 3 more users can be served within one cell. Obviously, the data rate-based utility function yields better performance compared to the delay-based utility function. This observation is confirmed if the unsatisfied users are considered in more detail. In Fig. 5.17 the number of users is depicted whose the QoS parameters are not fulfilled. Again, the percentage of those users is much lower if reallocation is performed, while the performance of the data rate based utility function is slightly better than that of the delay based utility function.

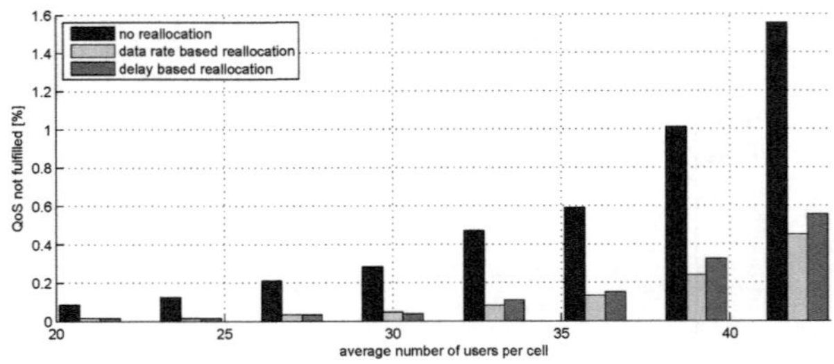

Figure 5.17: QoS not fulfilled, Fixed Beams

5.2.4 Summary

A new system concept has been proposed for MIMO-OFDM-based self-organizing networks. The key idea of the system concept is that the allocation of new resources is performed in such a way that the interference in the system is predictable. Therefore, interference can be kept on a low level by measurement at the base stations as well as the mobile terminals. The results show that the amount of users who can be served within a cell is significantly increased if beamforming techniques are applied, compared to single antenna systems. The reasons are the increased receive power at the mobile terminal due to the steering of the transmit power, SDMA gains, and a better interference situation within the system. The system performance can be further increased if reallocation of resources is applied. It was shown by simulation results that not only the number of satisfied users in the system can be increased, but also the number of unsatisfied users due to violation of the QoS demands is reduced.

Bibliography

[1] R. Grünheid, B. Chen and H. Rohling, "Joint Layer Design for an Adaptive OFDM Transmission System," *IEEE International Conference on Communications*, vol. 1, pp. 542-546, 2005

[2] H. Rohling and R. Grünheid, "Cross Layer Considerations for an Adaptive OFDM-based Wireless Communication System", *Wireless Personal Communications*, pp. 43 - 57, Springer, 2005

[3] D. Galda, D., N. Meier, H. Rohling and M. Weckerle, "System Concept for a Self-Organized Cellular Single Frequency OFDM Network," *8th International OFDM-Workshop*, Hamburg, Germany, 2003

[4] T. Yoo and A. Goldsmith, "Optimality of Zero-Forcing Beamforming with Multiuser Diversity," *IEEE International Conference on Communications*, vol. 1, pp. 542-546, 2005

[5] C. Fellenberg, A. Tassoudji, R. Grünheid and H. Rohling, "QoS aware Scheduling in Combination with Zero-Forcing Beamforming Techniques in MIMO-OFDM based Transmission Systems," *14th International OFDM-Workshop*, Hamburg, Germany, 2009

[6] WINNER "System Concept Descritpion", *Deliverable D7.6*, IST-2003-507581 WINNER, 2005

[7] R. Grünheid, B. Chen and H. Rohling, "Joint Layer Design for an Adaptive OFDM Transmission System," *IEEE International Conference on Communications*, vol. 1, pp. 542-546, 2005

[8] G. Song and Y. Li, "Utility-Based Resource Allocation and Scheduling in OFDM-Based Wireless Broadband Networks," *IEEE Communications Magazine*, vol. 43, no. 12, pp. 127-134, 2005

[9] H. Kim and Y. Han, "A Proportional Fair Scheduling for Multicarrier Transmission Systems," *IEEE Communications Letters*, vol. 9, no. 3, 2005

[10] M. Andrews, K. Kumaran, K. Ramanan, A. Stolyar, P. Whiting and R. Vijayakumar, "Providing Quality of Service over a Shared Wireless Link," *IEEE Communications Magazine*, pp. 150-154, 2001

[11] G. Song, Y. Li, L. J. Cimini, Jr., H. Zheng,"Joint Channel-Aware and Queue-Aware Data Scheduling in Multiple Shared Wireless Channels," *IEEE Wireless Communications and Networking Conference*, vol. 3, pp. 1939-1944, 2004

5.3 Pricing Algorithms for Power Control, Beamformer Design, and Interference Alignment in Interference Limited Networks

D. Schmidt, W. Utschick, Technische Universität München, Germany

5.3.1 Introduction

In this chapter, we examine the problem of finding good resource allocations in networks of many interfering transmitter-receiver pairs, where the receivers are not able to decode the signals from any but the desired transmitter and thus must treat interference as noise. While one approach in such systems is to allocate resources orthogonally (e. g., by assigning separate time slots or frequency bands to the transmitters), it will often be advantageous to allow users to share the resources to some extent. The transmit strategies must then, however, be optimized in order to reduce the negative effects of interference as much as possible.

The resulting optimization problems turn out to have undesirable properties: multiple (locally optimal) solutions to the necessary optimality conditions may exist, and these solutions in general cannot be explicitly computed. Therefore, it is necessary to rely on iterative algorithms to determine the transmit strategies. Also, due to the decentralized nature of the underlying system model, special attention must be payed to the distributed implementability of the algorithms.

Multiple antennas at the transmitters or receivers allow for spatial interference avoidance, where, e. g., a transmitter can focus its beam in the direction of its intended receiver and away from the unintended receivers. In MIMO systems, furthermore, the issue of *interference alignment* arises, leading to a whole new class of high-SNR optimal strategies.

5.3.2 System Model

We examine a system with K transmitter-receiver pairs (synonymously called users), where each receiver is only interested in the signal from its associated transmitter and all interference is treated as additional noise. Each receiver (transmitter) has M (N) antennas, respectively. The received signal vector of user k is

$$\boldsymbol{y}_k = \underbrace{\boldsymbol{H}_{kk}\boldsymbol{x}_k}_{\text{desired signal}} + \underbrace{\sum_{j\neq k} \boldsymbol{H}_{kj}\boldsymbol{x}_j}_{\text{interference}} + \underbrace{\boldsymbol{n}_k}_{\text{noise}}, \tag{5.6}$$

where $\boldsymbol{H}_{kj} \in \mathbb{C}^{M\times N}$ is the matrix of channel coefficients between transmitter j and receiver k, $\boldsymbol{x}_j \in \mathbb{C}^N$ is the vector of symbols transmitted by transmitter j, and $\boldsymbol{n}_k \in \mathbb{C}^M$ is the noise experienced at the M antennas of receiver k.

We assume the noise vector $\boldsymbol{n}_k$ to be uncorrelated with variance σ^2

$$\mathrm{E}\left[\boldsymbol{n}_k\boldsymbol{n}_k^{\mathsf{H}}\right] = \sigma^2\mathbf{I}, \tag{5.7}$$

and impose a unit power constraint on the transmit vector at each transmitter:

$$\mathrm{E}\left[\|\boldsymbol{x}_k\|_2^2\right] \leq 1. \tag{5.8}$$

The receive and transmit processing of user k is performed by the filter vectors $\boldsymbol{g}_k \in \mathbb{C}^M$ and $\boldsymbol{v}_k \in \mathbb{C}^N$, respectively. Note that for the sake of notational simplicity we limit ourselves to one data stream per user. The transmit vector $\boldsymbol{x}_k$ is obtained by multiplying the data symbol s_k with the beamformer $\boldsymbol{v}_k$:

$$\boldsymbol{x}_k = \boldsymbol{v}_k s_k. \tag{5.9}$$

For unit variance data symbols, this implies that the transmit power constraint (5.8) can be written as

$$\|\boldsymbol{v}_k\|_2^2 \leq 1. \tag{5.10}$$

After processing the received signal $\boldsymbol{y}_k$ with the receive filter vector $\boldsymbol{g}_k$, the estimate $\hat{s}_k$ of the data symbol s_k is

$$\hat{s}_k = \boldsymbol{g}_k^{\mathsf{H}} \boldsymbol{y}_k = \underbrace{\boldsymbol{g}_k^{\mathsf{H}} \boldsymbol{H}_{kk} \boldsymbol{v}_k s_k}_{\text{desired signal}} + \underbrace{\sum_{j \neq k} \boldsymbol{g}_k^{\mathsf{H}} \boldsymbol{H}_{kj} \boldsymbol{v}_j s_j}_{\text{interference}} + \underbrace{\boldsymbol{g}_k^{\mathsf{H}} \boldsymbol{n}_k}_{\text{noise}}. \tag{5.11}$$

The favorability of a certain situation to user k can be measured by its signal to interference-plus-noise ratio (SINR)

$$\gamma_k = \frac{|\boldsymbol{g}_k^{\mathsf{H}} \boldsymbol{H}_{kk} \boldsymbol{v}_k|^2}{\sum_{j \neq k} |\boldsymbol{g}_k^{\mathsf{H}} \boldsymbol{H}_{kj} \boldsymbol{v}_j|^2 + \sigma^2 \|\boldsymbol{g}_k\|_2^2} = \frac{S_k}{I_k + N_k}, \tag{5.12}$$

where S_k, I_k, and N_k are the desired signal power, interference power, and noise power after the receive filter, respectively. The goal is to maximize the system-wide sum utility, where each user's utility $u_k(\gamma_k)$ is an increasing function of its SINR:

$$\max_{\substack{\boldsymbol{v}_1, \ldots, \boldsymbol{v}_K \\ \boldsymbol{g}_1, \ldots, \boldsymbol{g}_K}} \sum_{k=1}^{K} u_k(\gamma_k) \qquad \text{s.t.:} \quad \|\boldsymbol{v}_k\|_2^2 \leq 1 \quad \forall k \in \{1, \ldots, K\}. \tag{5.13}$$

For many relevant utility functions (such as the 'rate' utility $u_k(\gamma_k) = \log(1 + \gamma_k)$, which is of special relevance since it can be interpreted as the achievable rate with Gaussian signaling), this problem is non-convex and requires a global optimization algorithm.

While suitable global optimization techniques exist (cf. [8]), they quickly become prohibitively complex with increasing system dimensions. In this work, we instead focus on finding good local optima by means of iterative algorithms.

5.3.3 Distributed Interference Pricing

Power Control in SISO Systems

We first examine the single-input single-output (SISO) case, where each transmitting or receiving terminal has a single antenna, i. e., $N = M = 1$. In this case, transmit filters, channel coefficients, and receivers are all scalars, and we write them as g_k, h_{kj}, and v_k, respectively. Since it is clear from (5.12) that the value of g_k is irrelevant for the SINR, we assume $g_k = 1\ \forall k$, w. l. o. g. Likewise, the SINR only depends on the squared magnitude of v_k, so we use the abbreviation $p_k = |v_k|^2$.

We define the *interference price* π_k of user k as the marginal decrease of the utility $u_k(\gamma_k)$ with the increase in received interference power I_k, evaluated at the current operating point:

$$\pi_k = -\frac{\partial u_k(\gamma_k)}{\partial I_k} = \frac{\partial u_k(\gamma_k)}{\partial \gamma_k} \cdot \frac{S_k}{(I_k + N_k)^2}. \tag{5.14}$$

In our iterative algorithm, each transmitter updates its transmit power p_k taking into account the interference prices of all other users:

$$p_k^{\text{new}} = \underset{p_k}{\operatorname{argmax}}\ u_k(\gamma_k) - \sum_{j \neq k} \pi_j |h_{jk}|^2 p_k \qquad \text{s. t.:} \quad 0 \leq p_k \leq 1. \tag{5.15}$$

The rationale behind this update rule is that instead of selfishly maximizing its own utility $u_k(\gamma_k)$, which would lead to every user always transmitting with full power $p_k = 1$, the user should maximize the own utility minus the 'cost' of causing interference to others. We can also interpret (5.15) as maximizing an approximated sum utility (cf. (5.13)), where all utility terms of the other users are linearized around the current operating point:

$$u_j(\gamma_j) \approx \frac{\partial u_j(\gamma_j)}{\partial I_j} \cdot \left.\frac{\partial I_j}{\partial p_k}\right|_{\text{Op. Point}} \cdot p_k + c = -\pi_j |h_{jk}|^2 p_k + c. \tag{5.16}$$

The constant c does not depend on p_k and is therefore irrelevant to the optimization problem (5.15).

The *distributed interference pricing algorithm* works as follows:

1. All transmit powers p_k are initialized (e. g., with unit power).
2. Each user k computes the interference price π_k according to (5.14) and announces it to all other users.
3. The users update their transmit strategies (powers, in the SISO case) according to (5.15).
4. Repeat from 2. until convergence.

It can be easily shown that, once this algorithm has converged, the Karush-Kuhn-Tucker (KKT) conditions of the original sum utility problem (5.13) are fulfilled, i. e., a local optimum has been found. Further convergence properties depend on the utility functions; for a detailed analysis, cf. [3].

Beamformer Design in MISO Systems

Next, we examine the multiple-input single-output (MISO) scenario, where all transmitters have $N > 1$ antennas while the receivers have $M = 1$ antennas each. The

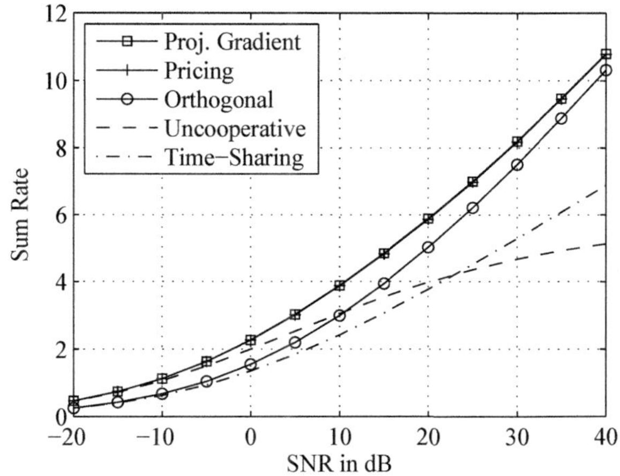

Figure 5.18: Average performance of different strategies in a two-user MISO interference channel with two antennas at each transmitter, in terms of sum rate utility

channel coefficients now form row vectors $\boldsymbol{h}_{kj}^{\mathsf{H}} \in \mathbb{C}^{1\times N}$; the transmit filters $\boldsymbol{v}_k$ are column vectors, whereas the scalar receivers g_k again are not relevant for the SINR.

With the same arguments as in the SISO case we obtain the following transmitter update:

$$\boldsymbol{v}_k^{\text{new}} = \underset{\boldsymbol{v}_k}{\operatorname{argmax}}\ u_k(\gamma_k) - \sum_{j\neq k} \pi_j |\boldsymbol{h}_{jk}^{\mathsf{H}} \boldsymbol{v}_k|^2 \qquad \text{s.t.:} \quad \|\boldsymbol{v}_k\|_2^2 \leq 1. \tag{5.17}$$

It is also possible to additionally linearize the own utility term $u_k(\gamma_k)$ around the current operating point w. r. t. $\boldsymbol{v}_k$ (or $|\boldsymbol{h}_{kk}^{\mathsf{H}}\boldsymbol{v}_k|^2$, which has the same effect), in order to simplify the update procedure. Again, convergence of the the pricing algorithm implies local optimality in the sum utility problem (5.13). For a detailed convergence analysis of the MISO case, cf. [6].

As can be seen for an exemplary scenario in Fig. 5.18, the distributed pricing algorithm performs as well as a generic gradient algorithm and clearly outperforms a number of straightforward non-iterative schemes for determining transmit strategies for a two-user MISO interference channel, regardless of the SNR.

Power Allocation in OFDM Systems

In order to accommodate multi-carrier scenarios, we must adjust our system model: we have $N = M = 1$ antennas at each terminal, with symbols transmitted on $Q > 1$ non-interfering carriers. The channel coefficient between transmitter j and receiver k on carrier q is h_{kj}^q; we will similarly use the superscript to denote the carrier index henceforth. Again, the receivers are irrelevant for the SINR, and user k's transmit strategy on carrier q, the allocated power, is $p_k^q = |v_k^q|^2$. The total transmit power of user k is the sum of powers allocated to its subcarriers, therefore the power constraint is

$$\sum_{q=1}^{Q} p_k^q \leq 1 \quad \forall k. \tag{5.18}$$

With similarly defined SINR γ_k^q, utility $u_k^q(\gamma_k^q)$, and interference price π_k^q for carrier q of user k, the update rule for user k in a multi-carrier system is:

$$\begin{aligned} \{p_k^{1,\text{new}}, \ldots, p_k^{Q,\text{new}}\} = \underset{\{p_k^1,\ldots,p_k^Q\}}{\text{argmax}} \sum_{q=1}^{Q} u_k^q(\gamma_k^q) - \sum_{j\neq k}\sum_{q=1}^{Q} \pi_j^q |h_{jk}^q|^2 p_k^q \\ \text{s. t.:} \quad \sum_{q=1}^{Q} p_k^q \leq 1 \quad \text{and} \quad p_k^q \geq 0 \;\forall q. \end{aligned} \tag{5.19}$$

The algorithm is proposed and analyzed in detail in [3].

5.3.4 MIMO Interference Networks and Interference Alignment

The distributed pricing algorithm can also be extended to multiple-input multiple-output (MIMO) scenarios, where both $N > 1$ and $M > 1$. In contrast to all previously discussed scenarios, the receive filters $\boldsymbol{g}_k$ are now relevant for the SINR. They must be continually updated in order to maximize the received SINR and must also be communicated to the other users alongside the interference prices. For a more detailed description of the algorithm, cf. [7].

In some cases, however, the full potential of the system is not achieved with the pricing algorithm. Let us consider, for example, a scenario with $N = M = 2$ antennas at all terminals, $K = 3$ users, and high SNR, where the goal is to maximize the sum rate utility $\sum_k \log(1+\gamma_k)$. The pricing algorithm in general will converge to a solution in which two users are active and the interference power after their receive filters $\boldsymbol{g}_k$ is close to zero. It is, however, possible to transmit three interference-free streams (one for each user) in this setting by ensuring that at each receiver the interference from both undesired transmitters is collinear and thus separable from the desired signal, i. e., by fulfilling the three conditions

$$\boldsymbol{H}_{12}\boldsymbol{v}_2 \parallel \boldsymbol{H}_{13}\boldsymbol{v}_3, \quad \boldsymbol{H}_{21}\boldsymbol{v}_1 \parallel \boldsymbol{H}_{23}\boldsymbol{v}_3, \quad \text{and} \quad \boldsymbol{H}_{31}\boldsymbol{v}_1 \parallel \boldsymbol{H}_{32}\boldsymbol{v}_2, \tag{5.20}$$

while using all available power at each transmitter, i. e., $\|\boldsymbol{v}_k\|_2^2 = 1 \;\forall k$. Note that for these conditions the number of equations equals the number of variables and it can be shown that they are always feasible. Clearly, for sufficiently high SNR, three separate data streams will lead to a higher sum rate utility than two.

These high-SNR optimal solutions are called *interference aligned* [1]. Due to the properties of the sum rate utility, interference aligned solutions are not easy to find by performing gradient based algorithms on the utility function. Two distributed algorithms specifically aimed at finding aligned solutions have been proposed in [2]: the 'min leakage' and 'max SINR' algorithms, both based on a heuristic of repeatedly exchanging the roles of transmitters and receivers. The former, while its convergence is guaranteed, is clearly suboptimal for moderate and lower SNR, as it attempts to

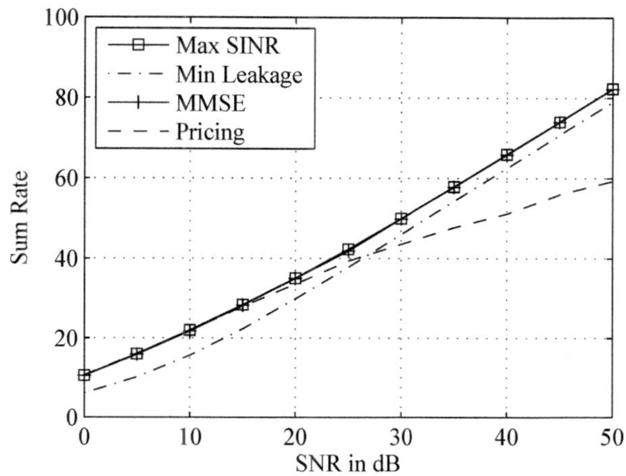

Figure 5.19: Average performance of different distributed algorithms in a five-user MIMO interference channel with three antennas at each terminal, in terms of sum rate utility

completely remove interference regardless of the noise power. The latter is not proven to converge, but appears to generally perform very well in terms of sum rate utility.

We propose an improved algorithm based on alternatingly updating transmitters and receivers to minimize the sum mean square error (MSE), i. e., $\sum_k \mathrm{E}[|s_k - \hat{s}_k|^2]$. The receiver update reads as

$$\boldsymbol{g}_k^{\mathrm{new}} = \left(\sum_j \boldsymbol{H}_{kj} \boldsymbol{v}_j \boldsymbol{v}_j^{\mathsf{H}} \boldsymbol{H}_{kj}^{\mathsf{H}} + \sigma^2 \mathbf{I} \right)^{-1} \boldsymbol{H}_{kk} \boldsymbol{v}_k, \tag{5.21}$$

and the transmitter update is

$$\boldsymbol{v}_k^{\mathrm{new}} = \left(\sum_j \boldsymbol{H}_{jk}^{\mathsf{H}} \boldsymbol{g}_j \boldsymbol{g}_j^{\mathsf{H}} \boldsymbol{H}_{jk} + \lambda_k \mathbf{I} \right)^{-1} \boldsymbol{H}_{kk}^{\mathsf{H}} \boldsymbol{g}_k, \tag{5.22}$$

where $\lambda_k \geq 0$ is chosen in order to fulfill $\|\boldsymbol{v}_k^{\mathrm{new}}\|^2 \leq 1$, e. g., by means of a line search with the Newton algorithm. These update rules turn out to be very similar to the updates of the 'max SINR' algorithm in [2], and also succeed in finding aligned solutions at high SNR. However, as the sum MSE is decreased after each update, convergence in terms of the sum MSE is ensured. Furthermore, the algorithm can be easily extended to minimizing a weighted sum of MSEs, thus assigning different priorities to the users. Finally, by adapting the weights of the weighted sum MSE minimization according to the current operating point, local optima of arbitrary utility functions can be found [5].

In the scenario shown in Fig. 5.19, we observe that the pricing algorithm performs well for moderate and low SNR, but does not achieve the full high-SNR slope (i. e., number of interference-free data streams). The 'min leakage' algorithm [2] does achieve the maximum slope, but is clearly suboptimal to the 'max SINR' algorithm [2] as well as the MMSE approach discussed above, which both achieve a similar sum utility in this case.

5.3.5 Summary

We have presented distributed algorithms for different types of interference networks, in which determining the globally optimal transmit strategy is not feasible. While for SISO, MISO, and OFDM systems the distributed interference pricing algorithm finds good locally optimal configurations, MIMO systems that allow for interference alignment require a different class of algorithm in order to achieve full system potential at high SNR. Here, minimizing the weighted sum MSE is a promising approach in addition to the algorithms in [2].

A comprehensive overview of pricing algorithms for interference networks can be found in [4].

Bibliography

[1] V. R. Cadambe and S. A. Jafar, "Interference alignment and degrees of freedom of the K-user interference channel," 54(8):3425–3441, August 2008.

[2] K. Gomadam, V. R. Cadambe, and S. A. Jafar, "Approaching the capacity of wireless networks through distributed interference alignment," in *Proc. IEEE Global Commun. Conf. (GLOBECOM)*, New Orleans, LA, November 2008.

[3] J. Huang, R. A. Berry, and M. L. Honig, "Distributed interference compensation for wireless networks," 24(5):1074–1084, May 2006.

[4] D. A. Schmidt, C. Shi, R. A. Berry, M. L. Honig, and W. Utschick, "Distributed resource allocation schemes," 26(5):53–63, September 2009.

[5] D. A. Schmidt, C. Shi, R. A. Berry, M. L. Honig, and W. Utschick, "Minimum mean squared error interference alignment," in *Proc. 43rd Asilomar Conf. Signals, Syst., Comput.*, Pacific Grove, CA, November 2009.

[6] C. Shi, R. A. Berry, and M. L. Honig, "Distributed interference pricing with MISO channels," in *Proc. 46th Annu. Allerton Conf. Commun., Control, Computing*, Monticello, IL, September 2008.

[7] C. Shi, D. A. Schmidt, R. A. Berry, M. L. Honig, and W. Utschick, "Distributed interference pricing for the MIMO interference channel," in *Proc. IEEE Int. Conf. Commun.*, Dresden, Germany, June 2009.

[8] H. Tuy, "Monotonic optimization: Problems and solution approaches," *SIAM J. Optimization*, 11(2):464–494, 2000.

5.4 Interference Reduction: Cooperative Communication with Partial CSI in Mobile Radio Cellular Networks

X. Wei, T. Weber, University of Rostock, Germany

5.4.1 Introduction

Future mobile radio wireless communication systems aim to provide high quality of service (QoS) with high spectrum efficiency [1, 2]. The remarkable capacity potential of wireless communication systems applying the multiple-input multiple-output (MIMO) technique has been indicated by pioneering work [3, 4]. In interference-limited cellular systems which can be considered as multiuser MIMO systems, interference management has already become the central task to achieve spectrally efficient communications [5]. Applying the orthogonal frequency-division multiplexing (OFDM) transmission technique [6], interference management is performed based on available channel-state information (CSI) of the investigated multiuser MIMO-OFDM system. A system concept based on the perfect full CSI is almost infeasible in realistic cellular systems due to the implementation complexity and the limited ability to track the CSI. Considering the above practical issues, we have proposed a cooperative communication scheme based on partial CSI with respect to significant CSI and imperfect CSI as a promising candidate for interference management. In this scheme, base stations (BSs) cooperate with each other to perform joint detection (JD) in the uplink (UL) and joint transmission (JT) in the downlink (DL). As compared to the state of the art JD and JT techniques, our contributions are shown as follows.

JD and JT can be performed based on the CSI in subsystems of the cellular system such as service areas (SAs) or group cells [7–9] selected according to the static geographic architecture. However, the MSs close to the boundary of the subsystem still suffer strong interference from the MSs in other subsystems in the UL and cause strong interference to the MSs in other subsystems in the DL. We have dynamically selected the significant CSI from the point of view of each MS in the form of significant useful channels and significant interference channels according to the functionality of the channels. In order to make full use of the selected significant CSI and to implement the cooperative communication in an efficient way, an iterative algorithm for JD and JT with significant CSI following the ideas of the zero-forcing (ZF) algorithm [10, 11] is implemented in a decentralized way. The implementation complexity is reduced since no CU is needed and only local significant CSI is required at each BS. Additionally, alternative BS antenna layouts with omnidirectional or sector antennas in combination with different transmission techniques are also investigated.

Although a lot of research work has been done on investigating the influence of imperfect CSI on multiuser MIMO systems considering full cooperation [12], the impact of imperfect CSI on multiuser MIMO systems considering partial cooperation with only significant CSI has rarely been mentioned in the literature. Our work has contributed to this point. The performance degradation caused by the imperfectness of the CSI used in the significant channel selection and in the JD/JT with partial CSI has been investigated. The question about how much CSI should be considered in JD/JT to obtain optimum system performance has been answered. Additionally, we have proposed an advanced detection algorithm based on the statistical knowledge of the imperfect CSI, from which optimum detection results can be obtained following the rationale of maximum-likelihood (ML) [13].

5.4.2 System Model and Reference Scenario

We have considered a realistic multicell cellular system with one BS in each cell. Applying the OFDM technique, it is sufficient to assume that one active MS is contained in each cell in the considered subcarrier and time slot. The active MSs can be selected through adaptive scheduling techniques which have been intensively investigated [14,15]. Applying multiple antennas at the BSs, the cellular system can be considered as a multiuser MIMO system with K_{A} antennas at the BS side and K_{M} antennas at the MS side in total. The UL channel matrix $\underline{\mathbf{H}}_{\mathrm{UL}}$ and the DL channel matrix $\underline{\mathbf{H}}_{\mathrm{DL}}$ can be easily obtained through the relation

$$\underline{\mathbf{H}}_{\mathrm{UL}}^{\mathrm{T}} = \underline{\mathbf{H}}_{\mathrm{DL}} = \underline{\mathbf{H}} = \begin{pmatrix} \underline{H}^{(1,1)} & \dots & \underline{H}^{(1,K_{\mathrm{A}})} \\ \vdots & & \vdots \\ \underline{H}^{(K_{\mathrm{M}},1)} & \dots & \underline{H}^{(K_{\mathrm{M}},K_{\mathrm{A}})} \end{pmatrix} \tag{5.23}$$

resulting from the reciprocity of the radio channels in the considered TDD systems. In realistic systems considering imperfect CSI, the estimated channel matrix

$$\hat{\underline{\mathbf{H}}} = \underline{\mathbf{H}} + \underline{\boldsymbol{\Delta}} \tag{5.24}$$

is obtained with the channel-error matrix $\underline{\boldsymbol{\Delta}}$, and the elements of $\underline{\boldsymbol{\Delta}}$ are assumed to be independent and identically Gaussian distributed with variance $\sigma_{\Delta}^2/2$ of real and imaginary parts. The estimated significant useful channel matrix $\hat{\underline{\mathbf{H}}}_{\mathrm{U}}$ and the estimated MS specific significant interference channel matrices $\hat{\underline{\mathbf{H}}}_{\mathrm{I},k_{\mathrm{M}}}$ obtained through the significant channel selection from the estimated channel matrix $\hat{\underline{\mathbf{H}}}$ can be applied in both JD and JT. With the data vector $\underline{\mathbf{d}}$, the transmitted vector $\underline{\mathbf{s}}$, the noise vector $\underline{\mathbf{n}}$, the received vector $\underline{\mathbf{e}}$ and the estimated data vector $\hat{\underline{\mathbf{d}}}$, the system model based on cooperative communication considering JD in the UL and JT in the DL with different types of available CSI is described in Fig. 5.20.

In the UL, the transmitted vector $\underline{\mathbf{s}}$ is obtained from the data vector $\underline{\mathbf{d}}$ in the simple OFDM transmitters at the MSs considering a scaling of the transmitted energy if it is required. For simplicity, we assume that a unit scaling factor is applied, and therefore

$$\underline{\mathbf{s}} = \underline{\mathbf{d}} = (\underline{d}^{(1)} \dots \underline{d}^{(K_{\mathrm{M}})})^{\mathrm{T}} \tag{5.25}$$

is obtained. From the received vector

$$\underline{\mathbf{e}} = \underline{\mathbf{H}}^{\mathrm{T}} \cdot \underline{\mathbf{s}} + \underline{\mathbf{n}} \ , \tag{5.26}$$

the estimated data vector $\hat{\underline{\mathbf{d}}}$ is obtained through JD at the BSs performed based on $\hat{\underline{\mathbf{H}}}_{\mathrm{U}}$ and $\hat{\underline{\mathbf{H}}}_{\mathrm{I},k_{\mathrm{M}}}$.

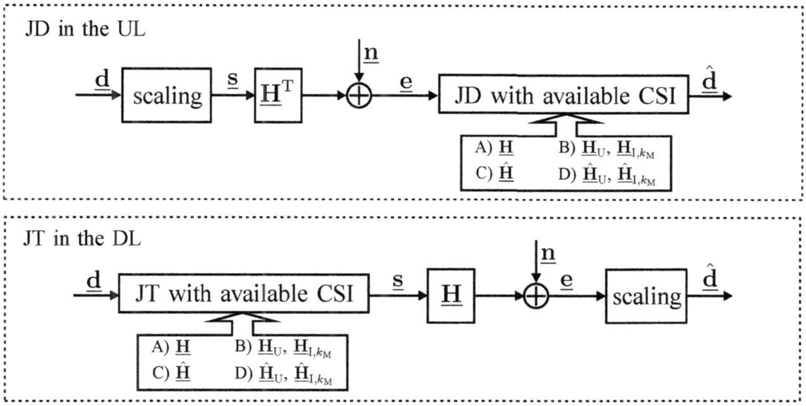

Figure 5.20: system model based on cooperative communication

In the DL, the transmitted vector $\underline{\mathbf{s}}$ is obtained from the data vector $\underline{\mathbf{d}}$ through JT at the BSs performed based on $\underline{\hat{\mathbf{H}}}_{\mathrm{U}}$ and $\underline{\hat{\mathbf{H}}}_{\mathrm{I},k_{\mathrm{M}}}$. In the simple OFDM receivers at the MSs from the received vector

$$\underline{\mathbf{e}} = \underline{\mathbf{H}} \cdot \underline{\mathbf{s}} + \underline{\mathbf{n}} \ , \tag{5.27}$$

we obtain the estimated data vector $\hat{\underline{\mathbf{d}}}$ through a scaling of the received energy to keep the received energy per data symbol unmodified as compared to the data symbol energy.

For intercell interference management in cellular systems, on one side we can apply an intelligent signal processing technique based on JD and JT with partial CSI as discussed above, on the other side we can consider a smart BS antenna layout suitable for cooperative communication. In general, a mobile radio communication system which is equipped with multiple antennas with significant distance from each other and where a joint signal processing based on these antennas is applied can be named as distributed antenna system (DAS). Two alternative BS antenna layouts named as omni-DAS and sector-DAS, respectively, are applied in an exemplary scenario as shown in Fig. 5.21. In this scenario, a 7-cell system including one center cell and 6 interfering cells around the center cell is considered with the frequency reuse cluster size 3. Uniform antenna gain is assumed in both BS antenna layouts for the sake of simplicity. For fairly comparing the system performance of the above two BS antenna layouts, it is assumed that each BS is equipped with 3 antennas in both omni-DAS and sector-DAS. In the omni-DAS, 3 omnidirectional BS antennas are located in the center of each cell. In the sector-DAS, 3 distributed 120 degree sector antennas are located in 3 vertices of each cell and point towards the cell center. The transformation between the omni-DAS and the sector-DAS is visualized using a 3-cell scenario in Fig. 5.21. Namely, the sector-DAS can be easily obtained from the omni-DAS by shifting the cell layout of the omni-DAS by a distance of the cell radius R in the direction of the arrow and replacing the 3 omnidirectional antennas at each BS with 3 sectorized antennas separately pointing towards the centers of 3 neighboring cells in the original network. Although after the above transformation

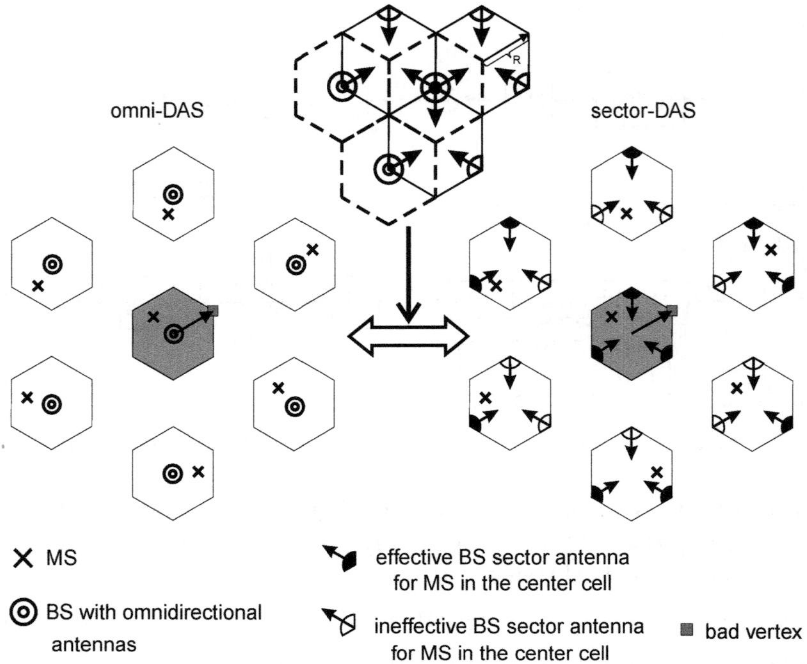

Figure 5.21: Two alternative BS antenna layouts: omni-DAS and sector-DAS, and their transformation

the BS antennas for each cell are served by 3 BSs in different physical locations, the total number of BSs and the total number of BS antennas are unmodified. In reverse, the omni-DAS can also be easily obtained from the sector-DAS. Under the assumption that antenna gains of sector antennas are zero at the directions more than 60 degrees with respect to the boresight, for each MS in the sector-DAS there are many ineffective antennas which have no influence on this MS. For example as shown in Fig. 5.21, all the effective sector antennas which have influence on the MS in the center cell and all the ineffective sector antennas which have no influence on the MS in the center are indicated. Therefore, less intercell interference inherently exists in the sector-DAS as compared to the omni-DAS due to the ineffective sectorized BS antennas in other cells for every MS in the sector-DAS. Obviously, the useful signals for the MSs close to their own BS antennas could be very strong. Therefore, comparing the above two BS-antenna-layouts, the omni-DAS is expected to offer a better system performance for the MSs close to the cell centers, while the sector-DAS is expected to offer a better system performance for the MSs close to the cell vertices where sectorized BS antennas could be located.

5.4.3 Significant CSI Selection Algorithm and Channel Matrix Formalism

In order to reduce the computational load of the JD/JT algorithms and the communication load between the BSs, we consider only the significant CSI corresponding to the channels which play a significant role in the system performance. Here we have distinguished the significant useful channels from the significant interference channels for each MS. Significant useful channels for one MS are the channels over which we get significant useful contributions when we estimate the data symbols transmitted from this MS in the UL, and the channels over which we generate significant useful contributions to the received signals for this MS in the DL. Significant interference channels for one MS are the channels over which we get significant interferences from other MSs when we receive the data symbols from this MS in the UL, and the channels over which we cause significant interferences to other MSs when we transmit the data symbol for this MS in the DL. According to the above definitions of significant channels, various mathematical criteria can be applied to determine the significant channels. A single channel indicator matrix $\widetilde{\mathbf{H}}_{\mathrm{U}}$ named the significant useful channel indicator matrix is sufficient to indicate the significant channels for all MSs. However, in general we have to use K_{M} individual channel indicator matrices $\widetilde{\mathbf{H}}_{\mathrm{I},k_{\mathrm{M}}}$ named the MS specific significant interference channel indicator matrices to represent the significant interference channels for the individual MSs k_{M}. Especially, in $\widetilde{\mathbf{H}}_{\mathrm{I},k_{\mathrm{M}}}$, there are some "don't care" elements corresponding to the channels which are irrelevant to the interference considered for MS k_{M} in our cooperative communication scheme. Two MSs have compatible significant interference channels if all the significant interference channels considered for one MS will never be considered as insignificant interference channels for the other MS. If all the MSs have compatible significant interference channels, we can obtain the combined significant interference channel indicator matrix $\widetilde{\mathbf{H}}_{\mathrm{I}}$. Details of the significant channel selection and the combination of significant interference channel indicator matrices have been described in [16, 18]. For simplicity, a 3-cell sector-DAS in Fig. 5.22 is used as an exemplary scenario to visualize the proposed significant channel selection algorithm and the corresponding channel matrix formalism. Especially, in the sector-DAS it is not necessary to consider the ineffective channels during the significant channel selection. We can just assign "0"s to the corresponding positions of these channels in $\widetilde{\mathbf{H}}_{\mathrm{U}}$ and $\widetilde{\mathbf{H}}_{\mathrm{I},k_{\mathrm{M}}}$.

The above selection algorithm is designed based on the perfect CSI in the channel matrix $\underline{\mathbf{H}}$. However, in realistic systems considering the imperfectness of the CSI, we can only perform a suboptimum selection of the significant channels based on the estimated channel matrix $\hat{\underline{\mathbf{H}}}$ in (5.24). From the mathematical point of view, we just replace $\underline{\mathbf{H}}$ with $\hat{\underline{\mathbf{H}}}$ in the above selection algorithm. The process to determine the significant channels is unmodified. Although only the magnitudes of the channel coefficients are required in our selection algorithm, imperfect channel estimation can still cause incorrect significant channel selection. The influence of imperfect CSI in the significant channel selection on the system performance will be investigated in Section 5.4.5.

Based on the above introduced significant channel indicator matrices $\widetilde{\mathbf{H}}_{\mathrm{U}}$, $\widetilde{\mathbf{H}}_{\mathrm{I},k_{\mathrm{M}}}$

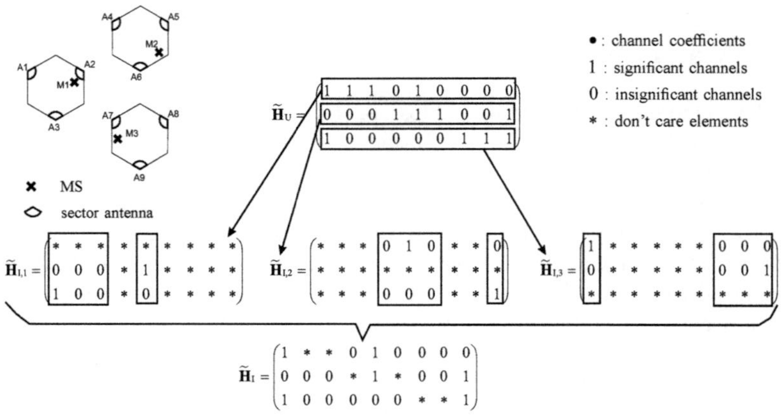

Figure 5.22: Example for channel selection and indicator matrix formalism in a sector-DAS

and $\widetilde{\mathbf{H}}_{\mathrm{I}}$, we obtain the estimated significant useful channel matrix and the estimated significant interference channel matrices as

$$\underline{\hat{\mathbf{H}}}_{\mathrm{U}} = \underline{\hat{\mathbf{H}}}^{\mathrm{T}}_{\mathrm{U,UL}} = \underline{\hat{\mathbf{H}}}_{\mathrm{U,DL}} = \underline{\hat{\mathbf{H}}} \odot \widetilde{\mathbf{H}}_{\mathrm{U}} = \underline{\mathbf{H}} \odot \widetilde{\mathbf{H}}_{\mathrm{U}} + \underline{\boldsymbol{\Delta}} \odot \widetilde{\mathbf{H}}_{\mathrm{U}}, \tag{5.28}$$

$$\underline{\hat{\mathbf{H}}}_{\mathrm{I},k_{\mathrm{M}}} = \underline{\hat{\mathbf{H}}}^{\mathrm{T}}_{\mathrm{I,UL},k_{\mathrm{M}}} = \underline{\hat{\mathbf{H}}}_{\mathrm{I,DL},k_{\mathrm{M}}} = \underline{\hat{\mathbf{H}}} \odot \widetilde{\mathbf{H}}_{\mathrm{I},k_{\mathrm{M}}} = \underline{\mathbf{H}} \odot \widetilde{\mathbf{H}}_{\mathrm{I},k_{\mathrm{M}}} + \underline{\boldsymbol{\Delta}} \odot \widetilde{\mathbf{H}}_{\mathrm{I},k_{\mathrm{M}}} , \tag{5.29}$$

$$\underline{\hat{\mathbf{H}}}_{\mathrm{I}} = \underline{\hat{\mathbf{H}}}^{\mathrm{T}}_{\mathrm{I,UL}} = \underline{\hat{\mathbf{H}}}_{\mathrm{I,DL}} = \underline{\hat{\mathbf{H}}} \odot \widetilde{\mathbf{H}}_{\mathrm{I}} = \underline{\mathbf{H}} \odot \widetilde{\mathbf{H}}_{\mathrm{I}} + \underline{\boldsymbol{\Delta}} \odot \widetilde{\mathbf{H}}_{\mathrm{I}} , \tag{5.30}$$

from the imperfect estimated channel matrix $\underline{\hat{\mathbf{H}}}$ as described in (5.24).

The selection algorithm mentioned in [16, 17] is based on the channel gain or the weighting factor magnitude of the interference which can be directly calculated from the channel coefficients. Now we take system performance criteria as the reference, and the significant CSI can be selected in the way to obtain the maximum SNIR indicating the maximum ratio of the received useful power to the received noise-plus-interference power, or to obtain the maximum average magnitude of log-likelihood ratios (LLRs) indicating the maximum reliability information of data detection. We take the model of a real-valued two user interference channel with the 2×2 channel matrix $\mathbf{H}$ including independent and identically distributed (i.i.d.) channel coefficients $h_{ij} \sim \mathcal{N}(0,1), i,j=1,2$ as an example. Applying BPSK modulation, equally distributed data symbols are included in the transmitted data vector $\mathbf{d} = (d_1, d_2)^{\mathrm{T}}$. The noise vector $\mathbf{n} = (n_1, n_2)^{\mathrm{T}}$ contains the Gaussian distributed noise signals, i.e., $n_k \sim \mathcal{N}(0, \sigma_{\mathrm{N}}^2), k=1,2, \sigma_{\mathrm{N}}^2 = 0.1$. The received vector $\mathbf{y} = (y_1, y_2)^{\mathrm{T}}$ is obtained as $\mathbf{y} = \mathbf{H} \cdot \mathbf{d} + \mathbf{n}$. As an example, we have investigated the loss of reliability information by comparing the selection algorithm I aiming at maximum channel gains, i.e., we select 3 out of 4 channel coefficients with largest channel gains, and the selection algorithm II aiming at maximum magnitude of LLR, i.e., we select 3 out of 4 channel coefficients which achieve the largest average magnitude of LLRs over random data symbols and noise signals. During the calculation, the known 3 channel coefficients and the limited statistical knowledge of the other channel coefficient are considered. For a snapshot of the channel matrix $\mathbf{H}$, we obtain

$$\overline{\left|LLR(\mathbf{H})\right|} = \underset{\{\mathbf{d},\mathbf{n}\}}{\mathrm{E}}\{(\left|L\left(d_1|\mathbf{y}\right)\right| + \left|L\left(d_2|\mathbf{y}\right)\right|)/2\} \ , \tag{5.31}$$

$$L\left(d_{\mathrm{i}}|\mathbf{y}\right) = \ln\frac{P(d_{\mathrm{i}}=+1|\mathbf{y})}{P(d_{\mathrm{i}}=-1|\mathbf{y})} = \ln\frac{P(\mathbf{y}|d_{\mathrm{i}}=+1,d_{\mathrm{j}}=+1)+P(\mathbf{y}|d_{\mathrm{i}}=+1,d_{\mathrm{j}}=-1)}{P(\mathbf{y}|d_{\mathrm{i}}=-1,d_{\mathrm{j}}=+1)+P(\mathbf{y}|d_{\mathrm{i}}=-1,d_{\mathrm{j}}=-1)} \ , \tag{5.32}$$

with i,j=1, 2, $i \neq j$. Assuming that h_{12} is selected as the insignificant channel, we would obtain

$$P(\mathbf{y}|d_1,d_2) = \int \frac{\exp\left(-\frac{(y_1-h_{11}d_1-\hat{h}_{12}d_2)^2}{2\sigma_{\mathrm{N}}^2}\right)}{\sqrt{2\pi\sigma_{\mathrm{N}}^2}} \cdot \frac{\exp\left(-\frac{(\hat{h}_{12})^2}{2}\right)}{\sqrt{2\pi}} \mathrm{d}\hat{h}_{12} \cdot \frac{\exp\left(-\frac{(y_2-h_{21}d_1-h_{22}d_2)^2}{2\sigma_{\mathrm{N}}^2}\right)}{\sqrt{2\pi\sigma_{\mathrm{N}}^2}}$$

$$= \frac{\exp\left(-\frac{(y_1-h_{11}d_1)^2}{2(d_2^2+\sigma_{\mathrm{N}}^2)} - \frac{(y_2-h_{21}d_1-h_{22}d_2)^2}{2\sigma_{\mathrm{N}}^2}\right)}{2\pi\sqrt{(d_2^2+\sigma_{\mathrm{N}}^2)\sigma_{\mathrm{N}}^2}} \tag{5.33}$$

The above formulae can be easily modified to the case that any other channel coefficient is selected as the insignificant one. Based on a sufficient number of snapshots of the channel matrix, the expectation $\underset{\{\mathbf{H}\}}{\mathrm{E}}\{\overline{\left|LLR(\mathbf{H})\right|}\} = \overline{\left|LLR\right|}$ can be calculated. The information loss of selection algorithm I as compared to selection algorithm II in this example is $(\overline{\left|LLR\right|}^{(\mathrm{II})} - \overline{\left|LLR\right|}^{(\mathrm{I})}) \approx 22.7 - 18.7 = 4$.

5.4.4 Decentralized JD/JT with Significant CSI for Interference Reduction

Based on the estimated significant channel matrices, we obtain the estimated channel energy scaling matrix

$$\hat{\mathbf{G}} = \hat{\mathbf{G}}_{\mathrm{UL}} = \hat{\mathbf{G}}_{\mathrm{DL}} = \mathrm{diag}\left(\underline{\hat{\mathbf{H}}}_{\mathrm{U}}\underline{\hat{\mathbf{H}}}_{\mathrm{U}}^{*\mathrm{T}}\right) \tag{5.34}$$

and the estimated channel correlation matrix

$$\underline{\hat{\mathbf{R}}} = \underline{\hat{\mathbf{R}}}_{\mathrm{UL}}^{\mathrm{T}} = \underline{\hat{\mathbf{R}}}_{\mathrm{DL}} = \left(\underline{\hat{\mathbf{H}}}_{\mathrm{I},1}\left[\underline{\hat{\mathbf{H}}}_{\mathrm{U}}^{*\mathrm{T}}\right]_1, \ldots, \underline{\hat{\mathbf{H}}}_{\mathrm{I},K_{\mathrm{M}}}\left[\underline{\hat{\mathbf{H}}}_{\mathrm{U}}^{*\mathrm{T}}\right]_{K_{\mathrm{M}}}\right) \tag{5.35}$$

of the multiuser MIMO system. In the UL, the iterative algorithm for JD with partial CSI can be described using matrix-vector-notation as

$$\underline{\hat{\mathbf{d}}}(i) = \hat{\mathbf{G}}^{-1}\left(\underline{\hat{\mathbf{H}}}_{\mathrm{U}}^{*} \cdot \underline{\mathbf{e}} - \overline{\mathrm{diag}}\left(\underline{\hat{\mathbf{R}}}^{\mathrm{T}}\right) \cdot \underline{\hat{\mathbf{d}}}(i-1)\right) \ . \tag{5.36}$$

The decentralized implementation of the iterative algorithm for each MS k_{M} can be described as

$$\underline{r}^{(k_{\mathrm{M}})} = \sum_{k_{\mathrm{A}}} \underline{\hat{H}}_{\mathrm{U}}^{(k_{\mathrm{M}},k_{\mathrm{A}})^*} \cdot \underline{e}^{(k_{\mathrm{A}})} \tag{5.37}$$

and

$$\underline{\hat{d}}^{(k_{\mathrm{M}})}(i) = \frac{1}{\hat{g}^{(k_{\mathrm{M}})}} \cdot \left(\underline{r}^{(k_{\mathrm{M}})} - \sum_{k_{\mathrm{A}}} \sum_{k'_{\mathrm{M}} \neq k_{\mathrm{M}}} \underline{\hat{H}}_{\mathrm{U}}^{(k_{\mathrm{M}},k_{\mathrm{A}})^*} \underline{\hat{H}}_{\mathrm{I},k_{\mathrm{M}}}^{(k'_{\mathrm{M}},k_{\mathrm{A}})} \cdot \underline{\hat{d}}^{(k'_{\mathrm{M}})}(i-1)\right) \ , \tag{5.38}$$

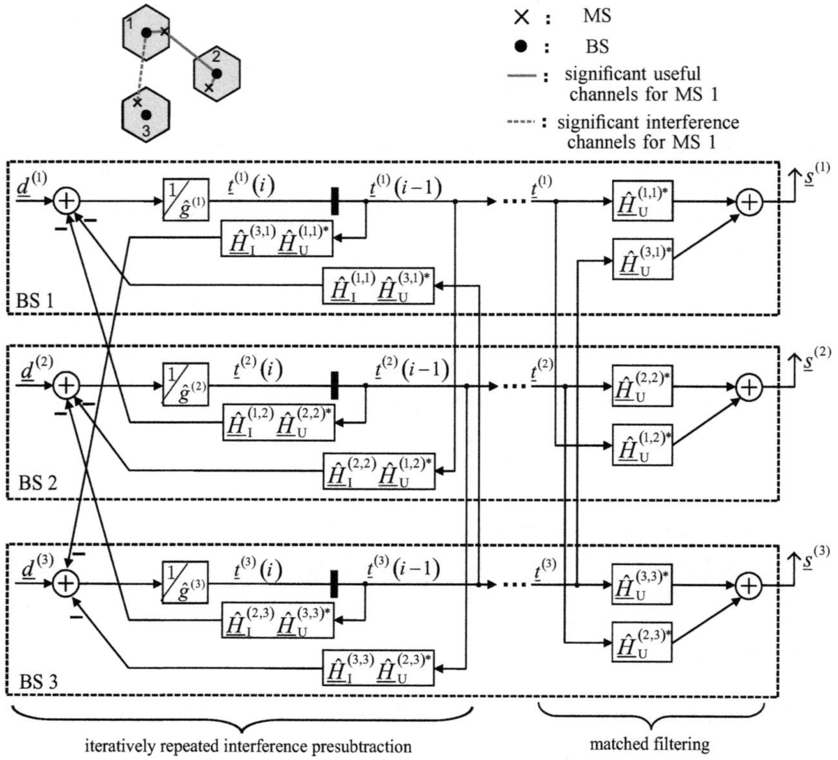

Figure 5.23: Decentralized implementation of JT with partial CSI in the DL of an exemplary 3-cell scenario

where the channel energy scaling factor is

$$\hat{g}^{(k_{\mathrm{M}})} = \sum_{k_{\mathrm{A}}} \underline{\hat{H}}_{\mathrm{U}}^{(k_{\mathrm{M}},k_{\mathrm{A}})} \underline{\hat{H}}_{\mathrm{U}}^{(k_{\mathrm{M}},k_{\mathrm{A}})*} \quad . \tag{5.39}$$

Firstly, at the BSs corresponding to the significant useful channels of the MS k_{M}, we calculate the matched filtering data estimates $\underline{r}^{(k_{\mathrm{M}})}$ considering only the significant useful CSI. Secondly, significant interferences from other MSs to the BSs antennas considered for MS k_{M} in the first step are reconstructed considering the significant interference CSI, and subtracted from the corresponding matched filtering data estimates $\underline{r}^{(k_{\mathrm{M}})}$. This step is implemented in an iterative way, and the preliminary detection results $\hat{\underline{d}}^{(k'_{\mathrm{M}})}(i-1)$ from other MSs are required.

In the DL, the matrix-vector-notation form of the iterative algorithm for JT with partial CSI can be described by

$$\underline{\mathbf{t}}(i) = \hat{\mathbf{G}}^{-1} \cdot \left(\underline{\mathbf{d}} - \overline{\mathrm{diag}}\left(\hat{\underline{\mathbf{R}}}\right) \cdot \underline{\mathbf{t}}(i-1)\right) \tag{5.40}$$

and

$$\underline{\mathbf{s}} = \hat{\underline{\mathbf{H}}}_{\mathrm{U}}^{*\mathrm{T}} \cdot \underline{\mathbf{t}} \ . \tag{5.41}$$

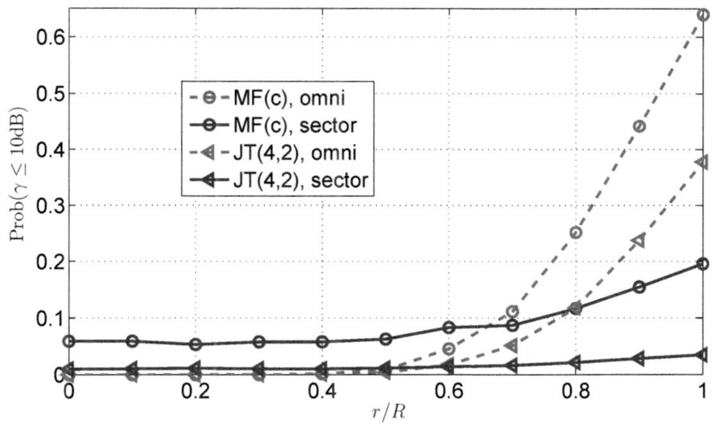

Figure 5.24: Outage probability of SIR for MS moving from cell center to the "bad vertex" in the center cell in the DL of 19 cell cellular system including one center cell and two tiers of cells around it with cluster size 3

The corresponding decentralized implementation of the iterative algorithm for each MS k_{M} can be described as

$$\underline{t}^{(k_{\mathrm{M}})}(i) = \frac{1}{\widehat{g}^{(k_{\mathrm{M}})}} \cdot \left(\underline{d}^{(k_{\mathrm{M}})} - \sum_{k_{\mathrm{A}}} \sum_{k'_{\mathrm{M}} \neq k_{\mathrm{M}}} \underline{\hat{H}}_{\mathrm{I},\mathrm{k}'_{\mathrm{M}}}^{(k_{\mathrm{M}},k_{\mathrm{A}})} \underline{\hat{H}}_{\mathrm{U}}^{(k'_{\mathrm{M}},k_{\mathrm{A}})^*} \cdot \underline{t}^{(k'_{\mathrm{M}})}(i-1) \right) \tag{5.42}$$

and

$$\underline{s}^{(k_{\mathrm{A}})} = \sum_{k_{\mathrm{M}}} \underline{\hat{H}}_{\mathrm{U}}^{(k_{\mathrm{M}},k_{\mathrm{A}})^*} \cdot \underline{t}^{(k_{\mathrm{M}})} \ . \tag{5.43}$$

Firstly, for each MS k_{M} at its considered BSs corresponding to its significant useful channels, the significant interferences from this MS to other MSs are predicted and compensated. Through this process, the predistorted data symbol $\underline{t}^{(k_{\mathrm{M}})}$ is obtained in an iterative way considering significant interference CSI, and the preliminary predistorted data symbol $\underline{t}^{(k'_{\mathrm{M}})}(i-1)$ from other MSs are required. Secondly, the transmitted signals $\underline{s}^{(k_{\mathrm{A}})}$ at the BS antennas corresponding to the significant useful CSI of the MSs are calculated from the predistorted data symbols $\underline{t}^{(k_{\mathrm{M}})}$ of all MSs.

Taking a 3-cell omni-DAS with a single antenna at each BS as an exemplary scenario, the above decentralized signal processing for JT with significant CSI is visualized in Fig. 5.23. For example, the signal processing for MS 1 is performed at its neighboring BS 1 and BS 2 corresponding to its two significant useful channels.

Not only the smart reception/transmission technique based on JD and JT with partial CSI as discussed above, we have also investigated and compared omni-DASs and sector-DASs which have been introduced in Section 5.4.2. The numerical results of the system performance in the DL with different combinations of transmission strategies and BS antenna layouts are shown in Fig. 5.24. It can be seen that iterative JT with N_{u} significant useful channels and N_{i} significant interference channels indicated by "JT $(N_{\mathrm{u}}, N_{\mathrm{i}})$" has much better performance as compared to intracell matched filtering (MF) indicated by "MF(c)", both in omni-DAS and sector-DAS.

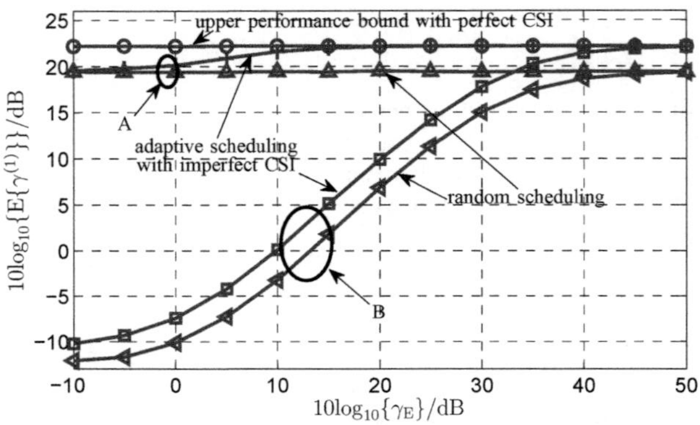

Figure 5.25: Average SNIR $E\{\gamma^{(1)}\}$ as a function of γ_E, applying JT(57, 1026), $10\log_{10}\{\gamma_N\} =$ 20dB, A: significant channel selection and JT are based on perfect CSI, B: significant channel selection and JT are based on imperfect CSI

The sector-DAS outperforms the omni-DAS even more when applying iterative ZF JT with significant CSI as compared to intracell MF.

5.4.5 Impact of Imperfect CSI on Cooperative Communication Based on JD/JT

The impact of imperfect CSI on cooperative communication scheme is investigated with the help of numerical results. We take a 19 cell cellular system including one cell in the center and two tiers of cochannel cells around it as the reference scenario. The performance of the MS in the center cell is assessed. Adaptive scheduling techniques have been intensively investigated in [14, 15]. In [16], the impact of imperfect CSI on the adaptive MS scheduling in the proposed cooperative communication scheme has been investigated. From Fig. 5.25, we can clearly see that with increasing imperfectness of the CSI, the system performance when applying adaptive scheduling approaches that of applying random scheduling. Under the realistic assumption that the significant channel selection and the JT with partial CSI are based on imperfect CSI, the system performance sharply decreases when the extent of imperfectness of CSI increases. The performance degradation caused by imperfect CSI used in the significant channel selection and JT is much larger than the performance degradation caused by replacing adaptive scheduling with random scheduling. In Fig. 5.26 and Fig. 5.27, the MS in the center cell is located at the fixed position of the "bad vertex", while the other MSs are randomly distributed in the other cells. In Fig. 5.26, taking the upper system performance bound obtained by applying perfect CSI in the whole cooperative transmission scheme as the reference, we can see the system performance degradation caused by the imperfect CSI used in the significant channel selection and JT with partial CSI, respectively. In Fig. 5.27, assuming imperfect CSI is applied in the whole cooperative transmission scheme, both omni-DAS and sector-DAS cooperative communication schemes considering different partial CSI selection strategies with different numbers of significant channels are investigated. As a func-

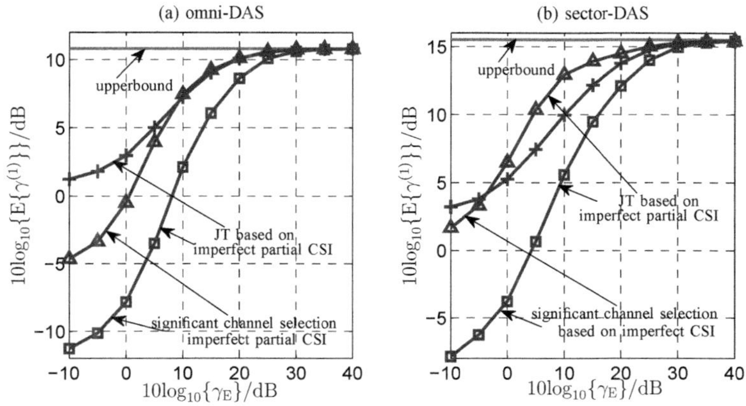

Figure 5.26: Average SNIR $E\{\gamma^{(1)}\}$ as a function of γ_E, applying JT(4, 2), $10\log_{10}\{\gamma_N\} = 20$dB

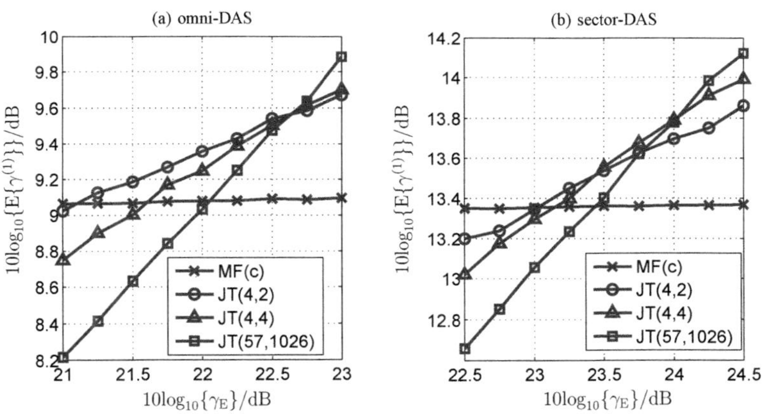

Figure 5.27: Average SNIR $E\{\gamma^{(1)}\}$ as a function of γ_E, $10\log_{10}\{\gamma_N\} = 20$dB

tion of the extent of imperfectness of the CSI, we can decide how much CSI should be considered in the cooperative communication scheme to obtain optimum system performance with moderate implementation complexity. Additionally, the influence of the imperfectness of CSI on the advanced JD/JT algorithms, i.e., SIC in the UL and THP in the DL with quantization, has been investigated. In Fig. 5.28 and Fig. 5.29 considering the exemplary channel matrix $\mathbf{H}_{UL} = \mathbf{H}_{DL}^T = (10\ 1\ 0.1; 10\ 2\ 0.1; 10\ 5\ 1)$, the input SNR has been adjusted in such a way that BER is always 10^{-3} when applying different JD/JT algorithms with perfect CSI. In Fig. 5.28, ZF, SIC, ML, and optimum MF with the ideal assumption of no multiuser interference are considered for JD in the receivers. In Fig. 5.29, ZF, THP, and optimum MF with the ideal assumption of no multiuser interference are considered for JT in the transmitters. We can see the sensitivity of different JD/JT algorithms to the imperfectness of CSI described by the variance σ_Δ^2 of the channel errors.

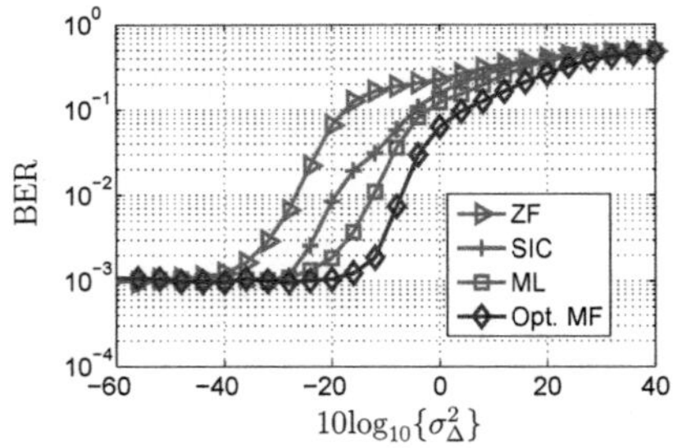

Figure 5.28: BER as a function of the extent of the imperfectness of CSI in the UL applying JD algorithms

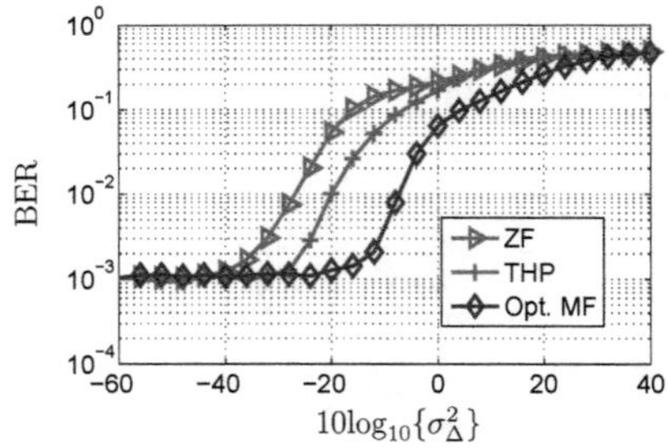

Figure 5.29: BER as a function of the extent of the imperfectness of CSI in the DL applying JT algorithms

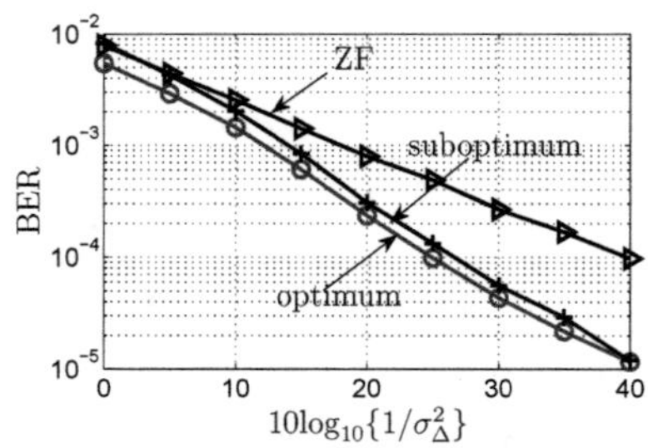

Figure 5.30: BER as a function of the extent of the perfectness of CSI with $10\log_{10}\{1/\sigma_\mathrm{N}^2\} = 40\mathrm{dB}$

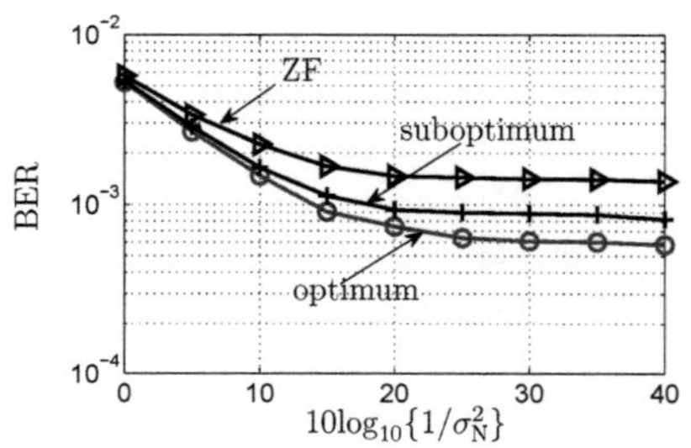

Figure 5.31: BER as a function of PSNR with $10\log_{10}\{1/\sigma_\Delta^2\} = 15\mathrm{dB}$

5.4.6 Advanced Algorithm Based on Statistical Knowledge of Imperfect CSI

Exploiting the statistical knowledge of imperfect CSI and the given alphabet $\mathbb{S}$ of the transmitted vector $\underline{\mathbf{s}}$, we propose an optimum detection scheme following the rationale of maximum likelihood (ML) as

$$\hat{\underline{\mathbf{s}}} = \underset{\mathbf{s}\in\mathbb{S}}{\operatorname{argmax}} \left\{ \mathrm{p}\left(\underline{\mathbf{e}}|\underline{\mathbf{s}}, \hat{\underline{\mathbf{h}}}\right) \right\} \ , \tag{5.44}$$

where $\mathrm{p}\left(\underline{\mathbf{e}}|\underline{\mathbf{s}}, \hat{\underline{\mathbf{h}}}\right)$ is calculated as the marginal PDF of $\mathrm{p}\left(\underline{\mathbf{e}}, \underline{\mathbf{h}}|\underline{\mathbf{s}}, \hat{\underline{\mathbf{h}}}\right)$ as

$$\mathrm{p}\left(\underline{\mathbf{e}}|\underline{\mathbf{s}}, \hat{\underline{\mathbf{h}}}\right) = \int \mathrm{p}\left(\underline{\mathbf{e}}, \underline{\mathbf{h}}|\underline{\mathbf{s}}, \hat{\underline{\mathbf{h}}}\right) \mathrm{d}\underline{\mathbf{h}} = \int \mathrm{p}\left(\underline{\mathbf{e}}|\underline{\mathbf{s}}, \underline{\mathbf{h}}\right) \mathrm{p}(\underline{\mathbf{h}}|\hat{\underline{\mathbf{h}}}) \mathrm{d}\underline{\mathbf{h}} \ . \tag{5.45}$$

In (5.45), $\mathrm{p}(\underline{\mathbf{h}}|\hat{\underline{\mathbf{h}}})$ can be calculated according to the Bayes' theorem as

$$\mathrm{p}(\underline{\mathbf{h}}|\hat{\underline{\mathbf{h}}}) = \frac{\mathrm{p}(\hat{\underline{\mathbf{h}}}|\underline{\mathbf{h}})\mathrm{p}(\underline{\mathbf{h}})}{\int \mathrm{p}(\hat{\underline{\mathbf{h}}}|\underline{\mathbf{h}})\mathrm{p}(\underline{\mathbf{h}})\mathrm{d}\underline{\mathbf{h}}} \ . \tag{5.46}$$

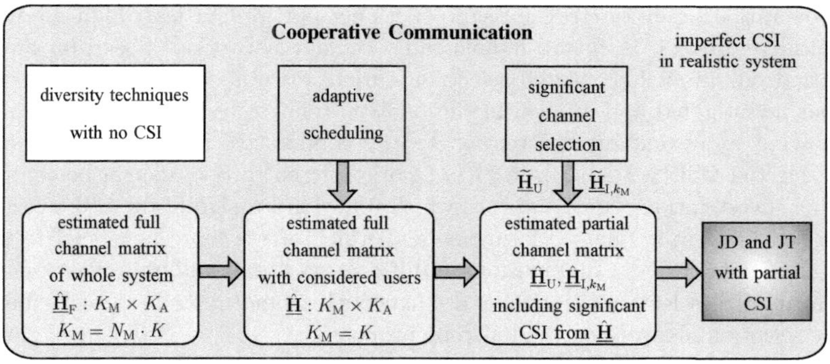

Figure 5.32: Cooperative communication with different levels of knowledge of CSI

Its application in two-user interference channels is presented. In Fig. 5.30 and Fig. 5.31, numerical results show the improvement of the system performance when applying optimum detection as compared to ZF and suboptimum detection where estimated channel coefficients are directly taken as perfect channel coefficients in ML detector.

5.4.7 System Concept Based on Different Levels of Knowledge of CSI

We aim at achieving high system performance in realistic mobile radio cellular systems. On one side, the system performance strongly depends on the available knowledge of CSI. On the other side, suitable communication strategies based on different levels of knowledge of CSI can be chosen. The system concept of cooperative communication in multiuser MIMO systems is proposed as shown in Fig. 5.32. Diversity techniques can be applied when no knowledge of CSI is available. Considering the influence of imperfectness of CSI in realistic systems, adaptive scheduling techniques are applied to select the users in each subcarrier. Applying the OFDM transmission technique, in the considered subcarrier significant CSI is selected and applied in JD and JT. In this way, cooperative communication with partial CSI concerning imperfect CSI and significant CSI can maintain significant useful contributions and eliminate significant interference with reduced communication load and computational load.

5.4.8 Outlook

The main focus of our work is the interference management in multiuser cellular systems from the interference cancellation signal processing point of view. Other interference management strategies such as resource allocation and power control can further reduce the influence of interference. In future work, various combinations of cooperative communication considering JD/JT, resource allocation, and power control can be investigated to jointly improve the performance of interference-limited cellular systems.

Cooperative reception/transmission with only partial CSI is of high interest for practical realization in future mobile radio cellular networks. Based on our contributions about significant CSI selection, various criteria of significant CSI can be further investigated and applied in the realistic cellular systems. Since researchers are paying more and more attention to the cooperative communication concept with selected CSI [8, 9, 19–23], in future work different proposals can be compared. Further investigations could indicate which criterion for significant CSI selection is more beneficial to system performance and which one is more beneficial to implementation complexity. Combinations of different proposals and improvements of the significant CSI selection algorithms are expected. Some concrete design guidelines of the selection algorithms could also be proposed.

Advanced nonlinear JD/JT algorithms, i.e., SIC with quantization in the UL and THP in the DL, with various significant CSI selection algorithms can be further investigated with respect to analytical and numerical results in future work. Distributed signal processing of these algorithms in the cooperative communication scheme could be implemented, and its implementation complexity can be assessed.

Statistical signal processing based on limited knowledge of CSI has also been investigated based on a real-valued two user interference channel model. Its application can be extended to realistic multiuser cellular systems in future work.

Bibliography

[1] X. You, G. Chen, M. Chen, and X. Gao, "Toward beyond 3G: The FuTURE project in China," *IEEE Communications Magazine*, vol. 43, no. 1, pp. 38–44, Jan. 2005.

[2] K. Tachikawa, "A perspective on the evolution of mobile communications," *IEEE Communications Magazine*, vol. 41, no. 1, pp. 66–73, Oct. 2003.

[3] G. J. Foschini and M. J. Gans, "On limits of wireless communications in a fading environment when using multiple antennas," *Wireless Personal Commun.*, vol. 6, pp. 311–335, Mar. 1998.

[4] E. Telatar, "Capacity of multi-antenna Gaussian channels," *European Transactions on Telecommunications*, vol. 10, no. 6, pp. 585–595, 1999.

[5] J. G. Andrews, "Interference cancellation for cellular systems: A contemporary overview," *IEEE Wireless Communications*, vol. 12, pp. 19–29, April 2005.

[6] R. D. J. van Nee and R. Prasad, *OFDM for Wireless Multimedia Communications*, Artech House, Boston, 2000.

[7] T. Weber, I. Maniatis, A. Sklavos, and Y. Liu, "Joint transmission and detection integrated network (JOINT), a generic proposal for beyond 3G systems," in *Proc. 9th International Conference on Telecommunications (ICT'02)*, Beijing, June 2002, vol. 3, pp. 479–483.

[8] P. Zhang, X. Tao, J. Zhang, Y. Wang, L. Li, and Y. Wang, "A vision from the future: Beyond 3G TDD," *IEEE Communications Magazine*, vol. 43, no. 1, pp. 38–44, January 2005.

[9] X. Tao, J. Xu, X. Xu, C. Tang, and P. Zhang, "Group cell FuTURE B3G TDD system," in *Proc. IEEE International Symposium on Personal, Indoor and Mobile Radio Communications (PIMRC'05).*

[10] A. Klein, G. K. Kaleh, and P. W. Baier, "Zero forcing and minimum mean-square-error equalization for multiuser detection in code-division multiple access channels," *IEEE Transactions on Vehicular Technology*, vol. 45, no. 2, pp. 276–287, May 1996.

[11] M. Meurer, P. W. Baier, T. Weber, Y. Lu, and A. Papathanassiou, "Joint transmission: Advantageous downlink concept for CDMA mobile radio systems using time division duplexing," *Electronics Letters*, vol. 36, no. 10, pp. 900–901, May 2000.

[12] T. Weber, A. Sklavos, and Meurer M., "Imperfect channel state information in MIMO transmission," *IEEE Transactions on Communications*, vol. 54, no. 3, pp. 543–552, March 2006.

[13] S. Verdu, *Multiuser Detection*, University Press, Cambridge, 1998.

[14] A. Kühne and A. Klein, "Adaptive MIMO-OFDM using OSTBC with imperfect CQI feedback," in *Proc. International ITG/IEEE Workshop on Smart Antennas (WSA'08)*, Darmstadt, Feb. 2008.

[15] A. Kühne and A. Klein, "Throughput analysis of multi-user OFDMA-systems using imperfect CQI feedback and diversity techniques," *IEEE Journal on Selected Areas in Communications*, vol. 26, no. 8, pp. 1440–1451, October 2008.

[16] X. Wei, T. Weber, A. Kühne, and A. Klein, "Joint transmission with imperfect partial channel state information," in *Proc. IEEE Vehicular Technology Conference (VTC'09-Spring)*, Barcelona, April 2009.

[17] X. Wei, T. Weber, A. Ahrens, and S. Deng, "Decentralized interference management in mobile radio networks," *Frequenz*, vol. 61, no. 11/12, pp. 259–269, 2007.

[18] X. Wei, and T. Weber, "Cooperative communication with partial channel-state information in multiuser MIMO systems," *AEÜ*, 2010.

[19] A. Papadogiannis, H.J. Bang, D. Gesbert, and E. Hardouin, "Downlink overhead reduction for multicell cooperative processing enabled wireless networks," in *Proc. IEEE International Symposium on Personal, Indoor, Mobile Radio Communications (PIMRC'08)*, Cannes, September 2008.

[20] A. Papadogiannis, D. Gesbert, and E. Hardoin, "A dynamic clustering approach in wireless networks with multi-cell cooperative processing," in *Proc. IEEE International Conference on Communications (ICC'08)*, Beijing, May 2008.

[21] P. Marsch and G. Fettweis, "A framework for optimizing the uplink performance of distributed antenna systems under a constrained backhaul," in *Proc. IEEE International Conference on Communications (ICC'07)*, Glasgow, June 2007.

[22] P. Marsch and G. Fettweis, "On multicell cooperative transmission in backhaul-constrained cellular systems," *Annals of Telecommunications*, vol. 63, no. 5–6, pp. 253–269, June 2008.

[23] S. Khattak and G. Fettweis, "Low backhaul distributed detection strategies for an interference limited uplink cellular system," in *Proc. IEEE 67th Vehicular Technology Conference (VTC'08-Spring)*, Singapore, May 2008, pp. 693–697.

6 OFDM/DMT for Wireline Communications

With wireline, we mean digital subscriber lines as well as optical transmission. Apart from the general aspects of multicarrier modulation, the channel properties are very different. It is thus unavoidable to go into some details of channel transfer characteristics and disturbances. In DSL, the channel characteristics are determined by the propagation constant, the characteristic impedance, and the loop structure, furthermore crosstalk from other loops and ingress (RFI, impulse noise) from outside. In copper loops, the frequency dependent transfer function of the twisted pairs together with possible reflections at splices lead to dispersion which we will also see in optical transmission. There, we distinguish chromatic dispersion and polarization mode dispersion. Optical transmission especially also suffers from non-linear characteristics (Kerr effect), which does not have a counterpart in copper cables. In copper transmission, non-linearities are especially due to the limitations of the D/A converters and power amplifiers, which may lead to clipping of the almost Gaussian distributed time-domain signal of OFDM. This holds for all OFDM transmission. Other non-linearities may be due to baluns (transformers) used in DSL, but practically, they are of minor relevance.

In the following two sections, we will study properties of the channel and specialties of the multicarrier transmission over twisted pairs and optical fibers. Thereafter, we consider a new approach for impulse noise cancellation in DSL and the simulation of optical multicarrier transmission.

6.1 Discrete MultiTone (DMT) and Wireline Channel Properties

W. Henkel, Jacobs University Bremen, Germany

This chapter describes basic properties of the wireline twister-pair channel and introduces DMT (Discrete MultiTone), the baseband variant of OFDM. Further information can, e.g., be found in a student book project [1], or [2–5].

6.1.1 Properties of the Twisted-Pair Channel

The twisted pair (TP) cable channel is characterized by a transfer function that is increasingly attenuating with frequency and cross-talk functions that increase with frequency, as well. Furthermore, ingress from radio interferers (RFI), e.g., amateur radio, and impulse noise have a strong impact. In the following, we describe the

H. Rohling (ed.), *OFDM: Concepts for Future Communication Systems*,
Signals and Communication Technology, DOI: 10.1007/978-3-642-17496-4_6,

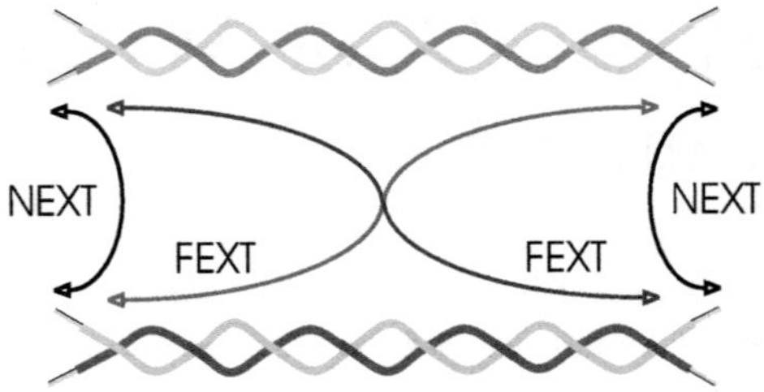

Figure 6.1: NEXT and FEXT

channel properties and interferences shortly, focusing a little more on impulse noise, especially in Section 6.3.

Transfer Characteristic

Two quantities determine the transfer function of a twisted pair: the propagation constant γ (together with the loop length l) and the characteristic impedance Z_0. The ABCD matrix of a loop is known to be

$$\mathbf{A} = \begin{bmatrix} \cosh(\gamma l) & Z_0 \cdot \sinh(\gamma l) \\ \frac{1}{Z_0} \cdot \sinh(\gamma l) & \cosh(\gamma l) \end{bmatrix} . \tag{6.1}$$

An ideally with Z_0 terminated line (or very long loop) has the transfer function

$$H(j\omega) = e^{-\gamma l} = e^{-\alpha l} e^{-j\beta l} . \tag{6.2}$$

Near-end Crosstalk (NEXT) and Far-end Crosstalk (FEXT)

NEXT results from coupling from other loops in the same cable from transmitters located at the same side as the own receiver (see Fig. 6.1). NEXT (as a power contribution) is modeled as

$$|H_{\text{NEXT}}(jf)|^2 = K_{\text{NEXT}} N^{0.6} f^{3/2} , \tag{6.3}$$

where N is the number of identical disturbers and the power of 0.6 is to halfways model the distribution of disturbers in cable.

FEXT results from coupling from other loops in the same cable from transmitters located at the opposite side of the own receiver (see Fig. 6.1). FEXT (as a power contribution) is modeled as

$$|H_{\text{FEXT}}(jf)|^2 = K_{\text{FEXT}} N^{0.6} \cdot l \cdot f^2 \cdot |H(jf,l)|^2 , \tag{6.4}$$

where N is the number of identical disturbers, l is the length of the coupling length, and the power of 0.6 is to halfways model the distribution of disturbers in the cable. The proportionality to l is intuitively obvious, since the longer the two loops are next to each other, the more power will be coupled. FEXT is influenced by the transfer function $H(jf,l)$, since it either traverses through it on the initial loop or after the coupling on the receiving loop. We ignore the slight differences in transfer functions on both sides which are mostly due to different twists. The so-called Equal-Level FEXT (EL-FEXT) is defined by eliminating the length dependency and the transfer function, i.e.,

$$|H_{\text{EL-FEXT}}(jf)|^2 = |H_{\text{FEXT}}(jf)| \cdot \frac{1}{l} \cdot \frac{1}{|H(jf,l)|^2} = K_{\text{FEXT}} N^{0.6} \cdot f^2 \,. \tag{6.5}$$

Measured NEXT and EL-FEXT functions of an 0.4-mm layered cable are shown in Fig. 6.2 and Fig. 6.3, respectively.

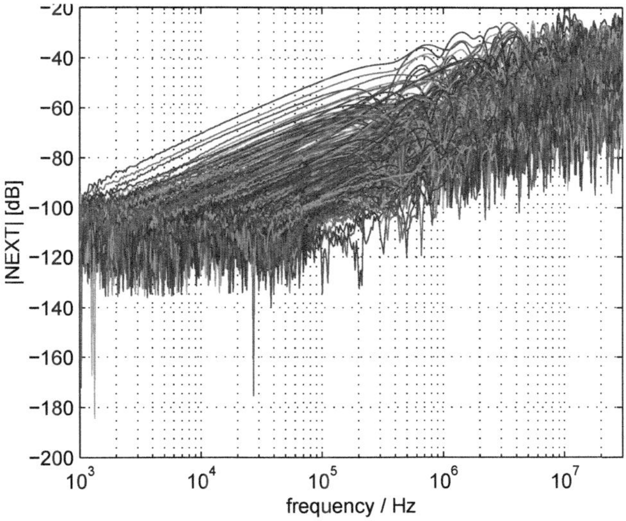

Figure 6.2: Measured NEXT of an 0.4-mm layered cable

Radio-frequency Interference (RFI) and Impulse Noise

Both kinds of ingress are due to non-symmetries of the twisted pair. The pair together acts as an antenna, which becomes visible in the *Common Mode*, the average mean of the two wires against ground. Due to non-symmetries, also in *Differential Mode*, an attenuated version of the signal will be present. Non-symmetries are characterized by unbalance parameters like Longitudinal Conversion Loss (LCL), Transverse Conversion Loss (TCL), Longitudinal Conversion Transfer Loss (LCTL), and Transverse Conversion Transfer Loss (TCTL). We will not define them in here, instead refer to [6].

Since **RFI** from amateur radio stations and others are very narrow-band compared to DSL services, they appear as a spike in the spectrum, i.e., almost a sinusoid in time domain. Countermeasures have been taken in VDSL by pulse shaping at the transmitter and windowing at the receiver. Transmitter pulse shaping reduced egress to radio receivers, windowing at the receiver reduces ingress. Otherwise, a

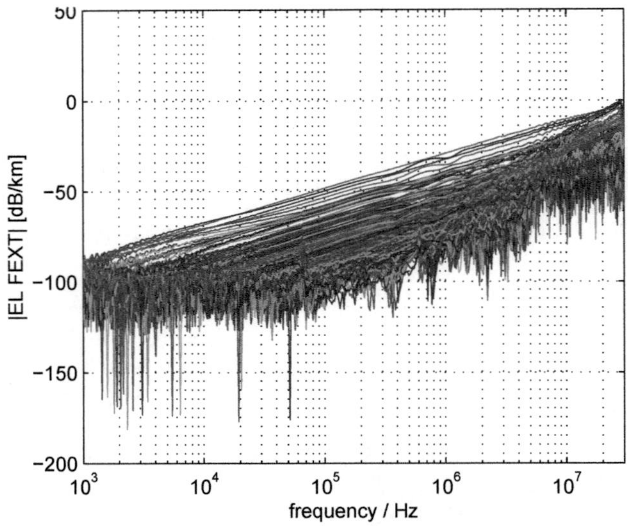

Figure 6.3: Measured EL-FEXT of an 0.4-mm layered cable

rectangular window will lead to a stronger leakage into neighboring frequencies. Combination of neighboring carriers are also a possible measure. Also cancellation methods based on the common mode have been proposed. These will be discussed in Section 6.3. RFI cancellation is necessary, since otherwise the desired receive signal would be hidden in the RFI signal at the analog interface, leading to saturation of the A/D-converter.

Impulse noise is another disturbance that can be seen as ingress, at least when central offices are digitalized. In earlier times, additional strong impulse noise was caused by relay switching in the central offices, which is not present in digitalized central offices any more. Nowadays, impulse noise is only caused by electrical home appliances, trains and trams, fluorescent tubes, ignition, lightning, and electrical machines, especially, e.g., welding. Figure 6.4 shows an impulse, measured in a workshop with a welding equipment and at another place caused by switching fluorescent tubes.
The coupling mechanism for impulse noise ingress is the same as for RFI disturbances, which means that common mode will also show a stronger interference than the differential mode, thereby offering possibilities for cancellation described in Section 6.3. The amplitude of impulse disturbances depends a lot on the cable infrastructure. Especially, aerial cables carry order of magnitudes higher impulse noise than buried cables.

In here, we shortly describe impulse-noise properties. Thorough studies on modeling and impulse noise generation can be found in [31–33]. The voltage follows a density similar to Weibull, given by

$$f_i(u) = \frac{1}{240u_0} e^{|-u/u_0|^{1/5}} \tag{6.6}$$

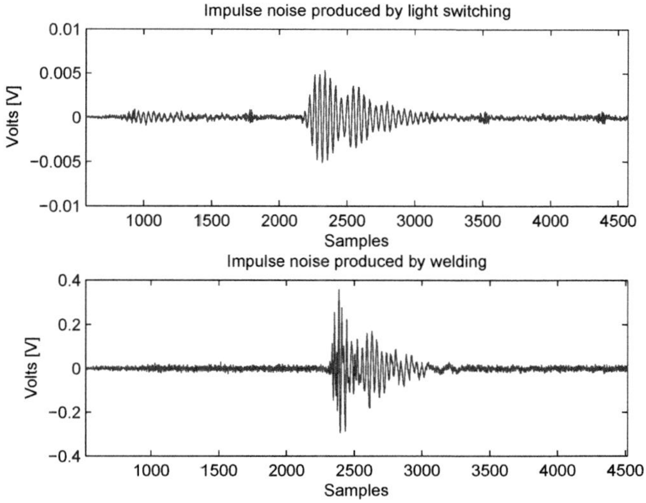

Figure 6.4: Impulses resulting from welding and fluorescent tubes, measured at a telephone socket

and the duration of an impulse follows roughly follows a combination of two log-normal densities in the form

$$f_l(t) = B\frac{1}{\sqrt{2\pi}s_1 t}e^{-\frac{1}{2s_1^2}\ln^2(t/t_1)} + (1-B)\frac{1}{\sqrt{2\pi}s_2 t}e^{-\frac{1}{2s_2^2}\ln^2(t/t_2)}. \quad (6.7)$$

In [33], furthermore, the inter-arrival times are determined by a Markov model where the states itself define a Poisson process. Additionally, a procedure is proposed, ensuring spectral properties and the voltage density at the same time.

6.1.2 Discrete MultiTone (DMT)

Twisted pairs have been designed for telephone services with only a bandwidth of 3.4 or 4 kHz. Necessarily, their transmission quality deteriorates with increasing frequency, determined by the attenuation and crosstalk functions, increasing with frequency. Thus, a baseband variant of OFDM is a must. We define DMT as

$$\begin{aligned} u(t) &= \sum_{i=0}^{N-1} F_i e^{j2\pi\frac{t}{T}i} = \sum_{i=1}^{N/2-1} F_i e^{j2\pi\frac{t}{T}i} + F_i^* e^{-j2\pi\frac{t}{T}i} \\ u_k &= \sum_{i=0}^{N-1} F_i e^{j\frac{2\pi}{N}ik} = \sum_{i=1}^{N/2-1} F_i e^{j\frac{2\pi}{N}ik} + F_i^* e^{-j\frac{2\pi}{N}ik} \end{aligned} \quad (6.8)$$

To make it a real time-domain signal, only half of the DFT components $i = 1, \ldots, N/2-1$ are available. The components $i = N-1, \ldots, N/2+1$ are conjugates and the components at 0 and $N/2$ are not used. They would otherwise have to be real, anyhow. Dependent on the underlying system, carriers up to number 5 or 32 may not be used, assuming the carrier spacing of 4.3125 kHz of ADSL and VDSL.

In contrast to wireless applications, where the cyclic prefix is typically chosen to

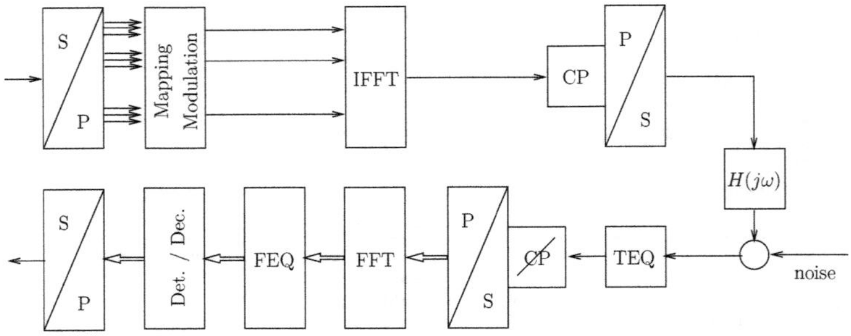

Figure 6.5: Components of DMT transmission

be a quarter of the symbol duration, DSL works with a very short cyclic prefix. In ADSL a cyclic prefix of 32 at a symbol duration of 512 samples is used, making it only 6.25 %. Nevertheless, the impulse response duration may significantly exceed the cyclic prefix (CP), which then requires a pre-equalization, called time-domain equalization (TEQ), to (roughly) shorten the impulse response to the length of the CP plus one. In wireless, this TEQ realization is not too feasible, due to a quickly time-variant channel. In wireline, the channel is almost stationary, apart from temperature changes and crosstalk variations due to switching of modems. The structure of the DMT transmission is given in Fig. 6.5. Additionally to the TEQ, the frequency-domain equalization is as in OFDM, except that, of course, only half of the carriers need to be equalized. There are quite some algorithms for time-domain equalization. In here, we may only mention a substitute-system approach, where the TEQ is adapted together with a virtual substitute system that represents channel plus equalizer. This reference system realizes the required short overall impulse response. The solution leads to an eigenvalue problem as shown, e.g., in [34]. Another option is to maximize the capacity taking into account the leakage effect of the DFT and external noise together with multidimensional search algorithms like downhill simplex or differential evolution [35].

Another specialty of DMT, due to the stationarity of the channel, is bit-loading in different forms. The oldest algorithm by Hughes-Hartogs [7] is a greedy approach, allocating bit-by-bit, where it costs the least incremental power at a given requested bit-error ratio (BER). Chow, Cioffi, and Bingham [8] use a modification of Shannon's capacity for the Gaussian channel, but introducing an additional margin, which is modified iteratively. Fisher and Huber [9] use the symbol-error ratio as a criterion, which leads to results tightly related to Chow et al.'s algorithm. George and Amrani [10] use a greedy approach not based on the incremental power, but on the increment in BER, assuming a constant power profile. Campello [11] uses a grouping of carriers with a similar (equal after quantization) capacity (after omitting the +1 in the argument). Finally, additional bits are placed in a Hughes-Hartogs/Levin- [12] -like fashion. Levin [12] introduced a bit-swapping argument when allocating bits.

There are manifold further algorithms, and they all lead to similar performances and differ mostly in their complexities. Especially, the oldest one by Hughes-Hartogs is the most complex of all. For these algorithms, we have worked out modifications that realize unequal error protection for different QoS requirements, e.g., [13–15]. For example, the Chow et al. algorithm is simply modified to realize UEP by introducing different margins for the different QoS classes. For robustness against impulse noise, one may put the most important information onto the worst carriers [13].

DMT, of course, has a the same drawback of a high peak-to-average power ratio (PAR) as OFDM. DMT has an almost Gaussian distribution in time domain, whereas OFDM leads to a Raleigh distribution of the amplitude due to Gaussian inphase and quadrature components. Many proposals for PAR reduction have been made. However, only the so-called Tone-Reservation method by Tellado [16] and oversampled versions thereof e.g., in [17] have made it into practical implementations due to the very low complexity. A further approach with acceptable complexity may be Partial Transmit Sequences based on a trellis search [18].

Further aspects of wireline transmission are upstream power back-off [19–21] to take care of the near-far problem, furthermore dynamic spectral management to allocate the available resources in a more dynamic way (e.g., iterative water filling [22, 23], NRIA [24], a game-theoretic approach [25]], and finally, MIMO approaches in the form of two-sided and one-sided processing. We shortly introduce these approaches which make use of the FEXT functions additionally to the transfer function. Note however, that such systems do only make sense for shorter distances below around 500 m. FEXT is attenuated by the transfer function of the cable and thus does not contribute significantly for longer loops. There are, of course, echo and NEXT cancelers, too, but we do not discuss them in here. In the following, we shortly discuss the MIMO approaches based on Singular-Value (SVD) or QR decomposition.

Two-Sided Processing for MIMO Based on SVD

The singular-value decomposition [26] rephrases the channel matrix in DFT domain $\mathbf{H}(\mathbf{n})$ at carrier number n as

$$\mathbf{Q}(n) \cdot \mathbf{\Lambda}(n) \cdot \mathbf{P}^H(n) \tag{6.9}$$

$\mathbf{\Lambda}(n)$ is a diagonal matrix. $\mathbf{P}(n)$ and $\mathbf{Q}(n)$ are unitary matrices.

Let $\mathbf{t}(n)$ and $\mathbf{r}(n)$ be input and output vectors, respectively.
At the transmitter side, we multiply the signal $\mathbf{t}(n)$ by $\mathbf{P}(n)$. Whereas, the signal at the receiver is multiplied by $\mathbf{Q}^H(n)$ to obtain the output $\mathbf{r}(n)$ (see Fig. 6.6).

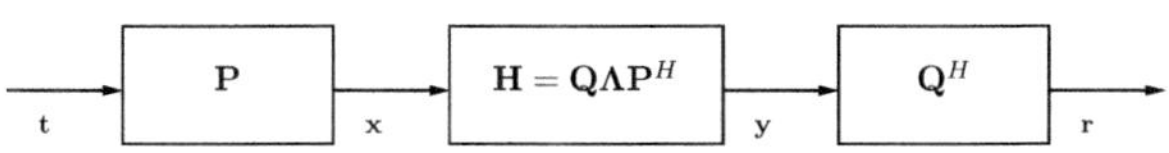

Figure 6.6: SVD MIMO diagonalization

$$\mathbf{r}(n) = \mathbf{Q}^H(n) \cdot \mathbf{y}(n) \tag{6.10}$$

Using the SVD in Eq. (6.9), we obtain

$$\mathbf{r}(n) = \mathbf{Q}^H(n) \cdot \mathbf{Q}(n) \cdot \mathbf{\Lambda} \cdot \mathbf{P}^H(n) \cdot \mathbf{x}(n) \tag{6.11}$$

$\mathbf{x}(n)$ is the product of $\mathbf{P}(n)$ and $\mathbf{t}(n)$,

$$\mathbf{r}(n) = \underbrace{\mathbf{Q}^H(n) \cdot \mathbf{Q}(n)}_{\mathbf{I}} \cdot \mathbf{\Lambda} \cdot \underbrace{\mathbf{P}^H(n) \cdot \mathbf{P}(n)}_{\mathbf{I}} \cdot \mathbf{t}(n) \tag{6.12}$$

OFDM and SVD as Reed-Solomon or RS-like codes RS codes are commonly defined as follows.

Definition 6.1.1. *A* **Reed-Solomon (RS) code** *of length N and minimum Hamming distance d_{Hm} is a set of vectors, whose components are the values of a polynomial $C(x) = x^l \cdot C'(x)$ of degree$\{C'(x)\} \leq K - 1 = N - d_{Hm}$ at positions z^k, with z being an element of order N from an arbitrary number field.*

$$c = (c_0, \ldots, c_{N-1}) \;, \quad c_i = C(x = z^i) \tag{6.13}$$

w_{Hm} and d_{Hm}, the minimum Hamming weight and distance, respectively, are known to be

$$w_{Hm} = \min ||c||_0 = d_{Hm} = N - (K-1) = N - K + 1 \;. \tag{6.14}$$

Since the samples are chosen to be powers of an element of order N, i.e., z^k, where z would be $e^{j2\pi/N}$ in the complex case, the equivalent description is known to be

$$c_i = z^{il} \cdot \sum_{k=0}^{N-1} C_k z^{ik} \;, \quad i = 0, \ldots, N-1 \;, \tag{6.15}$$

with $C_k = 0$ for $K \leq k \leq N-1$, Eq. (6.15) is nothing else than an IDFT with some consecutive DFT components unused. This means, OFDM inherently represents an RS code, when cyclically consecutive carriers are not used. In the case of DMT, these unused carrier positions need to be symmetric due to the conjugate symmetry.

Let us now compare the structures of OFDM and SVD.

$$\text{OFDM:} \quad \mathbf{W}^H \cdot \overbrace{\mathbf{W \Lambda W}^H}^{\text{channel}} \cdot \mathbf{W} \;, \tag{6.16}$$

$$\text{SVD:} \quad \mathbf{V}^H \cdot \underbrace{\mathbf{V \Lambda U}^H}_{\text{channel}} \cdot \mathbf{U} \;. \tag{6.17}$$

We observe that both make use of a pre- and post-processing by unitary matrices. IDFT and DFT diagonalize the channel Toeplitz matrix representing the convolution with the channel impulse response. SVD diagonalizes arbitrary channel matrices. The similarities lead to a more general concept of RS-like codes. The pre-processing matrix of the SVD offers a code with the same Hamming distance as an RS code for continuous and random channel matrices.

One-sided Processing for MIMO Based on QR Decomposition

Unfortunately, most of the time, we do not have the situation of a bundled connection of only a few hundred meters where SVD would be applicable. Distances between the central office and the cabinet are usually too long to profit from FEXT due to the attenuation of the loop. Short loops usually exist form the cabinet to customer premises or directly from the central office to the customer. These loops allow only for one-sided precessing at the cabinet (central office). This means post-processing for upstream and pre-processing for downstream. The linear algebra tool ideal for this task is the QR decomposition leading to a spatial decision feedback equalizer structure or Tomlinson-Harashima precoder, respectively [27–30].

QR decomposition for upstream processing For upstream processing we write the $L \times L$ channel matrix as (In the following, we omit the carrier index n.)

$$\mathbf{H} = \mathbf{QR} \,, \tag{6.18}$$

with a unitary matrix $\mathbf{Q}$ and an upper triangular matrix $\mathbf{R}$. Working with column vectors for information and received values, a post-processing with $\mathbf{Q}^H$ leads to[1]

$$\mathbf{r}^Q = \underbrace{\mathbf{Q}^H\mathbf{Q}}_{\mathbf{I}}\,\mathbf{Rt} + \mathbf{n} = \mathbf{Rt} + \mathbf{n} \,, \tag{6.19}$$

Due to the triangular matrix, a back-substitution approach can be applied to finally obtain an estimate of the information symbols $\hat{t}_k$, which may be further refined by channel decoding. The equation for $\hat{t}_k$

$$\hat{t}_k = \text{decode}\left[\frac{1}{R_{k,k}} \cdot r_k^Q - \sum_{j=k+1}^{L} \frac{R_{k,j}}{R_{k,k}} \cdot \hat{t}_j\right] \tag{6.20}$$

unveils the decision-feedback structure shown in Fig. 6.7.

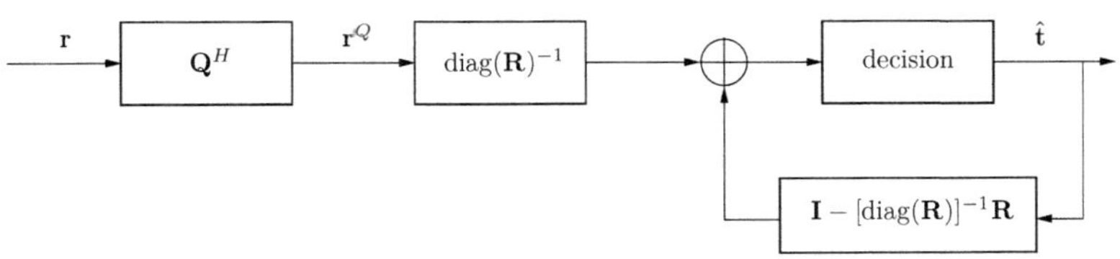

Figure 6.7: Spatial DFE structure resulting from QR decomposition

QR decomposition for downstream processing For downstream processing, the idea is to apply a QR decomposition to the transpose[2] of the channel matrix. This enables us to do a pre-processing instead of the post-processing of the previous paragraph. We hence obtain

$$\mathbf{H}^T = \mathbf{QR} \text{ , i.e., } \mathbf{H} = \mathbf{R}^T\mathbf{Q}^T \,. \tag{6.21}$$

[1] H stands for conjugate transpose.

[2] One may of course also take the conjugate transpose.

Pre-multiplying with $\mathbf{Q}^*$ leads to[3]

$$r^{Q_2} = \mathbf{H} \cdot \mathbf{Q}^* \cdot \mathbf{t}' = \mathbf{R}^T \cdot \underbrace{\mathbf{Q}^T \cdot \mathbf{Q}^*}_{=\mathbf{I}} \cdot \mathbf{t}' = \mathbf{R}^T \cdot \mathbf{t}' \tag{6.22}$$

Let us choose $\mathbf{t}'$ as

$$\mathbf{t}' = \mathbf{R}^{-T} \cdot \mathrm{diag}(\mathbf{R}^T) \cdot \mathbf{t} \tag{6.23}$$

The inverse of $\mathbf{R}^T$ already ensures the desired decoupling. The diagonal matrix is added to obtain a similar structure in the following formula as in (6.20). Equation (6.23) can equivalently be rephrased as

$$t'_k = \Gamma_M \left[t_k - \sum_{j=1}^{k-1} \frac{R_{j,k}}{R_{k,k}} \cdot t'_j \right] \tag{6.24}$$

Γ_M represents the modulo operation of the Tomlinson-Harashima precoding. Figure 6.8 shows the precoder structure.

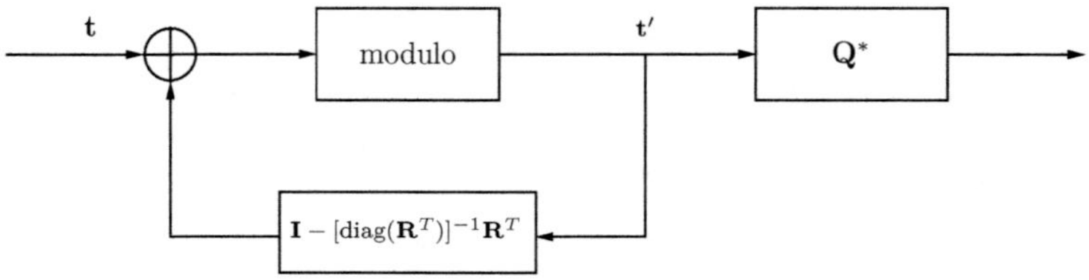

Figure 6.8: Spatial precoder structure resulting from QR decomposition

[3] * means conjugate

6.2 Optical OFDM Transmission and Optical Channel Properties

M. Mayrock, H. Haunstein, University Erlangen-Nürnberg, Germany

Commercially available systems for high bit-rate optical data transmission utilize on-off-keying or differential phase shift keying (DPSK) and reach bit-rates up to 40 Gbit/s. The transmission channel itself consists of single mode fibers exhibiting very low loss. In order to regenerate the power of the optical signal, optical amplifiers, e.g., erbium doped fiber amplifiers, are placed along the link. By this means the optical power is increased without the need of opto-electrical signal conversion. However, optical amplifiers introduce noise, which leads to a reduction of the optical signal-to-noise power ratio (OSNR). Depending on the signal powers and bandwidths along with other parameters, systems can be designed for transmission over hundreds to a couple of thousands kilometers.
When transmitting over long distances various kinds of signal distortion accumulate. Chromatic Dispersion (CD) describes the effect that different frequencies travel at different speeds along the optical wave-guide. Hence, pulses are broadened and finally inter-symbol interference (ISI) occurs. In today's systems CD-compensation is done with the help of dispersion compensating fibers or other optical devices like fiber gratings, which have to be adapted to the individual transmission scenario.
Polarization Mode dispersion (PMD) is a further performance limiting effect. Due to mechanical stress, imperfect fiber geometry, etc. the fiber becomes birefringent, i.e., signal contributions of orthogonal polarizations exhibit different speeds of propagation. The optical axes of different fiber sections are not aligned. Therefore, the "slow" and the "fast" modes of a certain piece of fiber couple into the optical axes of the next fiber section in a way which depends on their relative orientation. This effect occurs all along the fiber and eventually generates a fading-channel with a large number of echoes. There are methods for optical PMD compensation which usually compensate for first order differential group delay. Equalization techniques in the electrical domain could overcome this limitation. However, today's systems use direct detection receivers which convert the optical signal to the electrical domain via a non-linear operation and thus also electronic equalization shows limited performance.
It is expected that future systems will need higher signal bandwidth leading to strict requirements for dispersion compensation (CD and PMD). On the other hand more and more powerful signal processing resources can be implemented in today's integrated circuit fabrication processes. Therefore OFDM along with coherent detection is a candidate for future high bit-rate optical systems.

The demand for higher bit-rates can also be accounted for using techniques which operate at high spectral efficiency. OFDM is interesting for such systems, as it allows for dense wavelength multiplexing. Moreover the modulation alphabet for individual OFDM sub-carriers can be scaled easily and thus spectral efficiency can be adapted to given channel conditions. Recently, results of an OFDM transmission experiment have been published, where the authors report 5.6 bit/s/Hz spectral efficiency by transmitting eight times 66.8 Gbit/s over 640 km of uncompensated standard single-mode fiber [52]. Some modification in the set of system parameters allows for enhancement to 7.0 bit/s/Hz [53]. However, the signal-to-noise power ratio cannot be increased to arbitrarily high values by means of increasing the optical transmit power. This limitation is caused by the Kerr effect, which describes the variation of the refractive index of an optical wave-guide under variation of the optical signal power. As a consequence, distorting phase modulation is caused, which is a function of the signal power and finally leads to non-linear signal distortion (self phase modulation). Furthermore the effect causes non-linear crosstalk between signals of different wavelengths and between orthogonally polarized signals.

To overcome the drawbacks of the transmission impairments and to guarantee a bit error rate (BER) $< 10^{-16}$ Forward Error Correction (FEC) coding is inevitable. In classical non-coherent direct detection receivers processing at data rates up to 10 Gbit/s no channel capacity achieving FEC schemes were required, since a hard decision BER of already $< 10^{-3}$ could be attained. So, first-generation FEC schemes mainly relied on the (255, 239) Reed-Solomon (RS) code over the Galois field GF(256), with only 6.7% overhead. In particular, this code was recommended by the ITU for long-haul submarine transmissions [63]. Then, the development of Wavelength Division Multiplexing (WDM) technology provided the impetus for moving to second-generation FEC systems, based on concatenated codes with higher coding gains [64]. Nowadays, third-generation FEC schemes based on soft-decision decoding have become subject of interest since stronger FEC schemes are seen as a promising way to achieve performance close to channel capacity. Therefore a straightforward approach is bit-interleaved coded modulation with iterative decoding (BICM-ID) , which can be considered as the most simple approach to achieve high spectral efficiency while providing a low decoding complexity [68].

6.3 Impulse-Noise Cancellation

O. Graur, W. Henkel, Jacobs University Bremen, Germany

As previously discussed in Section 1.1, impulse noise can strongly affect transmission quality, occasionally even leading to DSL modem restart. In this section, a cancellation method is described which relies on the strong coupling of interference into Common-Mode [36].

6.3.1 Common Mode and Differential Mode

Differential-Mode (DM) signals have been the conventional approach of transmission over copper cables. The reason behind this is that they are less susceptible to strong interference such as impulse noise and RFI. Since DM signals appear as a voltage difference on two wires, any incident signals would couple equally, keeping the differential signal unchanged. Unlike DM, Common-Mode (CM) signals are taken as the arithmetic mean of the two signals measured with respect to ground, which makes them prone to interference. Both DM and CM signals are readily available on the receiver side.

$$x_{DM}(t) = x_1(t) - x_2(t) \tag{6.25}$$

$$x_{CM}(t) = \frac{x_1(t) + x_2(t)}{2} \tag{6.26}$$

CM signals consist mainly of ingress: independent noise, a component correlated with the desired signal from DM, and noise correlated with the noise in DM [41]. In the case of strong interference, such as in the case of impulse noise, there is a strong correlation between DM and CM signals. While the dominant component in CM will be the impulse, this might not necessarily be the case for DM. There, impulsive noise might be buried within the rest of the signal, making the detection less straightforward.

6.3.2 Coupling and Transfer Functions

For the rest of this section, channels were considered linear time invariant (LTI). It is also reasonable to assume that all the transmitted signals can be modeled as independent Gaussian random variables. Since neither statistical properties, nor coupling functions were previously defined for CM, our model relies on measurements. For the measurements in this section, a Swiss 0.4 mm cable with 50 pairs of length 100 m was used. As Fig. 6.9 shows, measurements revealed a −50 dB attenuation of signal coupling into CM, for frequencies below 2 MHz.

Figures 6.10 and 6.11 illustrate the NEXT and FEXT coupling functions obtained from measurements.

6.3.3 Common-Mode Reference-Based Canceler

The principle of impulse noise cancellation using the CM signal is illustrated in Fig. 6.12.

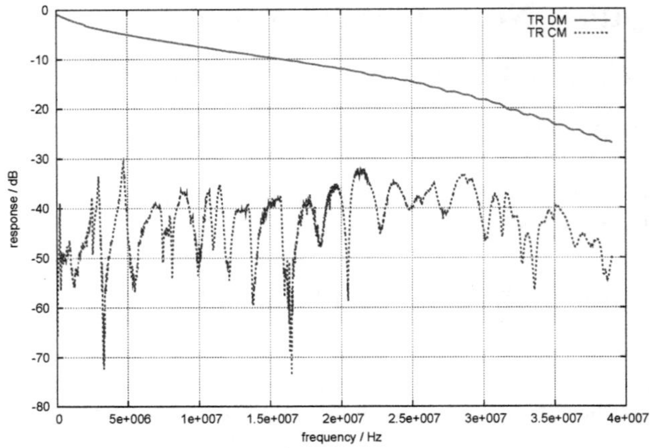

Figure 6.9: Transfer functions for DM and CM obtained from measurements of a 0.4 mm Swiss cable of length 100 m

For convenience, the impulses were received and saved into non overlapping blocks of length N, before attempting cancellation. Superscripts DM and CM refer to Differential-Mode and Common-Mode and the subscript in $\mathbf{H}_{j,i}^{DM}$ refers to the path from the ith pair into jth pair. The principle illustrated in Fig. 6.12 can be extended to a multipair cable with an arbitrary number of disturbers according to the Eq. (6.27). We assume L equal-length FEXT and K NEXT disturbers. We transmit signal $\mathbf{s}$ as a voltage difference at the transmitter side on pair j. At the receiver side, we measure two signals $\mathbf{y}_j^{DM}$ and $\mathbf{y}_j^{CM}$ which can be expressed as given in (6.27), where $\mathbf{s}_j$ is the transmitted signal of size Nx1 on pair j, $\mathbf{H}_{j,j}^{DM}$ denotes the NxN convolution matrix describing the DM to DM path on the jth pair. $\mathbf{w}^{DM}$ denotes uncorrelated AWGN in DM referred to as background noise, and $\mathbf{i}^{DM}$ represents the DM coupled impulse noise signal. Similar notation stands for CM signals.

$$
\begin{bmatrix} \mathbf{y}_j^{DM} \\ \mathbf{y}_j^{CM} \end{bmatrix} = \begin{bmatrix} \mathbf{H}_{j,j}^{DM} \\ \mathbf{H}_{j,j}^{CM} \end{bmatrix} \begin{bmatrix} \mathbf{s}_j \end{bmatrix} + \underbrace{\begin{bmatrix} \mathbf{H}_{j,1}^{DM} \cdots \mathbf{H}_{j,j-1}^{DM} \mathbf{H}_{j,j+1}^{DM} \cdots \mathbf{H}_{j,L}^{DM} \\ \mathbf{H}_{j,1}^{CM} \cdots \mathbf{H}_{j,j-1}^{CM} \mathbf{H}_{j,j+1}^{CM} \cdots \mathbf{H}_{j,L}^{CM} \end{bmatrix}}_{FEXT} \begin{bmatrix} \mathbf{s}_1 \\ \vdots \\ \mathbf{s}_{j-1} \\ \mathbf{s}_{j+1} \\ \vdots \\ \mathbf{s}_L \end{bmatrix}
$$

$$
+ \underbrace{\begin{bmatrix} \mathbf{H}_{j,L+1}^{DM} \cdots \mathbf{H}_{j,L+K}^{DM} \\ \mathbf{H}_{j,L+1}^{CM} \cdots \mathbf{H}_{j,L+K}^{CM} \end{bmatrix}}_{NEXT} \begin{bmatrix} \mathbf{v}_1 \\ \vdots \\ \mathbf{v}_K \end{bmatrix} + \underbrace{\begin{bmatrix} \mathbf{w}^{DM} \\ \mathbf{w}^{CM} \end{bmatrix}}_{AWGN} + \underbrace{\begin{bmatrix} \mathbf{i}^{DM} \\ \mathbf{i}^{CM} \end{bmatrix}}_{impulse\ noise} \tag{6.27}
$$

Measurements of impulse noise have been taken at inhouse phone outlets, both in DM and CM. Figure 6.13 presents an impulse measured both in DM and CM.

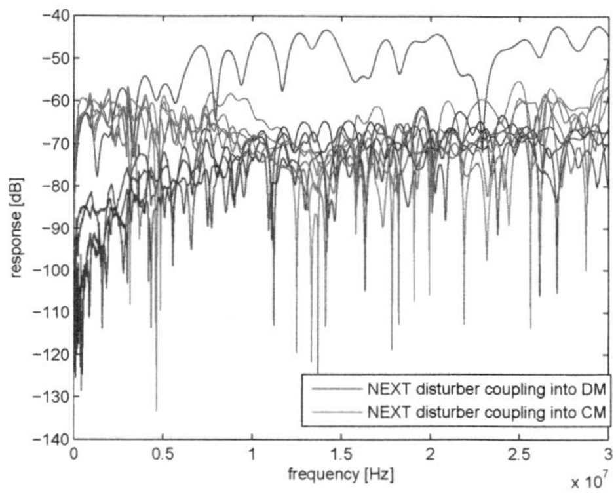

Figure 6.10: NEXT coupling functions, obtained from measurements of different TPs in the bundle. The outlier is due to measuring the other TP in the same star quad.

6.3.4 Impulse Noise Detection and Cancellation

Detection

The CM signal, besides providing a reference for most of the undesired interference in the system, comes with the advantage that its dominant component is impulse noise, which facilitates the detection of corrupted samples. Although many other detection methods for impulse noise have been successfully described in literature [47], the current section presents two simple methods. For the first method (6.14), in order to obtain the envelope, the CM is split into non overlapping frames of size M. Out of every frame, the maxim value is chosen and interpolation is performed among all local maxima. That is, after k distinct blocks of size M, $k-1$ values can be linearly interpolated, and from the corresponding $(k-1)M$ samples, the ones above a certain threshold τ can be flagged. Once flagged, the CM signal passes through the adaptive FIR filter which updates the coefficients only when a new flagged sample is detected. Under ideal assumptions, the resulting output would contain the DM signal, undistorted by impulse noise, and a term describing a minimum residual error.

A second method which can be easily implemented in the analog domain uses a rectifier and a low pass filter to detect the envelope of the CM signal (see Fig. 6.15).

Cancellation

For our simulations, the Normalized Least Mean Squares algorithm (NLMS) was used. NLMS is typically used due to its reduced computational complexity and robustness [46]. As illustrated in Eq. (6.27), the CM signal can be split into a component correlated with the DM noise, a component correlated with the DM signal and one uncorrelated component. This leakage of DM signal into CM poses

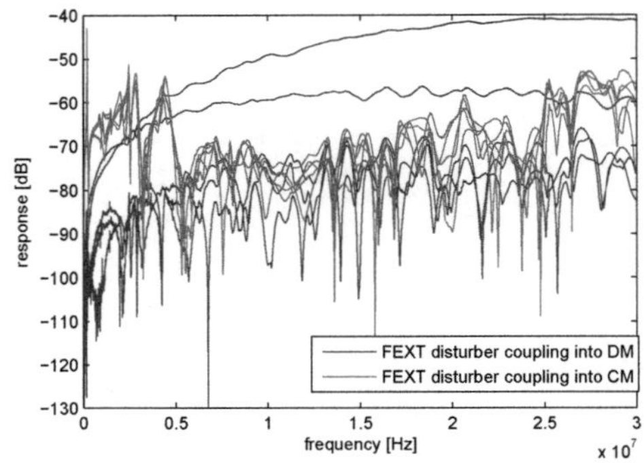

Figure 6.11: FEXT coupling functions, obtained from measurements of different TPs in the bundle. The outliers are due to measuring adjacent TPs

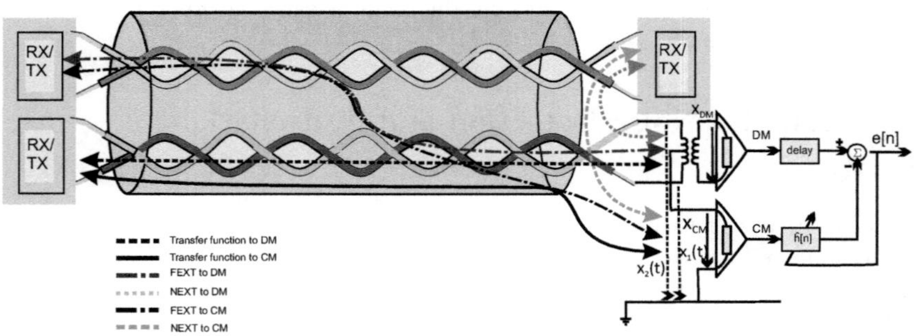

Figure 6.12: Coupling functions and canceler structure

the risk of canceling the useful component, which is much more likely for a high DM to CM coupling and a high SNR. One way to circumvent this problem is to update the filter coefficients only when the far-end transmitter is inactive [43].

Uncorrelated CM in-band noise induces the possibility that it will leak to the output of the adaptive filter, which will result in SNR loss. This undesired effect can be minimized by updating the filter coefficients only when impulse noise is detected in CM, which is what we went for in our simulations. Crosstalk is not canceled along with impulse noise since the total burst time is much smaller than the total transmission time, and the filter adaptation is performed sporadically.

6.3.5 Simulation Results

Impulse noise cancellation was investigated in the context of ADSL transmission. Coupling and transfer functions were measured for both DM and CM for a cable length of 100 m. Length-scaling for ADSL-specific loop distances was employed using the method described in Chapter 1.1. Note that the same length-scaling method was used both for DM and CM, although this might not necessarily be accurate in the case of CM. If was used, nevertheless, since no other model for a length dependency

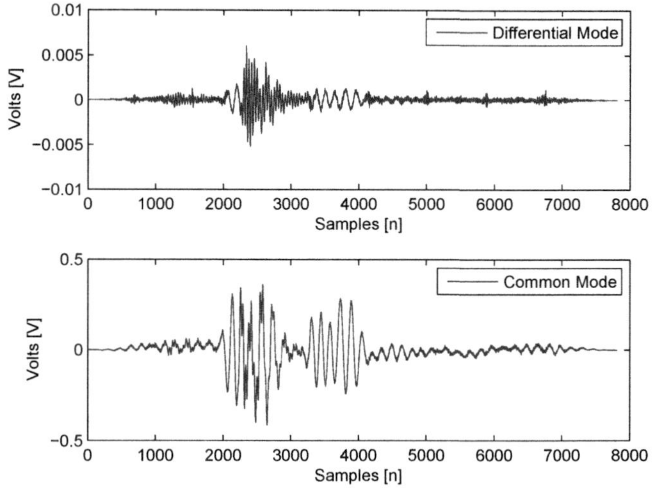

Figure 6.13: Impulse noise generated from light switching, both in DM and CM. Please note the different amplitudes of the waveforms.

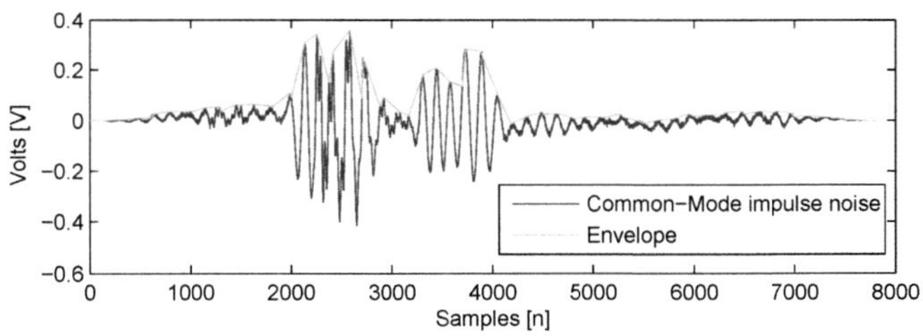

Figure 6.14: Envelope of CM signal (green)

could be found in literature[4] for CM transfer functions. Transmit signals were modeled according to the PSD of ADSL as specified in [49]. For NEXT modeling, the AslMx (German abbreviation for subscriber loop multiplexer) spectral mask [48] was used. Far-end crosstalk was generated as established in [49]. Simulations used sets of measured impulses generated in industrial settings (caused by welding), as well as in household environments (caused by light switching). ADSL transmission and reception was simulated, given the measured transfer and coupling functions, for different loop lengths and different number of NEXT and FEXT disturbers. Figure 6.16 depicts the canceler output (in blue) for an ADSL simulation employing 5 FEXT and 4 NEXT disturbers. For illustration purposes, since the amplitude of the measured impulse noise vectors was relatively small, the length of the loop was extended beyond ADSL-specific loop lengths, in order to achieve a lower SINR

[4]to the knowledge of the authors

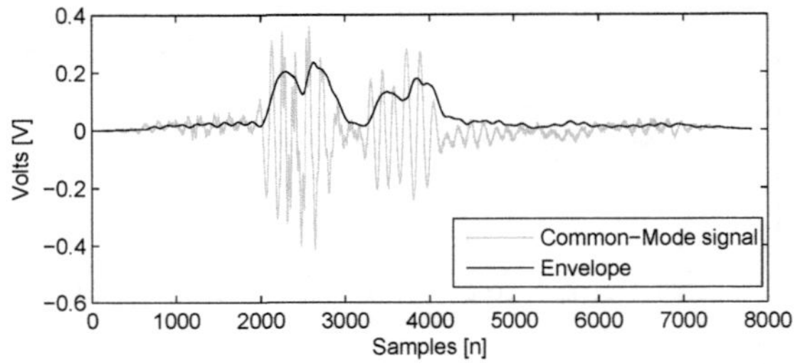

Figure 6.15: Second method for CM envelope detection

ratio. The green line in Fig. 6.16 depicts the overall received DM signal, which is corrupted by impulsive noise, while the black waveform illustrates the same DM signal, impulse noise free. As expected, the canceler produces a good estimate of the uncorrupted DM signal but does not suppress crosstalk. The red line presents the ideal transmitted signal, with no interference, only attenuated by the loop. For perfect impulse noise cancellation and no crosstalk cancellation, the blue line should resemble the black waveform as closely as possible, which is indeed the case.

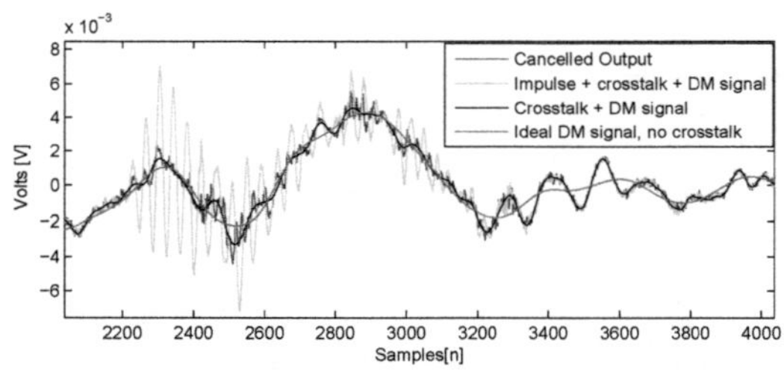

Figure 6.16: Output of the NLMS canceler (blue). The red line represents the ideal received signal, no crosstalk, no impulse noise, and no background noise. The black waveform denotes the sum of the ideal signal, 4 NEXT and 5 FEXT disturber signals, along with background noise. The green line illustrates the DM signal including impulse noise before cancellation. For perfect impulse noise cancellation and no crosstalk cancellation, the blue line should resemble the black signal as closely as possible.

6.4 Dual Polarization Optical OFDM Transmission

M. Mayrock, H. Haunstein, University Erlangen-Nürnberg, Germany

In this section we investigate an optical OFDM transmission system which deploys dual polarization transmission as well as wavelength division multiplexing. The system parameters are chosen according to the experimental setup of [52]. Our analysis is based on a system identification approach which treats the whole setup as a "weakly non-linear" system. This term reflects that the system characteristic is dominated by a linear transfer function. All kinds of distortion are treated as additive noise, i.e., besides actual noise, distortion due to non-linear effects is modeled as an additive noise-like contribution.

6.4.1 Setup

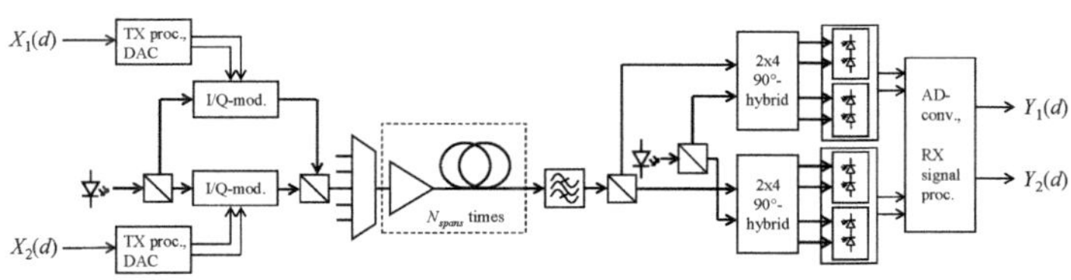

Figure 6.17: Investigated system: Dual polarization OFDM transmission, coherent detection.

Figure 6.17 depicts the investigated OFDM transmission system. Two independent baseband signals modulate the orthogonally polarized parts of the transmit laser signal. To achieve this, the signal of the TX laser source is split by a polarization beam splitter. Next, two external optical I/Q-modulators are applied before both signal contributions are recombined and launched into the optical waveguide. At the receiver, polarization diverse coherent detection is deployed. Once again polarization beam splitters are required to provide orthogonally polarized contributions of the received signal as well as the local laser to optical hybrids. Balanced photodetectors then convert their outputs to electrical representations of the inphase and quadrature components of both orthogonal RX signals. For our simulations the gross bit-rate per polarization is set to 30 Gbit/s. $Q = 108$ sub-carriers are modulated with symbols from a 16-QAM-alphabet. The cyclic prefix length equals 1/8 of the original OFDM symbol duration. This choice of system parameters reproduces a setup published in [52]. Eight WDM channels (8.4 GHz bandwidth each) on a 9 GHz grid are simulated; at the receiver there is a 12.5 GHz optical band-pass filter. Laser phase noise is accounted for by multiplication of the complex-valued receive signal with $\exp(j\phi(t))$. The random process $\phi(t)$ is obtained as an integral [55]

$$\phi(t) = \int_0^t \phi'(\tau)\,\mathrm{d}\tau. \qquad (6.28)$$

Here $\phi'(\tau)$ denotes zero-mean white Gaussian noise with power spectral density $2\pi\Delta\nu$. Finally, the parameter $\Delta\nu$ describes the laser line-width, which is set to 100 kHz in the sequel. The transmission link itself consisted of identical spans of standard single mode fiber (length: 80 km, chromatic dispersion coefficient: 17 ps/nm/km, attenuation: 0.2 dB/km). The simulation model for the optical channel considers the Kerr effect with the non-linear coefficient $\gamma = 1.33$/W/km [56]; polarization dependent loss is neglected. Optical amplifiers compensate for attenuation; their noise-figure is assumed to be 4 dB. It should be mentioned that there are no fibers or devices for optical dispersion compensation.

6.4.2 Noise Variance Estimation

In order to determine an estimate for the maximum achievable spectral efficiency, at first the signal distortion shall be quantified. As mentioned above, the analysis is based on a system identification approach which treats the whole setup as a "weakly non-linear" system, i.e., the system's characteristic is dominated by a linear transfer function. All kinds of distortion are treated as additive noise. The transmission of symbol vectors $[X_1(d)\ X_2(d)]^T$ on sub-carrier d can be written as

$$\begin{bmatrix} Y_1(d) \\ Y_2(d) \end{bmatrix} = \begin{bmatrix} H_{11}(d) & H_{12}(d) \\ H_{21}(d) & H_{22}(d) \end{bmatrix} \cdot \begin{bmatrix} X_1(d) \\ X_2(d) \end{bmatrix} + \begin{bmatrix} n_1(d) \\ n_2(d) \end{bmatrix}. \tag{6.29}$$

The samples $[n_1(d)\ n_2(d)]^T$ comprise noise which is added to the signal by optical amplification as well as distortion due to non-linear fiber effects modeled as an additive noise-like contribution. This assumption is not valid for arbitrary points of operation, but reasonable for values of optical powers where we expect best transmission performance.
After estimating the linear transfer characteristic (usually done with the help of pilot symbols) we subtract known data symbols (either further pilot symbols or data after decision) which have been affected by linear distortion through the channel from the received symbols. Then the relative noise variance for both receive branches can be determined [57]:

$$\frac{N_i}{S_i} = \frac{\sum_{d=1}^{Q} |n_i(d)|^2}{\sum_{d=1}^{Q} |H_{i1}(d)X_1(d) + H_{i2}(d)X_2(d)|^2}, \quad i \in \{1, 2\} \tag{6.30}$$

The summation in the nominator and denominator represents integration over the discrete frequency spectrum. In our simulations, the per-channel optical input power was varied from -12 to -3 dBm. The number of fiber spans ranges from 4 to 32. Based on the transmission of 100 OFDM symbols per polarization the inverse SNR of the orthogonal polarizations at the receiver is estimated according to (6.30). Figure 6.18 shows a contour plot which depicts N_1/S_1 in logarithmic scale. The relative noise power increases with longer transmission distances: from less than -20 dB for short links to more than -14 dB beyond 23 spans. Furthermore the plot shows the interrelation between estimated noise variance and optical input power: For low input powers, a power increment reduces the variance of additive noise at the receiver. At a certain power level, distortion due to fiber non-linearity comes

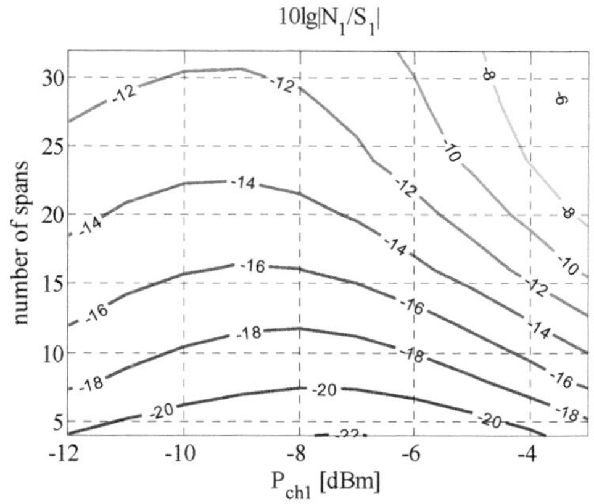

Figure 6.18: Estimated relative noise power.

into play and noise power increases.

Numerous simulations have been carried out in order to quantify these observations. Moreover, the influence of the launch powers of neighboring WDM channels on the noise variance has been studied. These results yield a method for separating the total noise variance into several contributions: Actual additive noise caused by optical amplifiers, noise-like distortion due to self phase modulation and contributions due to non-linear crosstalk between WDM channels [58].

Achievable Spectral Efficiency

According to Shannon the maximum information-rate which can be transmitted over a band-limited additive white Gaussian noise channel is

$$C = B \cdot \log_2\left(1 + \frac{S}{N}\right), \tag{6.31}$$

where S/N and B denote the signal-to-noise power ratio and the used bandwidth. Division by B results in the achievable spectral efficiency Γ. In order to obtain an estimate for the maximum achievable spectral efficiency of the simulated OFDM system we determine Γ for both receive branches on sub-carrier basis

$$\Gamma_i(d) = \log_2\left(1 + \frac{E\{|H_{i1}(d)X_1(d) + H_{i2}(d)X_2(d)|^2\}}{E\{|n_i(d)|^2\}}\right), \; i \in \{1,2\}. \tag{6.32}$$

The optical channel is assumed to be free of polarization dependent loss. Thus the contributions of the orthogonal polarizations are added up. Furthermore we average over the OFDM sub-carriers

$$\Gamma = \eta \cdot \left(\frac{1}{Q}\sum_{d=1}^{Q} \Gamma_1(d) + \frac{1}{Q}\sum_{d=1}^{Q} \Gamma_2(d)\right). \tag{6.33}$$

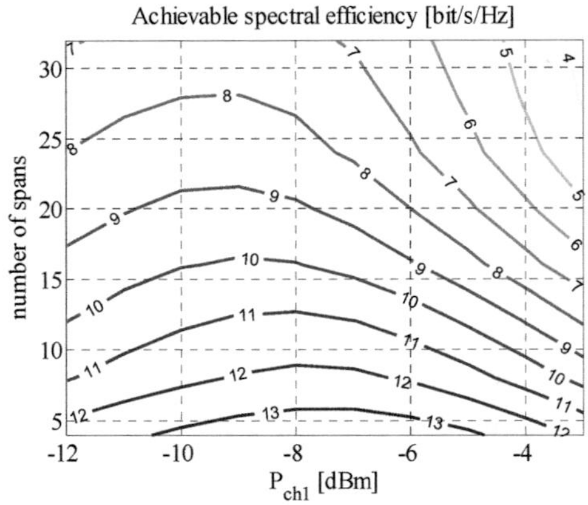

Figure 6.19: Achievable spectral efficiency versus distance and launch power.

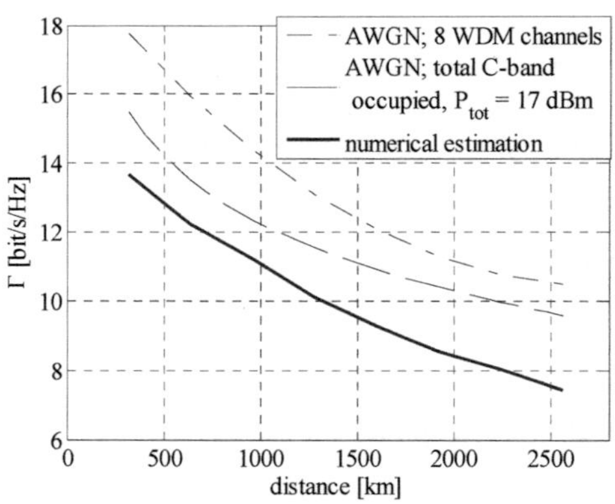

Figure 6.20: Achievable spectral efficiency vs. distance.

The coefficient η accounts for guard bands between WDM channels; in this scenario there is a fill factor $\eta = 8.4/9$. The simulation data obtained for distances of 4 to 32 spans and optical powers of -12 to -3 dBm was analyzed according to these equations. The results are summarized in the contour plot in Fig. 6.19. The simulated transmission scenario allows for spectral efficiency of slightly more than 13 bit/s/Hz for a distance of 4 fiber spans. For each transmission distance there is a distinct optical power level which maximizes the SNR in the sense of a weakly non-linear system and at the same time yields the corresponding capacity. The extraction of the associated maximum achievable spectral efficiency versus distance is shown in Fig. 6.20. We observe a capacity decay to about 7.5 bit/s/Hz for a distance of 2500 km. Due to the non-linear characteristic of the channel these results are a valid estimate for the given optical setup, powers, and WDM-parameters. The channel capacity is limited by actual additive noise along with non-linear signal distortion and non-linear cross-talk. For a comparison the diagram shows two curves which were obtained from approaches which consider linear AWGN channels. The investigated transmission system is assumed to consist of equally spaced, identical optical amplifiers. The optical SNR (OSNR) after N_{spans} amplifiers is given by [59]

$$\mathrm{OSNR} = \frac{P_{opt} \cdot \lambda_0}{G\,F_N \cdot h \cdot c \cdot N_{spans} \cdot B_{ref}}. \tag{6.34}$$

Here h and c denote Planck's constant and the speed of light, respectively. G is the amplifiers' gain which shall equal the loss of one fiber span. The noise figure is given by F_N. Finally λ_0 and B_{ref} denote the reference wavelength (1550 nm) and bandwidth for noise power measurement. In optical communications usually B_{ref} equals 12.5 GHz. However, insertion of the actual OFDM signal bandwidth leads to an accurate SNR estimate which can be applied to Shannon's equation. The upper curve in Fig. 6.20 was obtained by this calculation, whereby for each transmission distance the optimum optical launch power found in Fig. 6.19 was inserted. It can be observed that the numerical estimates, which consider the fiber non-linearity as an additive noise contribution, deviate from the pure AWGN channel by approximately 3 bit/s/Hz. Hence, at the optimum point of operation with respect to input power, the system still suffers from a distinct amount of distortion due to non-linear effects. An alternative way of finding a sensible value for the per-channel launch power is based on WDM transmission parameters. Commercially availably optical amplifiers operate in the so-called C-band which approximately ranges from 1525 to 1565 nm. Equivalently a total band-width of 5.0 THz can be used for parallel transmission. The maximum optical power typically reaches 17 dBm, e.g., used in [60]. In the case where the total bandwidth shall be occupied by OFDM bands on a 9 GHz grid, the per-channel launch power amounts to $P_{opt} = 10.4$ dBm. These considerations are the basis for the dashed line in Fig. 6.20, obviously leading to reduced capacity compared to the 8 channel AWGN scenario. Full occupation of the C-band results in penalty due to increased non-linear signal distortion. The numerical estimates for the achievable spectral efficiency in Fig. 6.20 exhibit a gap of approximately 6 bit/s/Hz compared to the reported values of transmission experiments [52] [53]. These results suggest that an increase of spectral efficiency is possible for the used fiber links. However, first the influence of quantization noise has to be incorporated into the simulation model.

6.4.3 ADC/DAC Resolution

High speed devices for analog-to-digital and digital-to-analog conversion (ADC and DAC) with sampling rates beyond 20 GS/s have been demonstrated [61] as well as the integration of four such devices, which are needed for sampling of inphase and quadrature components of the received signals of both polarizations. At such high sampling rates these devices have a more or less limited number of quantization levels. The authors of [61] report 6 bits resolution while state-of-the-art measurement devices provide higher accuracy. Thus, quantization noise as well as clipping has to be taken into consideration as performance limiting factors for high speed optical OFDM transmission. For this purpose the simulation models of the transmitter and receiver were extended by DAC and ADC device characteristics.

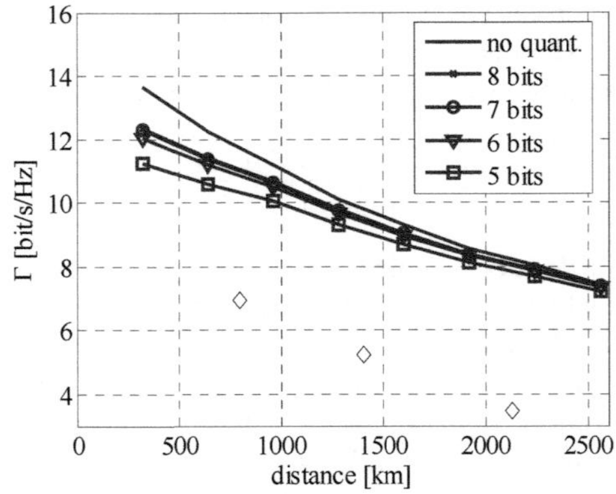

Figure 6.21: Achievable spectral efficiency versus distance considering quantization noise. Diamonds denote actual spectral efficiencies; pre-FEC BER: 10^{-3}; code rate 0.93.

In further simulations the number of quantization levels was varied. The clipping ratio (maximum amplitude over the signal's root-mean-square) was set to 10 dB. Fig. 6.21 depicts the resulting estimated achievable spectral efficiency. One can observe that there is marginal difference between the curves corresponding to 8, 7, and 6 bit resolution. In these cases there is a loss of approximately 1 bit/s/Hz for a distance of 500 km and just 0.5 bit/s/Hz for 1000 km. This loss is mainly attributed to distortion due to clipping. For longer transmission distances optical amplifier noise and fiber non-linear effects dominate. Usage of 5 bit DACs/ADCs introduces a distinct amount of quantization noise. Therefore, for actual implementations it will be desirable to use at least 6 bit quantization. In a further analysis of the simulation data of Fig. 6.18, transmission distances were determined, which lead to a bit error ratio of 10^{-3} when different QAM alphabets are used. This value is a typical maximum pre-FEC bit error ratio, which can be handled by forward error correction codes, which have been standardized for optical communications. These codes exhibit a rate of 0.93. Diamonds in Fig. 6.21 show respective distances along with the achieved spectral efficiency for 16QAM, 8QAM, and 4QAM. The calculation of the spectral efficiency considers the code rate as well as the fill factor of the WDM spectrum. Hence, such systems are assumed to transmit over 800 km at a spectral efficiency of 7 bit/s/Hz. Error-free transmission at 3.5 bit/s/Hz is supposed to work over 2100 km. These results imply that transmission performance of future systems may be enhanced by introducing more powerful FEC coding schemes.

6.5 Forward Error Correction

T. Lotz, W. Sauer-Greff, R. Urbansky, University of Kaiserslautern, Germany

Classical optical communication systems processing at data rates up to 10 Gbit/s employ non-coherent direct detection receivers that exhibit a pre-FEC BER $< 10^{-3}$. Therefore a (255,239) Reed Solomon (RS) code is sufficient to guarantee an overall BER $< 10^{-16}$. But due the increasing demand of higher signal bandwidth, more powerful FEC-schemes, such as turbo codes [65] or LDPC codes [66] have to be considered, in order to allow processing close to the theoretical Shannon limit. Hence we applied the concept of bit-interleaved coded modulation with iterative decoding (BICM-ID) to our system since it takes into account the multilevel characteristics of the signal as well as the code properties in an iterative decoding process. The basic idea of BICM-ID and the obtained simulation results are presented in the following.

6.5.1 BICM-ID System Model

Traditionally iterative decoding is applied with either a parallel or a serial concatenation of at least two convolutional codes [65]. But as noted in [67] the "Turbo principle" can be used not only with traditional concatenated coding schemes, but is more generally applicable to several other schemes that can be found in modern digital communications. Due to the use of a multilevel modulation scheme, we are able to adapt the Turbo Principle to iterative soft-demapping together with channel decoding, also referred to as BICM-ID [68].

The basic structure of a BICM encoder and iterative decoder is given in Fig. 6.22 [68].

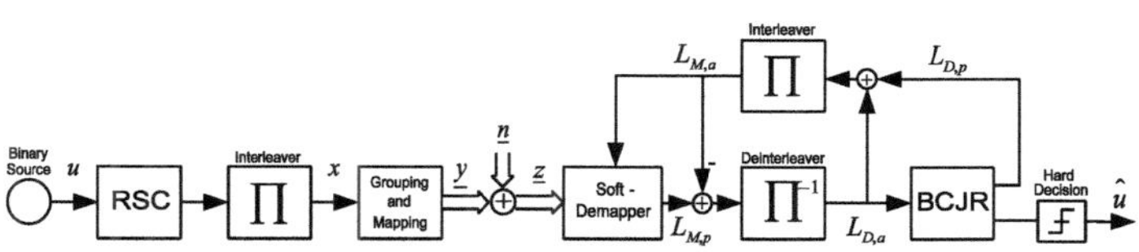

Figure 6.22: Iterative demapping and decoding, BICM-ID system configuration

The concept is as follows: The transmission is done on block basis where an info sequence of k bits is convolutionally encoded to a code word (CW) of $n > k$ bits with a recursive systematic convolutional (RSC) code. The generated CW is fed to a random bit interleaver which is a pre-requirement for an iterative receiver to guarantee statistical independence between adjacent bits. After interleaving the grouping operation takes M coded bits $x_{0,..,M-1} \in \{0,1\}$ to form the complex output symbol $\underline{y} = \text{map}(x_{0,..,M-1})$. This mapping operation map() is essential for the iterative demapping and decoding, since it links up the several bits $x_{0,..,M-1}$ to a constellation symbol and mutual dependencies arise between them. With traditional non-iterative decoding the mapping operation typically applies a Gray labeling for the lowest pre-FEC BER. With Gray mapping, neighboring signal points differ by only one binary digit in their binary decomposition. However, as is well known, Gray mapping performs worst for iterative demapping and decoding [68]. Hence different mapping strategies need to be considered, as discussed in Section 6.5.2.

After the mapping operation the complex symbols $\underline{y}$ are transmitted over the channel. For the assumption of an AWGN channel, which is valid for the investigated operation area of the optical OFDM system, the symbols $\underline{z}$ at the input of the soft-demapper are only corrupted by noise, so $\underline{z} = \underline{y} + \underline{n}$, where $\underline{n} = n_I + jn_Q$, with n_I, n_Q being realizations of two independent Gaussian random variables. Their variance is the double-sided noise power spectral density $\sigma^2 = \sigma_I^2 = \sigma_Q^2 = N_0/2$.

In the soft-demapper the channel symbols $\underline{z}$ are demapped and ungrouped to each M reliabilities in form of log-likelihood ratio (LLR) values $L_{M,p}$. In the first iteration the a-posteriori probabilities $L_{M,p}$, computed by the soft-demapper, are deinterleaved and soft-in/soft-out decoded in a symbol-by-symbol APP estimator, implemented by the BCJR algorithm [69]. At the output of the BCJR we find the APP LLR values $L_{D,p}$. For iterative decoding it is necessary to exchange only extrinsic information between the soft-demapper and decoder, so we have to remove statistically dependent information. Therefore the soft-input LLR values of the coded bits for the soft-demapper are the interleaved a-posteriori LLR values of the BCJR without the a-priori information, $L_{M,a} = \Pi\{L_{D,p} - L_{D,a}\}$. Respectively, the a-priori LLR values for the BCJR are computed as the deinterleaved extrinsic LLR values of the soft-demapper, $L_{D,a} = \Pi^{-1}\{L_{M,p} - L_{M,a}\}$.

The BCJR decoder performs bitwise soft-input processing, thus the demapper has to extract a soft value of each coded bit $x_{0,..,M-1}$ of a 2^M-ary complex channel symbol $\underline{z}$. When a-priori information is available, the APP soft information LLR value $L_{M,p}$ of bit k is computed as

$$L_{M,p}(x_k|\underline{z}) = \ln \frac{p(x_k = 1|\underline{z})}{p(x_k = 0|\underline{z})} =$$

$$L_{M,a}(x_k) + \ln \frac{\sum\limits_{i=0}^{2^{M-1}-1} p\left(\underline{z}|x_k{=}1, x_{j\neq k}{\equiv}\mathrm{bin}(i)\right)\cdot \exp \sum\limits_{\substack{j=0,j\neq k \\ \mathrm{btst}(i,h)=1}}^{M-1} L_{M,a}(x_j)}{\sum\limits_{i=0}^{2^{M-1}-1} p\left(\underline{z}|x_k{=}0, x_{j\neq k}{\equiv}\mathrm{bin}(i)\right)\cdot \exp \sum\limits_{\substack{j=0,j\neq k \\ \mathrm{btst}(i,h)=1}}^{M-1} L_{M,a}(x_j)}, \tag{6.35}$$

where $x_{j\neq k} \equiv \mathrm{bin}(i)$ denotes the joint event of the variables $x_{j\neq k}; k, j \in \{0..M-1\}$ having the values 0, 1 according to the binary decomposition of i. The function $\mathrm{btst}(i, h)$ takes the value '1' if bit number h is set in the binary decomposition of i, otherwise it is '0', where

$$h = \begin{cases} j & , j < k \\ j-1 & , j \geq k. \end{cases} \tag{6.36}$$

When the channel symbols $\underline{z}$ are corrupted by complex Gaussian noise, the conditional PDF calculates to

$$p(\underline{z}|\underline{y}) = \frac{1}{2\pi\sigma^2} \exp\left(-\frac{1}{2\sigma^2}|\underline{z} - \underline{y}|^2\right). \tag{6.37}$$

With (6.37) we can rewrite the soft-demapping algorithm of (6.35) for the complex

AWGN channel as

$$L_{M,p}(x_k|\underline{z}) = L_{M,a}(x_k) + \ln \frac{\sum\limits_{i=0}^{2^{M-1}-1} \exp\left[-\frac{1}{2\sigma^2} |\underline{z} - \underline{y}_{k,(1,i)}|^2 + \sum\limits_{\substack{j=0, j\neq k \\ \text{btst}(i,h)=1}}^{M-1} L_{M,a}(x_j) \right]}{\sum\limits_{i=0}^{2^{M-1}-1} \exp\left[-\frac{1}{2\sigma^2} |\underline{z} - \underline{y}_{k,(0,i)}|^2 + \sum\limits_{\substack{j=0, j\neq k \\ \text{btst}(i,h)=1}}^{M-1} L_{M,a}(x_j) \right]}, \quad (6.38)$$

where $\underline{y}_{k,(1,i)} = \text{map}(x_k{=}1, x_{j\neq k}{\equiv}\text{bin}(i))$ and $\underline{y}_{k,(0,i)} = \text{map}(x_k{=}0, x_{j\neq k}{\equiv}\text{bin}(i))$, $j \in \{0..M-1\}$.

6.5.2 Influence of the Applied Mapping

As noted before, the choice of the mapping is crucial to achieve a good performance of BICM-ID. If Gray mapping is applied the initial BER performance is acceptable, but the performance of the iteration loop is inferior, whereas for, e.g., an anti-Gray mapping its convergence behavior can be significantly improved for the price of a worse initial BER. So, for anti-Gray mapping a quite high SNR is required to reach the so called "turbo cliff", which is the required SNR at which the decoding process starts to deliver an iteration gain.

However, systems applying the Turbo Principle are known to run into a certain error floor. Hence, if a lower BER is intended outer coding is indispensable. When the necessary pre-FEC BER of the outer FEC scheme is known, it is possible to optimize the BICM-ID system to reach that specific error region by applying an irregular modulation at a lower SNR compared to the case of applying either a pure Gray or a pure anti-Gray mapping. So the symbols within a CW are mapped to a ratio of α according to Gray and to a ratio $1 - \alpha$ according to anti-Gray,

$$\alpha = \frac{N_{Gray}}{N_{Gray} + N_{Anti-Gray}} \quad (6.39)$$

with $N_{\{Gray,Anti-Gray\}}$ denoting the number of symbols within a CW owning a Gray or anti-Gray mapping, respectively [62].

6.5.3 Simulations on the Performance of Coded OFDM

In this section BICM-ID is optimized according to the properties of the optical OFDM system. The investigated coded OFDM transmission system with polarization multiplexing is depicted in Fig. 6.23.

Outer coding is implemented by the (255,239) RS-Code code of rate $R_{RS} = 0.937$ over the Galois field GF(256), specified by the ITU in [63]. This particular code has the property to guarantee a BER $< 10^{-16}$ when the RS decoder input BER is $< 10^{-3}$.

To enhance the error correction capabilities of the transmission system, inner coding is performed with a BICM encoder. So, the RS-encoder output sequence u is convolutionally encoded with a RSC encoder of memory 5 and punctured to result

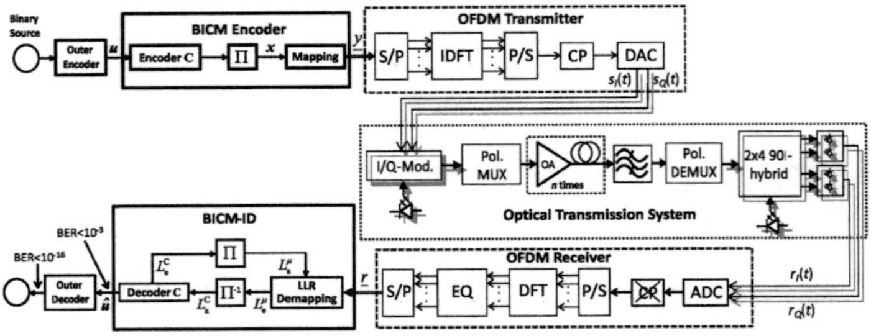

Figure 6.23: Complete optical COFDM system

in a rate of $R_{RSC} = 0.8$. Considering the particular outer code, the systems total code rate is $R = 0.75$. Since transmission is performed on a block basis the CW length was set to $N = 30720$, which is the number of bits in 10 OFDM symbols resulting from the used OFDM parameters. An analysis on the optimum mapping, specifically the ratio α between Gray and anti-Gray mapping, is given in below.

For our simulations the gross bit rate per polarization is set to 75 Gbit/s. $Q = 256$ sub-carriers are modulated with a constellation size of 64-QAM. The cyclic prefix length equals 1/12 of the original OFDM symbol duration.

In the optical transmission sub-system of Fig. 6.23 we find a simplified illustration of the system given in Fig. 6.17. Shadowing indicates that those units are available twice due to the separate use of both polarizations. The properties of the transmission link and the modulating laser are the same as in Section 6.4.1. In our simulations we included five WDM channels with 13.5 GHz bandwidth each on a 14 GHz grid. The pre-channel optical input power was varied from -12 to 0 dBm. The number of fiber spans was set to 12; the whole transmission link has a length of 960 km. The receiver includes a 18 GHz optical band-pass; sampling was performed at a rate of 24 GHz with a resolution of 6 bit. Taking into account the code and OFDM properties we achieve a spectral efficiency of 8 bit/s/Hz in both polarizations.

The resulting BERs after 7 iterations are shown for different channel input powers and different mappings in Fig. 6.24 .

Obviously, only an irregular modulation with $\alpha = 0.5$ exhibits after a finite number of iterations a BER $< 10^{-3}$, which is sufficient for the outer RS-code to obtain a total BER $< 10^{-16}$.

An explanation of BICM-ID to reach a turbo cliff only when applying an irregular modulation can be found by meas of extrinsic information transfer (EXIT) chart analysis, which is a powerful tool to visualize the flow of mutual information (MI) between decoder and demapper in the iterative decoding process. In the EXIT chart the MI of the decoder/demapper is plotted versus its a-priori input, i.e., the MI of the demapper/decoder. If an optimum demapper/decoder is used, the knowledge of the MI contained in the a-priori information is sufficient to derive the MI of the decoder/demapper. Denoting the encoder/demapper input by X_a and the corresponding extrinsic decoder/demapper output by X_e, the MI $I(X_e; X_a)$ calculates

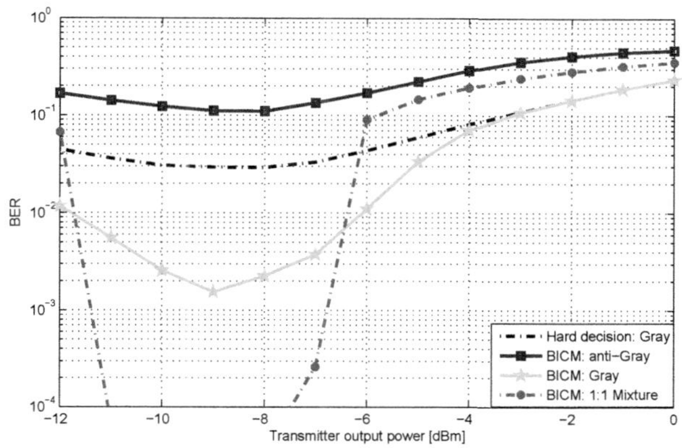

Figure 6.24: Simulations on systems performance; BER at BICM-ID output

to

$$I(X_e; X_a) = \int_{X_e} \int_{X_a} f(x_e, x_a) \log_2 \frac{f(x_e, x_a)}{f(x_e) f(x_a)} \delta x_a \delta x_e. \quad (6.40)$$

Figure 6.25 depicts the EXIT functions of decoder and demapper for different mapping schemes at the optimum optical input power set to −9 dBm. Obviously, when Gray mapping is applied the EXIT function is flat, so the performance of the iteration loop is inferior. Whereas anti-Gray seems to be more promising since the extrinsic MI $I_{\text{e,demapper}}$ of the demapper increases with an increasing a-priori MI $I_{\text{a,demapper}}$, but due to the fact that for low $I_{\text{a,demapper}}$ also $I_{\text{e,demapper}}$ is low, the EXIT functions of demapper and decoder intersect significantly before that of the Gray mapping. So its performance is even worse than the hard decision of Gray mapping. Only an irregular modulation combines the benefits of both mappings, which allows the EXIT functions to intersect at an MI higher than that necessary to obtained a BER $< 10^{-3}$, visualized by the dashed line.

6.6 Summary

This chapter gave an impression of some of the research issues related to the wireline use of multicarrier modulation. Many aspects are similar as in wireless, but the channels offer different possibilities or have other challenges, such as, e.g., more stationary behavior, an additional common mode in twisted-pair, or non-linearities in optical communication.

In the case of twisted-pair communication, we showed that the additionally available common-mode signal can serve as a reference for impulse-noise cancellation. The common mode predominantly shows external disturbances like radio-frequency interference and impulse noise. Crosstalk and the own signal are of less influence. The amplitude of impulse noise on common mode is significantly higher than on differential mode and provides an ideal reference input for a canceler to significantly

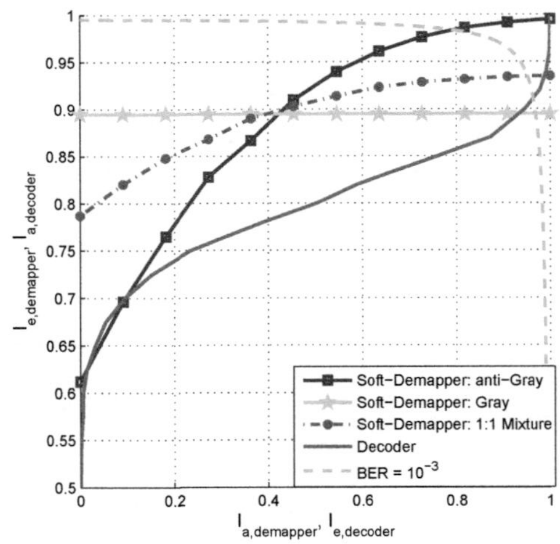

Figure 6.25: EXIT functions of decoder and soft-demapper for different mappings at optical input power of -9 dBm.

reduce impulse-noise in differential mode.

In optics, we were aiming at a maximum achievable spectral efficiency, which is limited by amplifier noise and non-linear fiber effects causing signal distortion and non-linear crosstalk between WDM channels. A weakly non-linear approach was applied which treats all these kinds of distortion as additive noise. Based on noise variance determination, estimates for the maximum achievable spectral efficiency are obtained, which amount to a range of 13 bit/s/Hz to 7.5 bit/s/Hz for distances from 320 km up to 2500 km. Realistic ADC/DAC characteristics turned out to slightly reduce the achievable spectral efficiency for short to medium distances. For long-haul transmission, quantization noise is not dominating. Forward error correction codes standardized for optical transmission leave a gap of approximately 4 bit/s/Hz to the estimated upper bound.

To overcome this gap, we adapted the principle of an iterative decoding scheme, namely BICM-ID, to our system. The obtained results for the investigated optical OFDM system promise to reach a spectral efficiency of 8 bit/s/Hz for a 960 km fiber link. However, we expect to enhance the performance beyond that by applying even more powerful FEC schemes, such as multi-level codes or LDPC codes.

Bibliography

[1] http://trsys.faculty.jacobs-university.de/files/DSL_book_project.pdf

[2] T. Starr, J.M. Cioffi, and P.J. Silverman, *Understanding Digital Subscriber Line Technology*, Prentice Hall, 1998.

[3] J.M. Cioffi, T. Starr, M. Sorbara, *DSL Advances*, Prentice Hall, 2002.

[4] W.Y. Chen, "Dsl: Simulation Techniques and Standards Development for Digital Subscriber Line Systems," Macmillan Technology Series, 1998.

[5] P. Golden, H. Debieu, K. Jacobsen, edts., *Fundamentals of DSL Technology*, Auerbach Publications, Boca Raton, FL, USA, 2006.

[6] ETSI TM 6 contribution, Luleå, Sweden, June 22-26, 1998, 970p02r3.

[7] D. Hughes-Hartogs, D., Ensemble Modem Structure for Imperfect Transmission Media, U.S. Patents, Nos. 4,679,227 (July 1987); 4,731,816 (March, 1988); and 4,833,706 (May 1989).

[8] P.S. Chow, J.M. Cioffi, J.A.C. Bingham, "A practical discrete multitone transceiver loading algorithm for data transmission over spectrally shaped channels," *IEEE Transactions on Communications*, Vol. 43, No. 234, pp. 773-775, Feb-Mar-Apr 1995.

[9] R.F.H. Fischer, J.B. Huber, "A New Loading Algorithm for Discrete Multitone Transmission," *proc. Globecom 1996*, Vol. 1, Nov. 18-22, 1996, pp. 724-728.

[10] Y. George, O. Amrani, "Bit loading algorithms for OFDM," *proc. International Symposium on Information Theory (ISIT) 2004*, p. 391, June 27 - July 2, 2004. 2004 Page(s):391 - 391

[11] J. Campello, "A Practical Bit Loading for DMT," *proc. ICC 1999*, Vancouver, Vol. 2, pp. 801-805, June 1999.

[12] H.E. Levin, "A complete and optimal data allocation method for practical discrete multitone systems," *GLOBECOM 2001*, San Antonio, TX, USA, Vol. 1, pp. 369-374, Nov. 25-29, 2001.

[13] W. Henkel and K. Hassan, "OFDM (DMT) Bit and Power Loading for Unequal Error Protection," *OFDM-Workshop 2006*, Hamburg, Aug. 30-31, 2006.

[14] K. Hassan and W. Henkel, "UEP with Adaptive Multilevel Embedded Modulation for MIMO-OFDM Systems," *OFDM-Workshop 2008*, Hamburg, Aug. 27-28, 2008.

[15] F. Yu, A. Willson, "A DMT Transceiver Loading Algorithm for Data Transmission with Unequal Priority over Band-limited Channels," *proc. Signals, Systems, and Computers*, 1999, Pacific Grove, CA, USA, Vol. 1, pp. 685-689, Oct. 24-27, 1999.

[16] J. Tellado, *Peak-to-Average Power Reduction for Multicarrier Modulation*, Ph.D. thesis, Stanford University, Sept. 1999.

[17] W. Henkel, V. Zrno, "PAR Reduction Revisited: An Extension to Tellado's Method," *6th International OFDM-Workshop*, Hamburg, Sept. 18-19, 2001.

[18] W. Henkel, V. Azis, "Partial Transmit Sequences and Trellis Shaping," *5th International ITG Conference on Source and Channel Coding (SCC)*, Erlangen, January 14-16, 2004.

[19] K.S. Jacobsen, "Methods of Upstream Power Backoff on Very High-Speed Digital Subscriber Lines," *IEEE Communications Magazine*, pp. 210-216, March 2001

[20] B. Wiese, K.S. Jacobsen, "Use of the Reference Noise Method Bounds the Performance Loss Due to Upstream Power Backoff," *IEEE JSAC*, Vol. 20, No. 5, pp. 1075-1084, June 2002.

[21] S. Schelstraete, "Define Upstream Power Backoff for VDSL," *IEEE JSAC*, Vol. 20, No. 5, pp. 1064-1074, June 2002.

[22] S.T. Chung and J.M. Cioffi, "Rate and Power Control in a Two-user Multicarrier Channel with no Coordination: the Optimal Scheme vs. a Suboptimal Method," *IEEE Transactions on Communications*, Vol. 51, No. 11, pp. 1768â€"2003, Nov 2003.

[23] S.T. Chung, S.J. Kim, J. Lee, and J.M. Cioffi, "A Game-theoretic Approach to Power Allocation in Frequency Selective Gaussian Interference Channels," in *Proc. of the IEEE International Symposium on Information Theory*, ISIT, Yokohama, Japan, p. 316. June-July 2003,

[24] D. Statovci, *Adaptive Resource Allocation for Multi-User Digital Subscriber Lines*, Ph.D. thesis, Vienna University of Technology, July 2005.

[25] A. Leshem and E. Zehavi, "Game Theory and the Frequency Selective Interference Channel," *IEEE Signal Processing Magazine*, Vol. 26, No. 5, pp. 28-40, September 2009.

[26] G. Tauböck, W. Henkel, "MIMO Systems in the Subscriber-Line Network," *5th International OFDM Workshop*, Hamburg, Sept. 12-13, 2000.

[27] G. Ginis, J.M. Cioffi, "Vectored Transmission for Digital Subscriber Line Systems," *IEEE JSAC*, Vol. 20, No. 5, pp. 1085-1104, June 2002.

[28] G. Ginis, J.M. Cioffi, "A multi-user precoding scheme achieving crosstalk cancellation with application to DSL systems," *proc. Thirty-Fourth Asilomar Conference on Signals, Systems and Computers*, 2000, Vol. 2, pp. 1627-1631, Oct. 29 - Nov. 1, 2000.

[29] G. Ginis, J.M. Cioffi, "On the relation between V-BLAST and the GDFE," *IEEE Communications Letters*, Vol. 5, No. 9, pp. 364-366, Sept. 2001.

[30] R.F.H. Fischer, C. Windpassinger, A. Lampe, and J.B. Huber, "Space-Time Transmission using Tomlinson-Harashima Precoding," *ITG Fachbericht*, 2002.

[31] W. Henkel, T. Kessler, "A Wideband Impulsive Noise Survey in the German Telephone Network: Statistical Description and Modeling," *AEÜ*, Vol. 48, No. 6, pp. 277-288, Nov. 1994.

[32] W. Henkel, T. Kessler, H.Y. Chung, "Coded 64-CAP ADSL in an Impulse-Noise Environment – Modeling of Impulse Noise and First Simulation Results," *IEEE JSAC*, Vol. 13, No. 9, Dec. 1995, pp. 1611-1621.

[33] I. Mann, S. McLaughlin, W. Henkel, R. Kirkby, and T. Kessler, "Impulse Generation wich Appropriate Amplitude, Length, Inter-arrival, and Spectral Characteristics," *IEEE JSAC*, Vol. 20, No. 5, pp. 901-912, June, 2002.

[34] M. van Bladel and M. Moeneclaey, "Time-domain equalization for multicarrier communication," in *Proc. Communication Theory Mini-Conf. (GLOBECOM 1995)*, pp. 167-171, Singapore, Nov. 13â€“17, 1995.

[35] W. Henkel and T. Kessler, "Maximizing the Channel Capacity of Multicarrier Transmission by Suitable Adaptation of the Time-Domain Equalizer," *IEEE Transactions on Communications*, Vol. 48, No. 12, pp. 2000-2004, Dec. 2000.

[36] W. Henkel and O. Graur, "Impulse Noise Cancellation based on the Common-Mode Signal," proc. *Sarnoff Symposium*, Princeton, NJ, USA, Apr. 12 -14, 2010, Princeton, NJ, USA.

[37] W. Henkel, T. Kessler,"A Wideband Impulsive Noise Survey in the German Telephone Network: Statistical Description and Modeling ," *AEÜ*, Vol. 48, No. 6, pp. 277-288, Nov./Dec. 1994.

[38] W. Henkel, T. Kessler, H.Y. Chung,"A Wideband Impulse-Noise Survey on Subscriber Lines and Inter-Office Trunks: Modeling and Simulation," *Globecom/CTMC '95*, Singapore, Nov. 13-17, 1995.

[39] I. Mann, S. McLaughlin, W. Henkel, R. Kirkby, T. Kessler,"Impulse Generation with Appropriate Amplitude, Length, Inter-arrival, and Spectral Characteristics," *IEEE J. Select. Area Commun.*, Vol. 20, No. 6, pp. 901-912, June 2002.

[40] G. Ginis, C. Peng,"Alien Crosstalk Cancellation for Multipair Digital Subscriber Line Systems," *EURASIP Journal on Applied Signal Processing*, Vol. 2006, 2006.

[41] T. Magesacher, P. Ödling, P. O. Börjesson and T. Nordström,"Exploiting the Common-Mode Signal in xDSL," *European Signal Processing Conference 2004*, Vienna, Austria, September 2004.

[42] T. Magesacher, P. Ödling, P. O. Börjesson,"Analysis of Adaptive Interference Cancellation Using Common-Mode Information in Wireline Communications," *EURASIP Journal on Applied Signal Processing*, Vol. 2007, 2007.

[43] T. Magesacher, P. Ödling, P. O. Börjesson,"Adaptive Interference Cancellation using Common-Mode Information in DSL," *Proc. European Signal Processing Conference 2005*, Antalya, Turkey, September 2005.

[44] P. Ödling, P. O. Börjesson, T. Magesacher and T. Nordström,"An Approach to Analog Mitigation of RFI," *IEEE J. Select. Area Commun.*, Vol. 20, No. 5, pp. 974-986, June, 2002.

[45] S. Haykin, Adaptive Filter Theory, Englewood Cliffs, Prentice Hall, 1986.

[46] B. Widrow, J. R. Glover, J. M. Mccool, J. Kaunitz, C. S. Williams, R. H. Hean, J. R. Zeidler, E. Dong, and R. C. Goodlin,"An Adaptive Noise Cancelling: Principles and Applications," *Proc. IEEE*, Vol. 63, No. 12, pp. 1692-1716, December, 1975.

[47] T. Chen, C. Tsai, T. Y. Chen,"An Intelligent Impulse Noise Detection Method by Adaptive Subband-Based Multi-State Median Filtering," *Second International Conference on Innovative Computing, Information and Control, ICICIC* 2007, 5-7 September, 2007.

[48] W. Henkel, T. Kessler,"Simulation of ADSL over ISDN on German Subscriber Lines," *IEEE Communications Letters*, Vol. 1, No. 5, pp. 124-126, September 1997.

[49] Draft ITU-T Recommendation G.996.1,"Test Procedures for Digital Subscriber Line (DSL) Transceivers," *International Telecommunication Union.*

[50] T. Nordström,"A model for an Austrian PE04 cable," *ETSI STC TM6*, Sophia-Antipolis, France, September 29 — October 3, 2003.

[51] L. Heylen, J. Musson,"Cable Models predict physically impossible Behaviour in Time Domain," *ETSI TM6 Meeting*, Amsterdam, Netherlands November 29 — December 1, 1999.

[52] H. Takahashi, A. A. Amin, S. L. Jansen, I. Morita, H. Tanaka, "8x66.8-Gbit/s Coherent PDM-OFDM Transmission over 640 km of SSMF at 5.6-bit/s/Hz Spectral Efficiency," *Proc. ECOC 2008*, Th.3.E.4, Sept.2008.

[53] H. Takahashi, A. A. Amin, S. L. Jansen, I. Morita, H. Tanaka, "DWDM Transmissi on with 7.0-bit/s/Hz Spectral Efficiency using 8x65.1-Gbit/s Coherent PDM-OFDM Signals," *Proc. OFC*, PDPB7, March 2009.

[54] R.-J. Essiambre, G. Foschini, P. Winzer, G. Kramer, "Capacity Limits of Fiber-Optic Communication Systems," *Proc. OFC*, OThL1, March 2009.

[55] L. Kazovsky, S. Benedetto, A. Willner, *Optical Fiber Communication Systems*, Artech House, 1996.

[56] G. P. Agrawal, *Nonlinear Fiber Optics*, Academic Press, 2nd edition, 1995.

[57] M. Mayrock, H. Haunstein, "Optical Monitoring for Non-Linearity Identification in CO-OFDM Transmission Systems," *Proc. OFC*, JThA58, Feb. 2008.

[58] M. Mayrock, H. Haunstein, "Monitoring of Linear and Non-linear Signal Distortion in Coherent Optical OFDM Transmission," *IEEE Journal of Lightwave Technology*, vol. 27, no. 16, pp. 3560-3566, Aug. 2009.

[59] I.P. Kaminow, T.L. Koch, *Optical Fiber Telecommunications IIIB*, Academic Press, 1997.

[60] X. Zhou, J. Yu, M.-F. Huang, Y. Shao, T. Wang, P. Magill, M. Cvijetic, L. Nelson, M. Birk, G. Zhang, S. Ten, H.B. Matthew, S.K. Mishra, "32Tb/s (320x114Gb/s) PDM-R Z-8QAM transmission over 580km of SMF-28 ultra-low-loss fiber," *Proc. OFC*, PD PB4, March 2009.

[61] B. Beggs, "Microelectronics Advancements to Support New Modulation Formats and DSP Techniques," *Proc. OFC*, OThE4, March 2009.

[62] T. Lotz, W. Sauer-Greff, R. Urbansky, "Iterative Demapping with Non-Euclidean Metric for Optical COFDM Systems," 14th *International OFDM-Workshop*, Hamburg, Germany, Sep. 2009

[63] "Interfaces for the Optical Transport Network," Telecommunication Standardization Sector, International Telecommunication Union, G. 709, Mar. 2003.

[64] "Forward error correction for high bit rate DWDM submarine systems," Telecommunication Standardization Sector, International Telecommunication Union, G.975.1, February 2004.

[65] C. Berrou , A. Glavieux, P. Thitimajshima, "Near Shannon limit error-correcting coding and decoding: Turbo-codes," *Proc. ICC* , Geneva, Switzerland, pp. 1064-1070, May 1993.

[66] R. G. Gallager, "Low-Density Parity-Check Codes", *IRE Trans. on Info. Theory*, Vol. IT-8, pp. 21-28, Jan. 1962.

[67] J. Hagenauer, "The Turbo Principle: Tutorial Introduction and State of the Art", *Symposium on Turbo Codes*, Brest, France, September 1997.

[68] S. ten Brink, J. Speidel, R.-H. Yan, "Iterative Demapping and Decoding for Multilevel Modulation," in *IEEE GLOBECOM*, Sydney, Australia, Vol. 1, pp. 579-584, July 1998.

[69] L. Bahl, J. Cocke, F Jelinek, J. Raviv, "Optimal Decoding of Linear Codes for Minimizing Symbol Error Rate," *IEEE Trans. Inform. Theory*, vol. 20, pp. 284-287, Mar. 1974.

Index

GPSR Compliance

The European Union's (EU) General Product Safety Regulation (GPSR) is a set of rules that requires consumer products to be safe and our obligations to ensure this.

If you have any concerns about our products, you can contact us on ProductSafety@springernature.com

In case Publisher is established outside the EU, the EU authorized representative is:

Springer Nature Customer Service Center GmbH
Europaplatz 3
69115 Heidelberg, Germany

Batch number: 10614744

Printed by Printforce, the Netherlands